S0-CDW-948

Aromatase Inhibition and Breast Cancer

Aromatase Inhibition and Breast Cancer

edited by

William R. Miller
Western General Hospital
Edinburgh, Scotland

Richard J. Santen
University of Virginia
Charlottesville, Virginia

MARCEL DEKKER, INC. NEW YORK • BASEL

Cover: Mammograms of the same breast before therapy (left) and after 3 months' therapy (right) with Femara®. Clear resolution of the cancer in the upper quadrant can be seen with therapy.

ISBN: 0-8247-0412-6

This book is printed on acid-free paper.

Headquarters
Marcel Dekker, Inc.
270 Madison Avenue, New York, NY 10016
tel: 212-696-9000; fax: 212-685-4540

Eastern Hemisphere Distribution
Marcel Dekker AG
Hutgasse 4, Postfach 812, CH-4001 Basel, Switzerland
tel: 41-61-261-8482; fax: 41-61-261-8896

World Wide Web
http://www.dekker.com

The publisher offers discounts on this book when ordered in bulk quantities. For more information, write to Special Sales/Professional Marketing at the headquarters address above.

Current printing (last digit):
10 9 8 7 6 5 4 3 2 1

PRINTED IN THE UNITED KINGDOM

Preface

Estrogens are key regulators of the normal development and growth of many tissues within the body, most notably the female reproductive system and secondary sexual organs. Given this central role, it is not surprising that excessive exposure to or inappropriate stimulation by estrogens may result in aberrant development and growth. In many instances, corrective procedures that reduce the production of estrogen may produce therapeutic benefits. Among such strategies is the use of drugs that inhibit the production of estrogen. In this respect, it is important to note that estrogens lie at the end of a biosynthetic sequence and that the blockade of any step in the pathway potentially inhibits estrogen production. However, the most specific method of blocking biosynthesis is to inhibit the final step in the sequence—the conversion of androgens to estrogens by the aromatase enzyme. Drugs that have the potential to inhibit aromatase have been available for some 30 years. In that time, the inhibitors have been used clinically to treat estrogen-dependent disease, most notably breast cancer, without making a major impact. Thus, even patients with breast cancer, which was suspected to be endocrine-sensitive, were unlikely to be given aromatase inhibitors as first-line hormone therapy, and then only after antiestrogens and progestins.

Given this background, a meeting entitled ''Aromatase Inhibitors into the Millennium'' would not have been imagined until a few years ago. What makes the concept highly attractive now is the development of new, third-generation drugs that can inhibit the aromatase enzyme with extraordinary potency and specificity. The testimony to this is that milligram doses of the inhibitors given orally and daily to postmenopausal women are capable of reducing circulating estrogens to undetectable levels without having apparent effects on any other class of steroid hormones.

The objectives of this volume are to (1) provide evidence of the endocrine effects of these novel aromatase inhibitors; (2) demonstrate how these effects translate into clinical benefits, using breast cancers as a primary example; (3) explore the other clinical indications for which aromatase inhibitors may be usefully employed; and (4) highlight recent research directed toward development of reagents, technologies, and models by which to optimize the use of aromatase inhibitors.

The scene is set by Mitch Dowsett and Harold Harvey, respectively, who review the drug development of aromatase inhibitors and the current management of advanced breast cancers (to date the major clinical setting in which inhibitors have been used). Steve Johnston and Ian Smith then go on to identify the place of aromatase inhibitors within the context of endocrine treatment of advanced breast cancers. They present evidence that new inhibitors such as letrozole, anastrozole, and exemestane all have proven efficacy when used as second-line therapy (after antiestrogens) and, indeed, in certain studies may produce benefits in terms of antitumor effects and less toxicity beyond older aromatase inhibitors or progestins. As a consequence, randomized trials are underway comparing the drugs with antiestrogens as first-line therapy and the question that is now seriously being asked is: ''Can aromatase inhibitors replace antiestrogens as first-choice endocrine therapy?'' Further and more direct evidence of powerful antitumor effects and clinical response can be elicited from studies in which aromatase inhibitors are given neoadjuvantly and the volume of the cancer within the breast is monitored. Mike Dixon presents the experience of the Edinburgh Breast Group.

These highly promising results have led to the use of aromatase inhibitors in earlier stages of the disease as an adjuvant to surgery. Henning Mouridsen reviews the major adjuvant trials that have been established to determine if the drugs may be used to treat micrometastatic disease and delay recurrence. However, it is clear that even in selected populations of patients, not all women will derive benefits from adjuvant treatment and there is a pressing need to identify accurately tumors that will respond to treatment. Manfred Wischnewsky explains how machine learning techniques may be used to address this issue. In terms of predicting and monitoring response, the ability to measure aromatase activity and expression may be important. Because levels of aromatase are low in peripheral tissues in postmenopausal women (in whom aromatase inhibitors are largely used), there is an immediate need to develop new reagents, technologies, and appropriate model systems. Hironobu Sasano and Urs Eppenberger provide state-of-the-art accounts of immunohistological assessments of aromatase protein and RT-PCR measurements at the level of mRNA.

While these reagents and technologies will be applied to relevant clinical material, there are limitations to patient-based studies, and these must be supplemented by experiments in appropriate model systems. The contributions by Bill Miller and Angela Brodie illustrate how model systems based on human material may be used to optimize the use of aromatase inhibitors in terms of differences between individual inhibitors and their combination with other forms of endocrine manipulations.

The adjuvant long-term use by breast cancer patients who are ostensibly disease-free will provide detailed information on the toxicity profiles of the new aromatase inhibitors and their long-term acceptability. This knowledge will be invaluable if the drugs are to be used in a preventative setting. Paul Goss is confident that aromatase inhibitors will be used to prevent breast cancer. If this occurs as a consequence of delaying the clinical appearance of occult disease, the antipromotional effects of aromatase inhibitors may be no greater than those of other forms of endocrine deprivation—for example, antiestrogens. On the other hand, if the estrogen molecules are carcinogenic initiators, the ability of aromatase inhibitors to reduce estrogen to exceptionally low levels (whereas antiestrogens such as tamoxifen do not) may provide a degree of protection beyond other hormonal agents.

The powerful endocrine properties of the new generation of aromatase inhibitors contraindicate widespread use in women without disease at this stage, and it is likely that they will be restricted to high-risk groups who will require careful monitoring to determine if effects at nontarget sites are associated with toxicity. Thus there is a need to appreciate health–economic issues in terms not only of the cost of side effects to the patient but also of financial constraints. This topic is reviewed by Suzanne Wait. Such considerations are particularly relevant if aromatase inhibitors are to be administered on a broader basis, for diseases in which the drugs are not classically indicated. The case for using aromatase inhibitors as therapy for pubertal gynecomastia is presented by Paul Kaplowitz, for prostate cancer by Matthew Smith, and for treatment of benign and malignant endometrial conditions by Serdar Bulun. The effective and optimal control of these and other hormone-dependent diseases still depends on the answers to basic questions, such as which inhibitor to use, at what dose, for how long, in what sequence, and for which patients? The next decade in the new millennium should provide many of the answers. The turn of the century is truly an exciting time for endocrinologists and oncologists interested in the enzyme called "aromatase."

William R. Miller
Richard J. Santen

Contents

PART IV NEW DIRECTIONS FOR AROMATASE RESEARCH

PART V NEW CLINICAL INDICATIONS FOR AROMATASE THERAPY

Contributors

Rainer Böhm Novartis Pharma AG, Basel, Switzerland

Angela M. H. Brodie, Ph.D. Professor, Department of Pharmacology and Experimental Therapeutics, University of Maryland School of Medicine, Baltimore, Maryland

Serdar E. Bulun, M.D. Director, Division of Reproductive Endocrinology, Department of Obstetrics and Gynecology, University of Illinois, Chicago, Illinois

Hillary A. Chaudry Novartis Pharma AG, Basel, Switzerland

J. M. Dixon, B.Sc., M.B.Ch.B., M.D., F.R.C.S., F.R.C.S.Ed. Senior Lecturer in Surgery, Consultant Breast Surgeon, Edinburgh Breast Unit Research Group, Western General Hospital, Edinburgh, Scotland

Mitch Dowsett, Ph.D. Professor, Academic Department of Biochemistry, The Royal Marsden Hospital, London, England

S. Eppenberger-Castori, Ph.D. Department of Research, Universitäts-Frauenklinik, Basel, Switzerland

Urs Eppenberger, Ph.D. Professor, Department of Research, Universitäts-Frauenklinik, Basel, Switzerland

D. Evans Novartis Pharma AG, Basel, Switzerland

Paul E. Goss, M.D., F.R.C.P.(UK), F.R.C.P.(C), Ph.D. Director, Breast Cancer Prevention Program, Department of Medical Oncology, Princess Margaret Hospital, University Health Network, Toronto, Ontario, Canada

Harold A. Harvey, M.D. Professor, Department of Medicine, Penn State College of Medicine, Hershey, Pennsylvania

Stephen R. D. Johnston, M.A., M.R.C.P., Ph.D. Consultant Medical Oncologist, Department of Medicine, The Royal Marsden Hospital, London, England

Paul B. Kaplowitz, M.D., Ph.D. Associate Professor, Department of Pediatrics, Virginia Commonwealth University School of Medicine, Richmond, Virginia

S. Levano Universitäts-Frauenklinik, Basel, Switzerland

Y. Liu University of Maryland School of Medicine, Baltimore, Maryland

B. Long University of Maryland School of Medicine, Baltimore, Maryland

Q. Lu University of Maryland School of Medicine, Baltimore, Maryland

William R. Miller, Ph.D., D.Sc. Professor of Experimental Oncology, Edinburgh Breast Unit Research Group, Western General Hospital, Edinburgh, Scotland

Takuya Moriya, M.D. Associate Professor, Department of Pathology, Tohoku University School of Medicine, Sendai, Japan

H. T. Mouridsen, M.D., M.Sci. Department of Oncology, Rigshopitalet, Copenhagen, Denmark

P. Mullen, B.Sc., M.Phil. Associate Scientific Officer, Western General Hospital, Edinburgh, Scotland

H. Müller, Ph.D. Universitäts-Frauenklinik, Basel, Switzerland

Kurt Possinger, M.D. Professor, Department of Oncology and Hematology, Medizinische Klinik II, Charité Campus Mitte, Humboldt University Berlin, Berlin, Germany

Kush Sachdeva Penn State College of Medicine, Hershey, Pennsylvania

Richard J. Santen, M.D. Professor, Division of Endocrinology and Metabolism, Department of Internal Medicine, University of Virginia, Charlottesville, Virginia

Hironobu Sasano, M.D., Ph.D. Professor and Chairman, Department of Pathology, Tohoku University School of Medicine, Sendai, Japan

Peter Schmid, M.D. Department of Oncology and Hematology, Medizinische Klinik II, Charité Campus Mitte, Humboldt University Berlin, Berlin, Germany

F. Schoumacher, Ph.D. Universitäts-Frauenklinik, Basel, Switzerland

Evan R. Simpson Professor and Director, Prince Henry's Institute of Medical Research, Monash Medical Centre, Clayton, Victoria, Australia

Ian E. Smith, M.D., F.R.C.P., F.R.C.P.E. Professor, Department of Medicine, The Royal Marsden Hospital, London, England

Matthew R. Smith, M.D., Ph.D. Department of Hematology/Oncology, Massachusetts General Hospital, Boston, Massachusetts

Takashi Suzuki, M.D. Assistant Professor, Department of Pathology, Tohoku University School of Medicine, Sendai, Japan

Kazuto Takayama Tohoku University School of Medicine, Sendai, Japan

R. Vidya, M.S., F.R.C.S. (Eng), F.R.C.S. (I) Clinical Research Fellow, Western General Hospital, Edinburgh, Scotland

Suzanne Wait Université Louis Pasteur, Strasbourg, France

Manfred B. Wischnewsky Center for Computing Technology (TZI), University of Bremen, Bremen, Germany

Khaled M. Zeitoun Columbia University College of Physicians and Surgeons, New York, New York

Part I

PLENARY LECTURE

1

Aromatase Inhibition: The Outcome of 20 Years of Drug Development

Mitch Dowsett
The Royal Marsden Hospital
London, England

I. ABSTRACT

The goal of developing highly effective, well-tolerated, orally active aromatase inhibitors was set about 20 years ago after the recognition that inhibition of aromatase was the principal mode of action of the breast cancer drug aminoglutethimide (AG). This goal has been realized over the last few years with the widespread availability of letrozole, anastrozole, and very recently exemestane. Clinical trials have confirmed the greater efficacy of these agents than the progestin megestrol acetate. Importantly, for the assessment of the gains made by these new drugs, letrozole has been found to provide survival advantages over AG. The eventual impact of these agents will be established by ongoing trials of their use in the adjuvant context and of pilot studies of their potential as prophylactic agents.

II. INTRODUCTION

Over the past 20 years the clinical development of aromatase inhibitors has led to the derivation of several highly specific, well-tolerated compounds which have

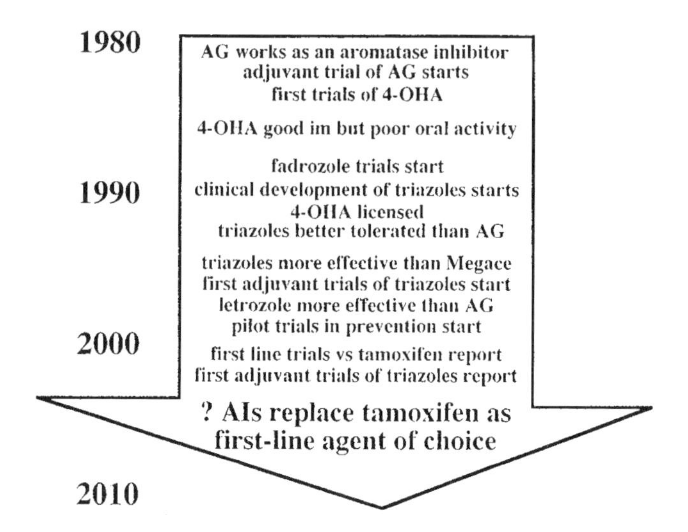

Figure 1 Time line of some of the major steps in the clinical advancement of aromatase inhibitors.

shown greater clinical efficacy than established agents. Large clinical trials are ongoing to establish the full role of these inhibitors in all stages of breast cancer therapy. The aim of this chapter is to highlight the main stages in their development over the past 20 years and to focus particularly on the central role which endocrine pharmacology has played in directing this development. The current positioning of new aromatase inhibitors is summarized and the impact of possible future results discussed. Particularly significant developments over the last 20 years are summarized in Figure 1.

III. AMINOGLUTETHIMIDE, THE PROTOTYPE AROMATASE INHIBITOR

Aminoglutethimide (AG) first entered preliminary trials in advanced breast cancer as a result of the observation that it inhibited adrenal steroidogenesis during its earlier investigation as an antiepileptic (1). The basis of the use of AG in this context was that adrenal androgens form the principal substrate for the synthesis of plasma estrogens by the enzyme aromatase in the peripheral tissues of postmenopausal women: Removal of these androgens would therefore be expected to elicit the attenuation of the estrogenic stimulus to the breast carcinoma by a process termed ‘‘medical adrenalectomy’’ (2). AG treatment was supplemented by replacement dosages of glucocorticoid to avoid potential problems of adrenal insufficiency.

It was soon established that AG plus glucocorticoid was a clinically effective treatment but with significant tolerability problems (3). During the early 1970s Siterii and colleagues (4) established that AG was an inhibitor of the aromatase enzyme, and a classic paper by Santen and colleagues (5) demonstrated that AG was an efficient suppressant of aromatase activity in postmenopausal breast cancer patients. This led to the development of a concept of a dual mode of action for AG; that is, suppression of adrenal androgens synthesis and inhibition of the conversion of any residual androgen to estrogen.

The pharmacology of AG and assessment of the role of its various activities in its clinical effectiveness is, however, complicated by the application of glucocorticoid. In a series of studies in which hydrocortisone or AG was given alone or in combination, it became clear that the effects of AG on adrenal steroidogenesis in fact result in a substantial increase in adrenal androgen secretion (rather than the decrease anticipated) (6–9). This results from its inhibition of the 21-hydroxylase enzyme, an effect which is only neutralized by the addition of hydrocortisone. Stuart-Harris et al. (9) therefore investigated the clinical efficacy of AG when used alone at a dose below that conventionally used. The clinical efficacy of this treatment with AG 125 mg twice a day established that the primary mode of action of AG was through aromatase inhibition.

Although this early work was important in establishing that aromatase inhibition was a viable method of treating postmenopausal patients with advanced breast cancer, it was clear that AG was far from an ideal agent. Endocrine studies indicated that it was only partially effective in suppressing plasma estrogen levels, and that its nonspecificity extended to the suppression of prostaglandin synthesis (10) and thyroxine synthesis (11) in addition to its adrenal effects which required the traditional use of glucocorticoid. Most significantly it had several marked side effects, including lethargy and somnolence extending to ataxia as well as nausea and vomiting (1). Thus, the scene was set for an explosion of pharmaceutical activity with the aim of deriving a specific, fully effective, well-tolerated aromatase inhibitor. These developments are conveniently considered according to the two main structural subdivisions; that is, steroidal and nonsteroidal compounds.

IV. STEROIDAL AROMATASE INHIBITORS

Two steroidal compounds have entered widespread clinical study: (1) 4-hydroxyandrostenedione (4-OHA, formestane) and (2) exemestane. Both of these compounds are analogues of the natural substrate androstenedione and act as competitive binders at the substrate binding site of the enzyme. However, in addition, both are converted by the aromatase enzyme to reactive intermediates which bind irreversibly to the enzyme and thereby permanently inactivating it.

A. 4-Hydroxyandrostenedione

4-OHA was one of about 200 compounds which were screened by Drs. Harry and Angela Brodie in the 1970s (12). A series of preclinical studies demonstrated its pharmacological effectiveness and clinical studies were initiated in the early 1980s. It was both expected and confirmed that the oral administration of 4-OHA had poor biological activity as measured by both estrogen suppression and inhibition of aromatization in vivo (13,14). This resulted from the glucuronidation of the critical 4-hydroxy modification through first-pass liver metabolism. Thus, further studies and clinical use focused on the intramuscular administration of the drug.

A series of dose-finding studies demonstrated that 4-OHA could maintain estrogen suppression on a schedule of once every 2 weeks. It was found that with a dose of 500 mg every 2 weeks there was no recovery of plasma estrogen levels prior to the next injection, but a small recovery was found with 250 mg every 2 weeks (13,15). Nonetheless, the 250-mg dose was chosen for widespread clinical use on the basis of a greater incidence of local tolerability problems with the higher dose (16).

Several Phase II studies documented the clinical efficacy of formestane (summarized in Ref. 16). In one Phase III study comparing 4-OHA to tamoxifen as first-line treatment of advanced breast cancer no difference in response rate or survival was recorded, but the median duration of response was significantly longer for tamoxifen (17). Another Phase III study compared 4-OHA as second-line treatment to megesterol acetate and no difference in response rate, time to progression, or survival was found (18). These data are somewhat disappointing, but it may be significant that the pharmacological effectiveness of 4-OHA, as measured by in vivo aromatase assays, is in fact less than that with AG at its full dose with mean inhibition of only approximately 85% (19).

B. Exemestane

Exemestane has the advantage of being orally active. As with other aromatase inhibitors, pharmacological studies have been important to demonstrate the efficacy of exemestane and to select an appropriate dose. However, the structure of exemestane required the development of special chromatographic separation methods for valid estrogen measurement, which was a complicating and delaying factor in its development. Nonetheless, Johannessen et al. (20) were able to select a dose of 25 mg per day given orally as that giving maximum estrogen suppression. More recently, it has been established that this results in an inhibition of aromatase in vivo by 98% (comparable with some of the modern nonsteroidal

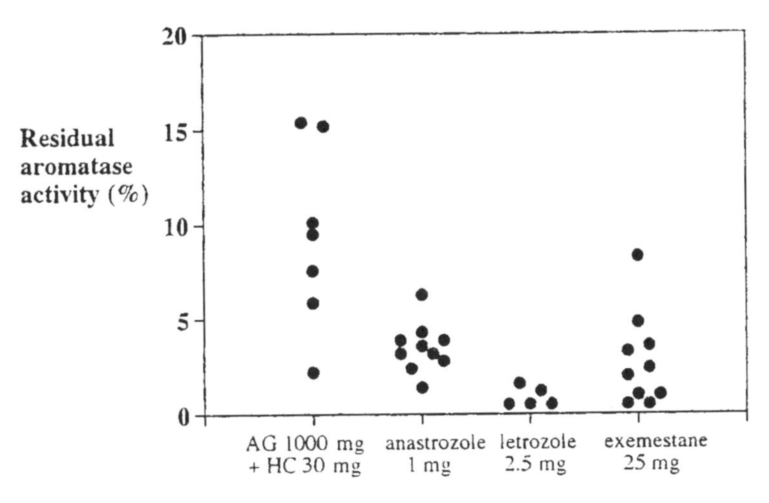

Figure 2 In vivo inhibition of whole body aromatase activity with anastrozole, letrozole, and exemestane at their clinically used dosages. Residual activity is shown in individual patients as a percentage of baseline activity.

inhibitors; see below) (21) (Fig. 2). The first Phase III study of exemestane in advanced breast cancer reported a survival advantage of this compound over megesterol acetate in addition to a significantly enhanced median time to progression (22). This orally active steroidal agent has recently been licensed for widespread use. It remains to be established whether its steroidal nature will be associated with any special activity which might have advantages over the more well-established nonsteroidal compounds.

V. NONSTEROIDAL INHIBITORS

Nonsteroidal inhibitors of aromatase act by virtue of their association with the haem prosthetic group of the aromatase enzyme (23). The specificity of this binding is determined by their fit into the substrate binding site of aromatase as opposed to that of the numerous other cytochrome P450 enzymes, which include many that are involved in steroidogenesis.

A. Fadrozole

Fadrozole is a highly potent compound in vitro, but it has a relatively short half-life which leads to its poorer in vivo activity compared with the more slowly cleared triazole inhibitors (24). Although the compound is effective clinically, the dose used has been limited by the effects that fadrozole has on aldosterone synthesis (25,26). The clinically used dose of 1 mg twice a day leads to a mean suppression of only about 85% of aromatase in vivo (24); that is, similar to 250 mg for 2 weeks of 4-OHA (14). At present, this compound is used widely only in Japan.

B. Triazole Inhibitors

Four triazole inhibitors have entered clinical studies for aromatase inhibition. Of these, vorozole and YM511 are no longer in substantial clinical trials. In contrast, anastrozole and letrozole are licensed for use in patients with advanced breast cancer after tamoxifen, and they are currently in large trials for both their first-line metastatic use and the possible use in the adjuvant scenario (see below). These compounds have been shown to lead to near complete inhibition of aromatase in vivo: Anastrozole at 1 mg/day inhibited aromatase by a mean 96.7% and letrozole by 98.9% (27,28) (see Fig. 2). Direct randomized comparisons of the effectiveness of these two agents both pharmacologically and clinically is currently underway. Both compounds have shown an improved tolerability over megesterol acetate and improved efficacy according to at least one clinical endpoint (29,30). Additionally, letrozole has been found to have greater efficacy than AG in terms of time to progression ($P = .008$) and overall survival ($P = .002$; median 28 vs. 20 months) (31). This last comparison emphasizes the improvement in efficacy that has occurred by virtue of the development of the new nonsteroidal aromatase inhibitors and also emphasizes the improvement in tolerability: Adverse events were 29% with letrozole versus 46% with AG. A detailed comparison of the clinical positioning of these compounds is provided in Chapters 3 and 4.

VI. ONGOING TRIALS WITH AROMATASE INHIBITORS

A. First-Line Treatment Versus Tamoxifen

A Phase III trial of tamoxifen versus 4-OHA indicated better efficacy overall for tamoxifen (17). Two trials have also been conducted with fadrozole versus tamoxifen. In both of these most endpoints were not significantly different, but in one (32) there was a significantly greater time to treatment failure with tamoxifen. In the other (33) response duration was longer with tamoxifen. The greater pharmacological effectiveness of the triazole inhibitors and the improved efficacy of letrozole in comparison with AG provide hope that they will be more effective than 4-OHA and fadrozole and may be at least as effective as tamoxifen. Whatever the result of these new trials, some lessons may be learned from examination of earlier results from studies of tamoxifen versus AG. Two things are clear: Some patients who do not respond to the aromatase inhibitor do respond to tamoxifen and vice versa; that is, there is not total cross resistance between these two approaches to therapy (34,35). Thus, it would be helpful to have a biological indicator which would predict the effectiveness of one of these agents rather than another in a specific patient. The only biomarker which is currently considered a candidate for this is intratumoral aromatase activity, and although some encouraging data have been reported (36,37), larger studies are required. Another approach that might be considered in choosing adjuvant therapy is the change in the proliferation marker Ki67 after short-term exposure to the medical agent(s) prior to surgery (38) (see Chap. 3).

These early studies with AG and tamoxifen also indicated that some patients would respond to AG after responding to tamoxifen and to first-line tamoxifen after first response to AG (34,35). However, one study (34) found that response to tamoxifen after AG was very infrequent, suggesting that resistance mechanisms to the aromatase inhibitor might preclude further endocrine responsiveness. This is an important issue that should be addressed in ongoing clinical studies as patients relapse from first-line comparisons between the aromatase inhibitor and tamoxifen as well as the adjuvant trials.

B. Adjuvant Therapy

The clear efficacy of aromatase inhibitors and their excellent tolerability has led to their comparative investigation with tamoxifen in a number of clinical trials in the adjuvant setting. The structure of a number of these trials, which involve international participation, is illustrated in Figure 3. It can be seen that the different design of the trials will enable questions of combination and sequencing with tamoxifen to be addressed in addition to a head-to-head comparison between the inhibitors and tamoxifen. Results from some of these may be expected from the year 2001 onward. A better tolerability or improved efficacy of the aromatase inhibitor compared with tamoxifen would lead to a very substantial increase in the use of aromatase inhibitors, as it would be likely to result in tamoxifen being replaced as the first-line agent of choice. This is a scenario which a few years ago would have seemed very unlikely. Aromatase inhibitors might even be more effective than tamoxifen, since it is now widely accepted that the acquired resistance that occurs in some breast cancer patients to tamoxifen is due to a sensitization to the agonist activity of the drug, which is expressed through its binding to the estrogen receptor (39). This mechanism cannot occur with aromatase inhibitors, since they do not interact directly with the estrogen receptor (ER).

C. Breast Cancer Prevention

The possible use of aromatase inhibitors in the prevention scenario is perhaps the most exciting possibility currently under investigation with these drugs (see

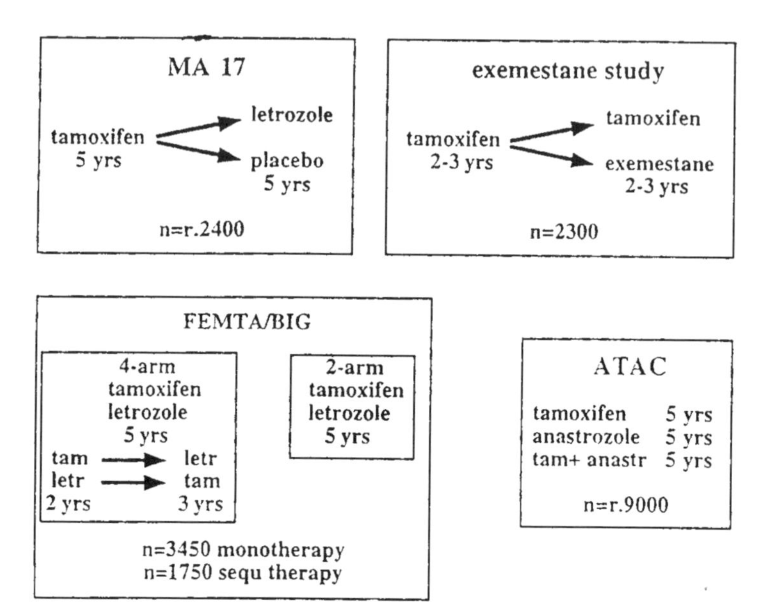

FIGURE 3 Design of the ongoing international adjuvant trials of aromatase inhibitors.

Chap. 9). There is a clear rationale for the use of these inhibitors, since there is a strong relationship between plasma estrogen levels in postmenopausal women and the eventual development of breast cancer (40). This is supported by the observation of an inverse relationship between bone mineral density and breast cancer risk (41). The eventual use of the inhibitors in this scenario, as opposed to one of the selective estrogen-receptor modulators (e.g., tamoxifen, raloxifene), is likely to depend in part on the effects of the inhibitors on bone mineral density, cholesterol levels, and the associated cardiac effects.

VII. OPTIMAL APPLICATION OF AROMATASE INHIBITORS

Although there is no doubt that the development of the new aromatase inhibitors over the past 20 years has been of major benefit in deriving better-tolerated and more effective compounds, it is not yet clear that their first-line application is more beneficial in terms of prolonged cancer control than the initial application of agents eliciting only partial suppression followed by one of the more potent compounds later. Studies which provoke consideration of this issue include the demonstration that patients will respond to the addition of an aromatase inhibitor

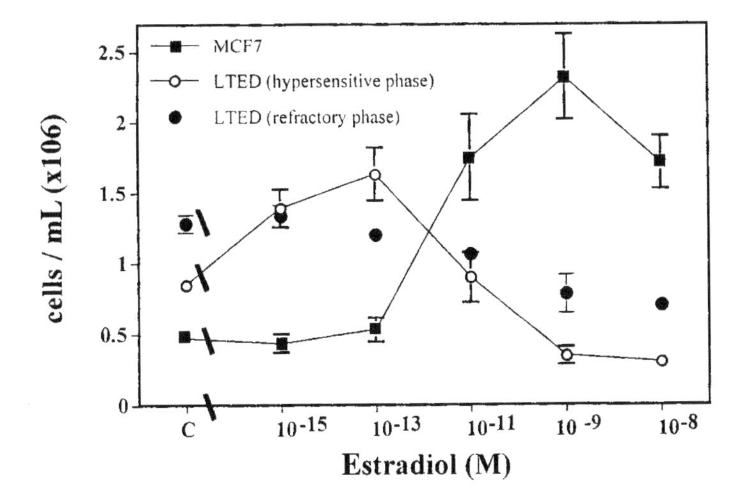

FIGURE 4 Growth response to estradiol of wild-type MCF7 cells and long-term estrogen-deprived MCF7 cells (LTED) after 15–50 weeks without estrogen (hypersensitive phase) or 50–80 weeks without estrogen (refractory phase) 43.

to gonadotropin-releasing hormone (GnRH) agonist treatment such as goserelin (premenopausal) after initial response to the GnRH agonist alone (42). Similarly, responses to the combination of AG and 4-OHA has been reported after initial treatment with 4-OHA alone (43). This indicates that breast cancer cells may become apparently resistant to treatment with an estrogen-depriving agent as a result of becoming hypersensitive to the stimulatory effects of residual estrogen. Laboratory studies by Santen's group et al. (42) and by our laboratory (45) (Fig. 4) recapitulate this effect of hypersensitivity with the long-term deprivation of human breast cancer cells in vitro. It remains to be established whether improved breast cancer control is achieved by immediate complete estrogen deprivation or by the stepwise approach using a partial estrogen deprivation initially. Randomized clinical trials are about to be undertaken to answer this important question.

VIII. CONCLUSIONS

The last 20 years have resulted in the development of a new mode of treatment of breast cancer with improved clinical efficacy having been demonstrated in the largest randomized clinical studies ever conducted in advanced breast cancer. The possible advantages of these new compounds over tamoxifen as a first-line agent will be established in the next few years in the metastatic and adjuvant setting. Only then will we know just how important these developments have been.

ACKNOWLEDGMENTS

A large number of collaborators have worked with me over the last 20 years contributing enormously to the results and thoughts discussed above. I specifically wish to acknowledge the central importance of Dr. Ian Smith, Professor Trevor Powles, Professor Charles Coombes, Professor Per Lønning, Dr. Stephen Johnston, and Professor Adrian Harris for their long-term support and collaboration. Many of the studies discussed above and the overall progress in the development of aromatase inhibitors would not have occurred without the energies of Dr. Stuart Hughes, who is sadly deceased.

REFERENCES

1. Hughes SWM, Burley DM. Aminoglutethimide. A 'side-effect' turned to therapeutic advantage. Postgrad Med J 1970; 46:409–416.
2. Griffiths CT, Hall TC, Saba Z, Barlow JJ, Nevinny HB. Preliminary trial of aminoglutethimide in breast cancer. Cancer Res 1973; 32:31–37.
3. Stuart-Harris RC, Smith IE. Aminoglutethimide in the treatment of advanced breast cancer. Cancer Treat Revs 1984; 11:189–204.
4. Thompson EA Jr, Siiteri PK. Utilization of oxygen and reduced nicotinamide adenine dinucleotide phosphate by human placental microsomes during aromatization of androstenedione. J Biol Chem 1974; 249:5364–5372.
5. Santen RJ, Santner S, Davis B, Veldhuis J, Samojlik E, Ruby E. Aminoglutethimide inhibits extraglandular estrogen production in postmenopausal women with breast carcinoma. J Clin Endocr Metab 1978; 46:1066–1074.
6. Harris AL, Dowsett M, Smith IE, Jeffcoate SL. Endocrine effects of low dose aminglutethimide alone in advanced postmenopausal breast cancer. Br J Cancer 1983; 47:621–627.
7. Harris AL, Dowsett M, Smith IE, Jeffcoate SL. Hydrocortisone alone versus hydrocortisone plus aminglutethimide: a comparison of the endocrine effects in postmenopausal breast cancer. Eur J Cancer 1984; 20:463–467.
8. Samojlik E, Velohuis JD, Wells SA, Santen RJ. Preservation of androgen secretion during estrogen suppression with aminoglutethimide in the treatment of metastatic breast carcinoma. J Clin Invest 1980; 65:602–612.
9. Stuart-Harris R, Smith IE, Dowsett M, Bozek T, McKinna JA, Gazet J-C, Jeffcoate SL, Kurkure A, Carr L. Low-dose aminoglutethimide in treatment of advanced breast cancer. Lancet 1984; 2:604–607.
10. Harris AL, Mitchell MD, Smith IE, Powles TJ. Suppression of plasma 6-keto-prostaglandin $F_1\alpha$ and 13,14-dihydro-15-keto-prostaglandin $F_2\alpha$ by aminoglutethimide in advanced breast cancer. J Cancer 1983; 48:595–598.
11. Santen RJ, Wells SA, Cohn N, Demers LM, Misbin RI, Foltz EL. Compensatory

increase in TSH secretion without effect on prolactin secretion in patients treated with aminoglutethimide. J Clin Endocr Metab 1977; 45:739–746.

12. Schwarzel WC, Kruggel W, Brodie HJ. Studies on the mechanisms of estrogen biosynthesis. VII. The development of inhibitors of the enzyme system in human placenta. Endocrinology 1973; 92:866–880.
13. Dowsett M, Cunningham DC, Stein RC, Evans S, Dehennin L, Hedley A, Coombes RC. Dose-related endocrine effects and pharmacokinetics of oral and intramuscular 4-hydroxyandrostenedione in postmenopausal breast cancer patients. Cancer Res 1989; 49:1306–1312.
14. MacNeill FA, Jacobs S, Dowsett M, Lonning PE. The effects of oral 4-hydroxyandrostenedione on peripheral aromatisation in post-menopausal breast cancer patients. Cancer Chemother Pharmacol 1995; 36:249–254.
15. Dowsett M, Goss PE, Powles TJ, Hutchinson G, Brodie AMH, Jeffcoate SL, Coombes RC. Use of the aromatase inhibitors 4-hydroxyandrostenedione in postmenopausal breast cancer: optimization of therapeutic dose and route. Cancer Res 1987; 47:1957–1961.
16. Coombes RC, Hughes SWM, Dowsett M. 4-Hydroxyandrostenedione: a new treatment for postmenopausal patients with breast cancer. Eur J Cancer 1992; 28A:1941–1945.
17. Carrión RP, Candel VA, Calabresi F, Michel RT, Santos R, Delozier T, Goss P, Mauriac L, Feuilhade F, Freue M, Pannuti F, van Belle S, Martinez J, Wehrle E, Royce CM. Comparison of the selective aromatase inhibitor formestane with tamoxifen as first-line hormonal therapy in postmenopausal women with advanced breast cancer. Ann Oncol 1994; 5:S19–S24.
18. Rose C, Freue M, Kjaer M, Boni C, Janicke F, Coombes C, Willemse PHB, van Belle S, Çarrion RP, Jolivet J, de Palacios PI. An open, comparative randomized trial comparing formestane vs oral megestrol acetate as a second-line therapy in postmenopausal advanced breast cancer patients. Eur J Cancer 1996; 32A:49.
19. Jones AL, MacNeill F, Jacobs S, Lønning PE, Dowsett M, Powles TJ. The influence of intramuscular 4-hydroxyandrostenedione on peripheral aromatisation in breast cancer patients. Eur J Cancer 1992; 28:1712–1716.
20. Johannessen DC, Engan T, di Salle E, Persiani S, Paonini J, Ornati G, Piscitelli G, Lønning PE. Endocrine and clinical effects of exemestane (FCE 24304) novel steroidal aromatase inhibitor in postmenopausal breast cancer patients: phase I study. Clin Cancer Res 1997; 3:1100–1108.
21. Geisler J, King N, Anker G, Omati G, Di Salle E, Lønning P, Dowsett M. In vivo inhibition of aromatization by exemestane, a novel irreversible aromatase inhibitor, in postmenopausal breast cancer patients. Clin Cancer Res 1998; 4:2089–2093.
22. Kaufmann M, Bajetta E, Dirix LY, et al. Survival advantage of exemestane (EXE, Aromasin) over megestrol acetate (MA) in postmenopausal women with advanced breast cancer (ABC) refractory to tamoxifen (TAM): results of a phase III randomised double-blind study. Proc Am Soc Clin Oncol 1999; 18:109a.
23. Kao Y-C, Cam LL, Laughton CA, Zhou D, Chen S. Binding characteristics of seven inhibitors of human aromatase: a site directed mutagenesis study. Cancer Res 1996; 56:3451–3461.
24. Lønning PE, Jacobs S, Jones A, Haynes B, Powles TJ, Dowsett M. The influence

of CGS 16949A on peripheral aromatisation in breast cancer patients. Br J Cancer 1991; 63:789–793.

25. Dowsett M, Stein RC, Mehta A, Coombes RC. Potency and selectivity of the nonsteroidal aromatase inhibitor CGS 16949A in postmenopausal breast cancer patients. Clin Endocrinol 1990; 32:623–634.
26. Santen RJ, Demers LM, Adlercreutz H, Harvey H, Santner S, Sanders S, Lipton A. Inhibition of aromatase with CGS16949A in postmenopausal women. J Clin Endocr Metab 1989; 68:99–106.
27. Geisler J, King N, Dowsett M, Ouestad L, Lundgren S, Walton P, Kormeset PO, Lønning PE. Influence of Anastrozole (Arimidex®) a non-steroidal aromatase inhibitor, on in vivo aromatisation and plasma estrogen levels in postmenopausal women with breast cancer. Br J Cancer 1996; 74:1286–1291.
28. Dowsett M, Jones A, Johnston, Jacobs S, Trunet P, Smith IE. In vivo measurement of aromatase inhibition by letrozole (CGS 20267) in postmenopausal patients with breast cancer. Clin Cancer Res 1995; 1:1511–1515.
29. Dombernowsky P, Smith I, Falkson G, Leonard R, Panasci L, Bellmunt J, Bezwoda W, Gardin G, Gudgeon A, Morgan M, Fornasiero A, Hoffman W, Michel J, Hatschek T, Tjabbes T, Chaudri HA, Hornberger U, Trunet PF for the Letrozole International Trial Group. Letrozole (Femara®), a new oral aromatase inhibitor for advanced breast cancer: double-blind randomized trial showing a dose effect and improved efficacy and tolerability compared with megestrol acetate. J Clin Oncol 1998; 6:453–461.
30. Buzdar A, Jonat W, Howell A, Jones SE, Blomqvist C, Vogel CL, Eiermann W, Wolter JM, Azab M, Webster A, Plourde PV. Anastrozole, a potent and selective aromatase inhibitor, versus megestrol acetate in postmenopausal women with advanced breast cancer: Results of overview analysis of two phase III trials. J Clin Oncol 1996; 14:2000–2011.
31. Gershanovich M, Chaudri HA, Campos D, et al. Letrozole, a new oral aromatase inhibitor: randomized trial comparing 2.5 mg daily, 0.5 mg daily and aminoglutethimide in postmenopausal women with advanced breast cancer. Ann Oncol 1998; 9: 639–645.
32. Thürlimann B, Beretta K, Bacchi M, Castiglione-Gertsch M, Goldhirsch A, Jungi WF, Cavalli F, Senn HJ, Fey M, Lohnert T. First-line fadrozole HC1 (CGS 16949A) versus tamoxifen in postmenopausal women with advanced breast cancer. Ann Oncol 1996; 7:471–479.
33. Falkson CI, Falkson HC. A randomised study of CGS 16949A (fadrozole) versus tamoxifen in previously untreated postmenopausal patients with metastatic breast cancer. Ann Oncol 1996; 7:465–469.
34. Smith IE, Harris AL, Morgan M, Gazet J-C, McKinna JA. Tamoxifen versus aminoglutethimide versus combined tamoxifen and aminoglutethimide in the treatment of advanced breast carcinoma. Cancer Res 1982; 42:3430s–3433s.
35. Harvey HA, Lipton A, White DS, et al. Cross-over comparison of tamoxifen and aminoglutethimide in advanced breast cancer. Cancer Res 1982; 44:3451s–3453s.
36. Shenton KC, Dowsett M, Lu Q, Brodie A, Sasano H, Sacks NPM, Rowlands MG. Comparison of biochemical aromatase activity with aromatase immunohistochemistry in human breast carcinomas. Breast Cancer Res Treat 1998; 49:S101–S107.

37. Miller WR, O'Neill J. The importance of local synthesis of estrogen within the breast. Steroids 1987; 50:537–548.
38. Makris A, Powles TJ, Allred DC, Ashley SE, Ormerod MG, Titley JC, Dowsett M. Changes in hormone receptors and proliferation markers in tamoxifen treated breast cancer patients and the relationship with response. Br Cancer Res Treat 1998; 48: 11–20.
39. De Friend DJ, Anderson E, Bell J, Wilks DP, West CML, Howell A. Effects of 4-hydroxtamoxifen and a novel pure antiestrogen (ICI 182780) on the clonogenic growth of human breast cancer cells in vitro. Br J Cancer 1994; 70:204–211.
40. Thomas HV, Key T, Allen D, Moore JW, Dowsett M, Fentiman IS, Wang DY. A prospective study of endogenous serum hormone concentrations and breast cancer risk in postmenopausal women on the island of Guernsey. Br J Cancer 1997; 76: 401–405.
41. Cauley JA, Lucas FL, Kuller LH, Vogt MT, Browner WS, Cummings SR. Bone mineral density and risk of breast cancer in older women: the study of osteoporotic fractures. JAMA 1996; 276: 1404–1408.
42. Stein RC, Dowsett M, Hedley A, Coombes RC. The clinical and endocrine effects of 4-hydroxyandrostenedione alone and in combination with goscrelin in premenopausal women with advanced breast cancer. Br J Cancer 1990; 62:679–683.
43. Lønning PE, Dowsett M, Jones A, Ekse D, Jacobs S, MacNeill F, Johannessen DC, Powles TJ. Influence of aminoglutethimide on plasma oestrogen levels in breast cancer patients on 4-hydroxyandrostenedione treatment. Br Cancer Res Treat 1992; 23:57–62.
44. Masamura S, Santner SJ, Heitjan DF, Santen RJ. Estrogen deprivation causes estradiol hypersensitivity in human breast cancer cells. J Clin Endocr Metab 1995; 80: 2918–2925.
45. Chan CMW, Dowsett M. Expression of RIP140, SUG-1, TIF-1 and SMRT in estrogen-hypersensitive MCF-7 cells. Proc Am Assoc Cancer Res 1999; 40:159.

Part II

ADVANCED BREAST CANCER

M. Dowsett, Chair
P. Goss, Chair

2

Role of Hormonal Therapy and Chemotherapy in Advanced Breast Cancer: An Overview

Harold A. Harvey and Kush Sachdeva
Penn State College of Medicine
Hershey, Pennsylvania

I. ABSTRACT

We have made recent significant advances in the management of early-stage breast cancer. However, advanced breast cancer is becoming an increasingly prevalent and challenging clinical problem. Nevertheless, with the introduction of new hormonal agents, including antiestrogens and selective aromatase inhibitors, and the availability of very active cytotoxic drugs, especially taxanes, a greater percentage of patients with advanced breast cancer are now experiencing a prolongation of survival. Several experimental chemotherapy and biological agents show great promise in the treatment of advanced breast cancer. The challenge in the next few years will be to integrate these new treatments into optimal management strategies. This chapter reviews the current status and potential expanded role of systemic hormonal therapy and chemotherapy in advanced breast cancer. It is important to stress the value of appropriate patient selection guided by individual patient preference, defined clinical criteria, and the use of new biochemical predictive factors. It is reasonable to anticipate higher complete response rates in advanced breast cancer and hence the possibility of curing a greater percentage of these patients.

II. INTRODUCTION

Breast cancer worldwide is the most common malignancy afflicting women and is the second most common cause of death from cancer. Despite an increasing incidence of the disease, cause-specific mortality has declined in the last decade in the United States and Europe (1). This encouraging trend is no doubt due to the more widespread and appropriate use of mammographic screening programs in developed countries. Perhaps this trend is also enhanced by the early use of effective systemic adjuvant therapies at the time of initial diagnosis. If the early promise of chemoprevention studies and the ability to screen for the genetic predisposition to develop breast cancer are fulfilled, then it is reasonable to anticipate even more significant decreases in breast cancer mortality. Despite the progress made in early disease, considerable challenges remain with respect to the management of advanced breast cancer. Approximately 20–40% of patients with breast cancer treated initially with curative intent will ultimately develop metastatic disease. Even in developed countries, too many patients still present for the first time with locally advanced disease (T_3, T_4). This clinical situation often reflects a knowledge deficit on the part of the patient or poor access to health care. On occasion, this type of clinical presentation is the result of aggressive and unfavorable tumor biology.

Physicians and patients traditionally regard advanced breast cancer as a uniformly fatal disease process. However, recent published experience from some large institutions and groups indicate an increasing survival rate in patients with advanced breast cancer when treated with systemic therapy. For example, at the MD Anderson Hospital and Tumor Institute, the 5-year survival rates for patients with advanced breast cancer have increased from approximately 5% in the 1950s to 10% in the 1960s and almost 20% in the 1970s (2). The EORTC has reported a 20% 5-year survival in patients who achieved a complete response after chemotherapy (3). The purpose of this chapter is to review the role of hormonal therapy and chemotherapy in the management of advanced breast cancer, to indicate areas of recent progress, and to suggest challenges that need to be solved in the new millennium.

III. HORMONAL THERAPY IN ADVANCED BREAST CANCER

Hormone-dependent breast cancers are estrogen receptor (ER) or progesterone receptor (PgR) positive. Other characteristics that distinguish hormone-dependent breast cancer from the hormone-independent phenotype include a more indolent tumor growth, metastases to favorable sites, and responsiveness to sequential endocrine therapies. In addition, hormone-dependent breast cancer is more prevalent among older patients. Recognition of the clinical and biological features of this subtype of breast cancer enables clinicians better to select patients who are most likely to benefit from endocrine therapy.

Recent advances in the endocrine therapy of breast cancer have been possible as a result of a better understanding of the biology of the ER and how various antiestrogenic compounds can modulate its expression and function. The therapeutic options in the hormonal management of advanced breast cancer depend on the age of the patient and response to prior therapy. Premenopausal women are best treated with ovarian ablation, luteinizing hormone–releasing hormone (LHRH) analogues, or antiestrogens. In postmenopausal women with advanced breast cancer, antiestrogens are the first line of therapy.

Pure antiestrogens are steroidal agents that are analogues of estradiol. They actively bind to the ER but dimerization with the receptor is incomplete and the formation of a functional transcriptional complex is disrupted. These compounds act as ''pure antiestrogens'' and are devoid of estrogenic activity in all tissues.

Among antiestrogenic drugs, tamoxifen has been most studied, but the newer antiestrogen toremifene appears to produce equivalent response rates and duration of clinical benefit (4). Most studies report complete cross resistance between tamoxifen and toremifene. Toxicity of the two agents is similar, and it remains to be proven whether toremifene will in fact be associated with less proliferative effects on endometrial tissue. Faslodex (ICI 182,780) is a parenterally administered pure antiestrogen that seems to be capable of overcoming tamoxifen resistance. Faslodex is currently under study in advanced breast cancer where it is being compared to tamoxifen or aromatase inhibitors (5). Orally administered SCH 57050 is another pure antiestrogen that is also under study.

LHRH analogues are potent agonists which produce a clinical benefit similar to oophorectomy when chronically administered to premenopausal women in the treatment of advanced breast cancer (6). The addition of tamoxifen to LHRH analogues, such as zoladex, may improve the response rate somewhat. New agents that are direct antagonists of LHRH are now being proposed for clinical trial. In postmenopausal women in whom antiestrogen therapy is no longer effective, progestational agents were the most commonly used second-line hormonal therapy. A more modern approach is the use of drugs designed to inhibit aromatase enzyme function, which is the major source of estrogen production in postmenopausal women. Letrozole and anastrazole are examples of highly potent, well-tolerated, orally administered nonsteroidal competitive inhibitors (type II) of aromatase. Formestane and exemestane are analogues of the androgenic substrate androstenedione and cause mechanism-based (type I) irreversible inhibition of the aromatase enzyme. The results of clinical trials which establish the value and positioning of the new aromatase inhibitors in the management of advanced breast cancer in postmenopausal women will be presented later in this book (Chap. 4). In general, and in the absence of direct head-to-head comparisons, the newer antiestrogens and aromatase inhibitors appear to have roughly equivalent efficacy and are well tolerated. Responses to these agents are quite durable, and patients who achieve disease stabilization have a survival that is similar to objective responders.

With the exception of the combination of LHRH analogues and tamoxifen, hormonal therapies are best used in sequence rather than in combination, since few studies show an improved clinical benefit with the latter approach. Patients with metastatic breast cancer who are best suited for hormonal therapy are those with both ER- and PgR-positive tumors, a long disease-free interval, an indolent disease course, and nonvisceral dominant metastases with prior response to an endocrine treatment. Hormonal therapy is generally preferred over chemotherapy in elderly patients and in individuals with significant comorbidities or organ dysfunction such as significant bone marrow, renal, or hepatic impairment. A major challenge for the future will be to understand the mechanisms of hormone resistance in breast cancer and thereby circumvent the problem.

IV. CHEMOTHERAPY IN ADVANCED BREAST CANCER

Compared to many other solid tumors, breast cancer is relatively sensitive to chemotherapy. Several drugs demonstrate clinically useful single-agent activity in advanced breast cancer. Until recently, most clinicians preferred to use combination regimens such as cyclophosphamide, methotrexate, 5-fluorouracil (CMF), cycophosphamide, adriamycin, 5-fluorouracil (CAF); or adriamycin, 5-fluorouracil (AF). Although responding patients clearly benefit from therapy, improvements in overall survival are generally modest. A recent meta-analysis of several reported randomized trials of chemotherapy and hormonal therapy in 31,510 women with advanced breast cancer suggested that multiagent regimens were associated with higher response rates than were seen with monotherapy. Chemotherapeutic regimens containing an anthracycline were generally more active than those that did not. In the meta-analysis, more intensive therapy tended to produce higher response rates but at the same time led to greater toxicity. The meta-analysis reported an improved survival rate with doxorubicin-containing regimens compared to older drug combinations (7).

Several new chemotherapeutic drugs have become available for the treatment of advanced breast cancer. Taxanes are a new class of agent with the greatest degree of activity in breast cancer since anthracyclines were introduced. Several studies show that the taxanes, paclitaxel and docitaxel, produce response rates that range from 20 to 60% in patients with anthracycline-resistant disease when used as single agents in advanced breast cancer. Taxanes are now being used as first-line single-agent therapy in metastatic breast cancer and as integral components of combination regimens with doxorubicin; doxorubicin and cyclophosphamide; and with herceptin (8).

Further issues related to the use and future development of taxanes as treatment for advanced breast cancer include the need to determine optimal dosing

TABLE 1 Strategies to Reduce Chemotherapy Toxicity

Doxorubicin vs. epirubicin, liposomal doxorubicin, continuous infusion
Oral thymidylate synthase inhibitors
Weekly taxanes
Vinorelbine vs. taxane
Oxaliplatin vs. cisplatin

schedules. It is possible that weekly dosing of taxanes would increase dose density, whereas possibly reducing both hematological toxicity and neurotoxicity (9). The therapeutic ratio of new taxanes could be further improved by the development of orally active agents with reduced toxicity and improved efficacy based on structural modifications, allowing the use of higher and prolonged intratumoral drug levels. The mechanisms of taxane drug resistance are poorly understood, and progress in this area will enhance the utility of these agents in breast cancer.

Several new drugs of different classes show promising activity in advanced breast cancer (Table 1), but their role as single agents or their use in polychemotherapeutic regimens will need to be established by clinical trials.

A. High-Dose Chemotherapy

One proposed strategy to overcome drug resistance and increase complete response rates in advanced breast cancer is the use of high-dose chemotherapy. There are several reported trials of dose-intense therapy with growth factor support or autologous bone marrow or peripheral stem cell support. Unfortunately, the benefit of such approaches remains controversial (10). Dose-intense therapy for advanced breast cancer is practical and has proved to be safer than when it was first proposed, but it remains a relatively toxic and expensive undertaking. The lack of a uniformly positive impact on survival indicates the need for more carefully conducted clinical research. Perhaps current strategies are limited by an ability to achieve truly effective, safe levels of dose intensification. Alternatively, it is possible that the currently adopted strategy is flawed, because it does not adequately account for breast cancer kinetics and DNA damage repair mechanisms.

V. PATIENT SELECTION

The treatment selection for an individual patient with advanced breast cancer is usually based on certain clinical criteria and practitioner experience. Factors that could reliably predict response to a particular treatment or class of agents would therefore be of enormous benefit to both physicians and patients. The best analogy is the use of the ER and PgR to predict response to endocrine therapy. There is a growing interest in the recognition and validation of other factors. Although progress is being made, much controversy still exists. For example, the expression of the Her-2/neu receptor protein in breast cancer probably does define a group of patients most likely to benefit from treatment with the monoclonal antibody trastuzumab (Herceptin). Some trials suggest that the expression of Her-2/neu in breast cancer may predict for response to anthracycline-based chemotherapy. Less certain are the claims that Her-2/neu expression predicts for reduced sensitivity to CMF chemotherapy or tamoxifen (11). Other possible predictive factors include p53 mutations in determining resistance to chemotherapy and hormonal therapy, and low thymidylate synthase (TS) tumor levels as a marker for sensitivity to therapy with fluoropyrimidines. Further advances in this area of research hold the promise of better therapy individualization for patients with advanced breast cancer.

Since bone metastases and the resultant skeletal morbidity is such a prominent feature of advanced breast cancer, tests that would predict the likelihood of their occurrence or the response to adjunctive therapies, such as the bisphosphanates, would be of enormous value. Future studies on the significance of PTH-related peptide (PTHrP) expression in breast cancer, blood levels of bone sialoprotein, or urinary levels of bone-specific collagen degradation products (e.g., N-teleopeptide levels) may provide useful predictive information.

Although biological characteristics are important in selecting appropriate therapy for patients with advanced breast cancer, other considerations may be equally important. Patient preferences and their perceptions of therapeutic goals must always be borne in mind and reconciled with the views of the treating physician. Some patients with advanced breast cancer, although aware of their overall prognosis, will accept aggressive therapeutic or experimental approaches for a long time in the hopes of increased longevity and reduction in tumor burden. For these patients, fairly small gains in response rates or duration of benefit are deemed important even at the cost of some toxicity. For other patients, the emphasis will be placed on achieving short-term goals such as surviving until an anniversary or other important personal events. These patients see treatments that relieve symptoms, have low toxicity, and are conveniently administered as a high priority. A recently published survey of 103 patients with advanced cancer indicated a preference for orally administered therapy over parenteral treatments provided there was no significant compromise in efficacy (12). Thus, in selecting treatment for patients with advanced breast cancer, the clinician must be sensitive to both biological and psychosocial issues.

A. Bone Metastases

Skeletal morbidity from bone metastases is a common clinical problem in advanced breast cancer. Large randomized clinical trials indicate that the amino-bisphosphonate pamidronate delays the time to the development of skeletal complications, reducing their rate (especially fractures), the need for radiation therapy, bone pain, and hypercalcemia. These studies also demonstrate that pamidronate can be given together with a variety of hormonal agents and chemotherapeutic regimens (13,14). These results suggest that it would seem prudent to add pamidronate to hormone treatments or chemotherapy once bone metastases are identified in breast cancer patients, and that this supportive measure should be continued for as long as the patient is at high risk of skeletal complications.

B. Advanced Breast Cancer in the Elderly

Elderly patients with advanced breast cancer generally benefit from hormonal therapy. They also tolerate standard doses of most cytotoxic chemotherapy regimens better than is generally supposed. Because of the greater incidence of comorbid conditions in the elderly, certain agents should be used with caution or not at all. For example, renal function is often compromised in the elderly so that greater toxicity might be anticipated with agents such as cisplatinum or methotrexate. Neurotoxicity tends to be more disabling in the elderly, so taxanes and vinca alkaloids may be less attractive in this population. Similarly, drugs such as anthracyclines or herceptin might more easily compromise cardiac function.

Fortunately, there are now good alternatives to several of these agents. Table 2 lists some strategies that could be used to reduce the toxicity of chemotherapy particularly in elderly patients. Frequently, elderly patients with advanced breast cancer have strong preferences to avoid toxic therapies, and their goals and expectations from treatment might be very different from those of younger women. Nevertheless, chronological age should not be the sole reason for withholding certain therapies with the exception of autologous bone marrow transplant. Rather biological or physiological function should help guide the choice of therapy in this age group. The algorithm in Figure 1 is a suggestion for the

TABLE 2 Advanced Breast Cancer: Promising New Agents

Antimetabolites
capecitabine
UFT; other TS inhibitors
gemcitabine
Topo-1 isomerase inhibitors
irinotecan, topotecan
Platinum analogues—oxaliplatin
Anthracyclines
doxil, oral idarubicin

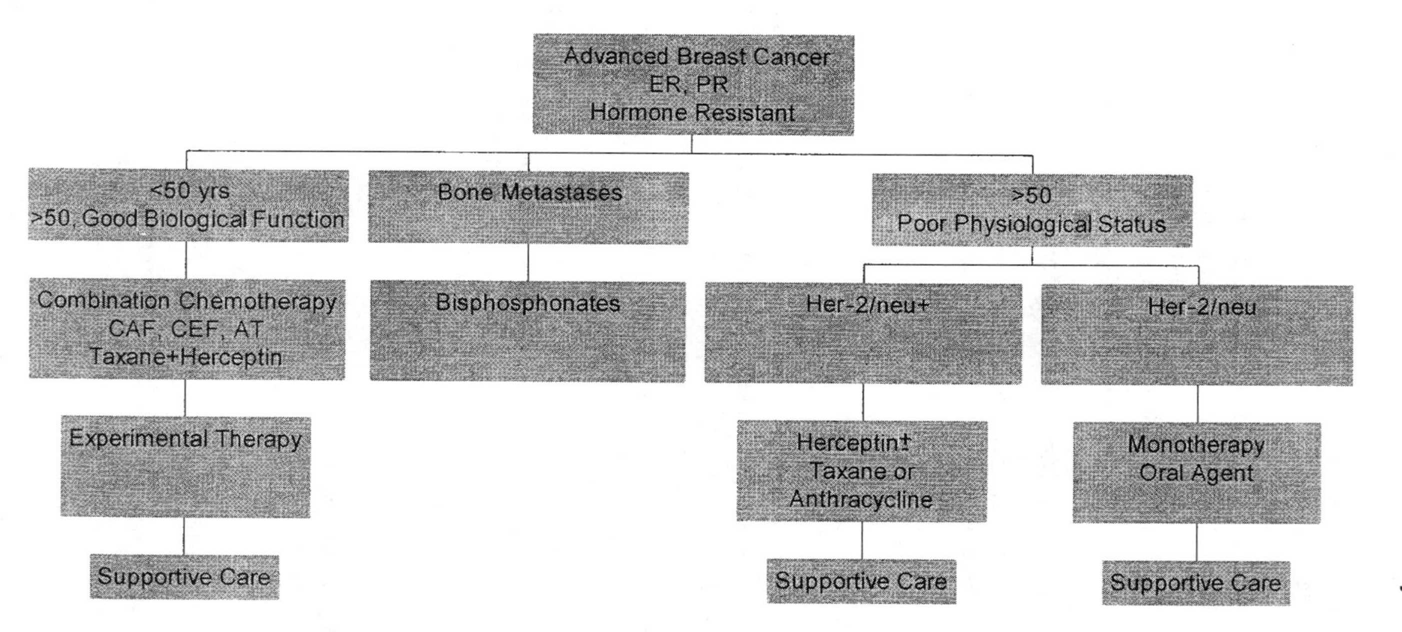

Figure 1 Algorithm for using chemotherapy in advanced breast cancer.

approach to the selection of chemotherapy in patients with advanced breast cancer.

C. Combined Therapy

The combined use of chemotherapy and hormonal therapy in treating advanced breast cancer does not confer a survival advantage. Therefore, this approach should probably be reserved for the rare circumstances of very bulky or life-threatening disease or bone marrow involvement when it is deemed that urgent tumor reduction is in the patient's best interest.

VI. NEW THERAPIES

Progress in the general understanding of tumor biology will eventually lead to the discovery of new potential therapeutic targets. Already some of these targets have been defined and examples include growth factor receptors and ligands, ras and other signal transduction pathways, angiogenesis and metalloproteinases, and cellular antigens suitable for antibody targeting or vaccine therapy. Clinicians in the first decades of the new millennium will be faced with the challenge of choosing therapeutic strategies from among a large number of options. Therefore, we must now develop practical approaches to keep abreast of these changes and integrate them into existing treatment strategies for the optimal management of patients.

VII. CONCLUSIONS

Advanced breast cancer is a prevalent and often chronic disease process. Patients with advanced breast cancer are living longer, with an improved quality of life, using a range of treatments currently available, which include hormonal agents, chemotherapeutic regimens, and adjunctive supportive measures. With the integration of new therapeutic approaches, based on rationally designed and carefully conducted clinical trials, it is indeed likely that prolonged survival in advanced breast cancer will become commonplace. And prolonged survival is but a short step from cure!

REFERENCES

1. Chu KC, Tarone RE, Kessler LG, Ries LAG, Hankey BF, Miller BA, Edwards BK. Recent trends in US breast cancer incidence, survival, and mortality rates. J Natl Cancer Inst 1996; 88:1571–1579.
2. Ross MB, Buzdar AU, Smith TL, Eckles N, Hortobagyi GN, Blumenschien GR, Freireich EJ, Gehan EA. Improved survival of patients with metastatic breast cancer

receiving combination chemotherapy. Comparison of consecutive series of patients in 1950s, 1960s, and 1970s. Cancer 1985; 55:341–346.

3. Tomiak E, Piccart M, Mignolet F, Sahmoud T, Paridaens R, Nooy M, Beex L, Fentiman IS, Muller A, van der Schueren E, Rubens RD. Characterisation of complete responders to combination chemotherapy for advanced breast cancer: a retrospective EORTC Breast Group study. Eur J Cancer 1996; 32A(11):1876–1887.
4. Hayes DF, Van Zyl JA, Hacking A, Goedhals L, Bezwoda WR, Mailliard JA, Jones SE, Vogel CL, Berris RF, Shemano I, Schoenfelder J. Randomized comparison of tamoxifen and two separate doses of toremifene in postmenopausal patients with metastattic breast cancer. J Clin Oncol 1995; 13:2556–2566.
5. Howell A, DeFriend D, Robertson J, Blamey R, Walton P. Response to a specific antiestrogen (ICI 182,780) in tamoxifen-resistant breast cancer. Lancet 1995; 345: 29–30.
6. Taylor CW, Green S, Dalton WS, Martino S, Rector D, Ingle JN, Robert NJ, Budd GT, Paradelo JC, Natale RB, Bearden JD, Mailliard JA, Osborne CK. Multicenter randomized clinical trial of goserelin versus surgical ovariectomy in premenopausal patients with receptor-positive metastatic breast cancer: an intergroup study. J Clin Oncol 1998; 16:994–999.
7. Fossati R, Confalonieri C, Torri V, Ghislandi E, Penna A, Pistotti V, Tinazzi A, Liberati A. Cytotoxic and hormonal treatment for metastatic breast cancer: A systematic review of published randomized trials involving 31,510 women. J Clin Oncol 1998; 16:3439–3460.
8. Nabholtz JM. The role of taxanes in the management of breast cancer. Semin Oncol 1999; 26(3 Suppl 8):1–3.
9. Seidman AD, Hudis CA, Albanel J, Jong W, Tepler I, Currie V, Moynahan ME, Theodoulou M, Gollub M, Baselga J, Norton L. Dose-dense therapy with weekly 1-hour paclitaxel infusions in the treatment of metastatic breast cancer. J Clin Oncol 1998; 16:3353–3361.
10. Rowlings PA, Williams SF, Antman KH, et al. Factors correlated with progression-free survival after high-dose chemotherapy and hematopoietic stem cell transplantation for metastatic breast cancer. J Am Med Assoc 1999; 282:1335–1343.
11. DiGiovanna MP. Clinical significance of HER-2/neu overexpression: Part II. PPO Updates Principles Practice Oncol 1999; 13:101–113.
12. Liu G, Franssen E, Fitch MI, Warner E. Patient preferences for oral versus intravenous palliative chemotherapy. J Clin Oncol 1997; 15:110–115.
13. Hortobagyi GN, Theriault RL, Lipton A, Porter L, Blayney D, Sino C, Wheeler H, Simeone JF, Seaman JJ, Knight RD, Hefferman M, Melk K, Reitsma DJ for the Protocol 19 Aredia Breast Cancer Study Group. Long-term prevention of skeletal complications of metastatic breast cancer with pamidronate. J Clin Oncol 1998; 16: 2038–2044.
14. Theriault RL, Lipton A, Hortobagyi GN, Leff R, Gluck S, Stewart JF, Costello S, Kennedy I, Simeone J, Seaman JJ, Knight RD, Mellars K, Heffernan M, Reitsma DJ. Pamidronate reduces skeletal morbidity in women with advanced breast cancer and lytic bone lesions: a randomized, placebo-controlled trial. Protocol 18 Aredia Breast Cancer Study Group. J Clin Oncol 1999; 17(3):846–854.

3

Place of Aromatase Inhibitors in the Endocrine Therapy of Breast Cancer

Stephen R. D. Johnston, Ian E. Smith, and Mitch Dowsett
The Royal Marsden Hospital
London, England

I. ABSTRACT

The development of orally active, potent, and selective third-generation aromatase inhibitors represents a major advance in the management of hormone-sensitive breast cancer. Randomized clinical trials have established their role as the treatment of choice following tamoxifen failure, with significant gains in clinical efficacy together with improved tolerability over the progestins. In advanced disease, aromatase inhibitor trials are ongoing in the first-line setting against tamoxifen. There are no data to favor combined use with other endocrine agents, although there is much interest in the correct sequencing of aromatase inhibitors (i.e., the possible merits of stepwise versus complete estrogen suppression and steroidal versus nonsteroidal inhibitors).

In terms of which patients benefit most from these new drugs, the enhanced efficacy of recent trials has shown that traditional clinical factors such as soft tissue versus visceral sites of disease and prior response to tamoxifen may be less discriminatory for second-line response to aromatase inhibitors than with previous agents. Expression of estrogen receptor (ER) remains the best determinant of endocrine response, although not all ER-positive tumors are known to be tamoxifen responsive and assessment of ER function/phenotype has been an area of research. Recent studies have shown that ER and progesterone receptor (PgR) expression is often maintained following tamoxifen failure, and that this may be an accurate predictor of response to aromatase inhibitors. It is clear that better patient selection will maximize the chance of response to new aromatase inhibitors. Current research in the primary medical (neoadjuvant) setting has led to analysis of early changes in biomarkers as predictors of subsequent response both to tamoxifen and aromatase inhibitors. In particular, changes in cell proliferation (Ki-67) can be easily measured in repeat fine-needle aspirates and may prove to be highly predictive. Ongoing primary endocrine therapy trials with the new aromatase inhibitors will determine whether biomarkers can accurately predict for response and thus allow us to optimize the use of this current generation of endocrine agents.

II. INTRODUCTION

A significant proportion of breast cancers are estrogen dependent and are therefore amenable to endocrine therapy. Since the original description of the therapeutic response to ovarian ablation over 100 years ago (1), several advances have been made within the last decade in developing new and effective hormone treatments and expanding their role from the management of advanced disease toward adjuvant therapy. Although tamoxifen, a competitive nonsteroidal antiestrogen, has been the mainstay of treatment for over 20 years, new agents have entered the clinic which have potentially superior activity together with an improved safety profile.

As described elsewhere in this book (Chap. 2), aromatase inhibition provides complete estrogen deprivation in postmenopausal women and has recently become established as the second-line endocrine treatment of choice following tamoxifen failure. Aminoglutethimide was a first-generation nonsteroidal aromatase inhibitor (2), and although plasma estrogens were suppressed, the drug had several side effects, including lethargy, skin rash, and reduction in plasma cortisol levels, which necessitated the concomitant use of hydrocortisone (3). Second-generation inhibitors included the nonsteroidal compound fadrozole and the steroidal agent formestane (Fig. 1), both of which were more potent than aminoglutethimide (4,5). However, clinical trials with these agents failed to show superiority over second-line progestin therapy (6,7), and further clinical development was limited by specific problems, including lack of selectivity due to inhibition of aldosterone production (fadrozole) or inconvenient intramuscular route of administration (formestane).

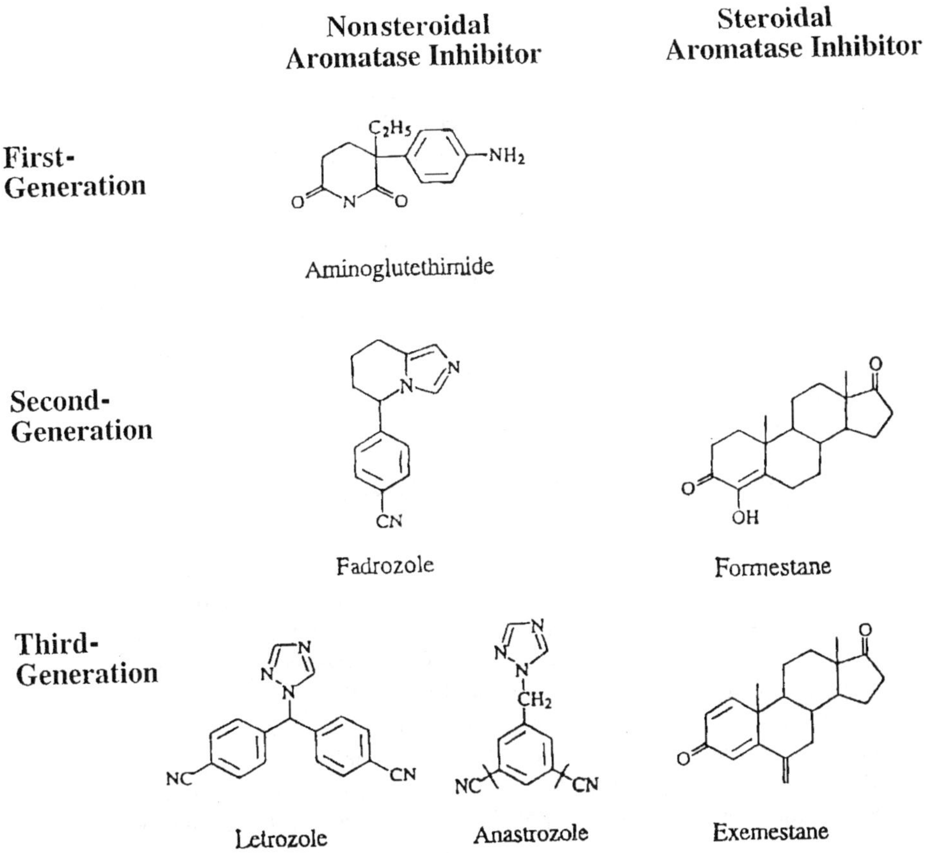

Nonsteroidal Aromatase Inhibitor

Steroidal Aromatase Inhibitor

First-Generation

Aminoglutethimide

Second-Generation

Fadrozole

Formestane

Third-Generation

Letrozole

Anastrozole

Exemestane

FIGURE 1 Structures of nonsteroidal and steroidal aromatase inhibitors.

Recently, considerable clinical progress has been made with the development of third-generation potent oral aromatase inhibitors, including the nonsteroidal inhibitors anastrozole and letrozole together with the steroidal inhibitor exemestane (Fig. 1). These novel agents are two to three orders of magnitude more potent than aminoglutethimide and are very effective in reducing serum estrogen levels in postmenopausal women (8–10). In addition, they are highly selective for the aromatase enzyme without affecting mineralocorticoid or glucocorticoid synthesis. The recent Phase III trials in patients with metastatic breast cancer who failed tamoxifen have compared these agents against previously used second-line therapies (i.e., progestins or aminoglutethimide). The recurring theme from all these trials has been not only superior clinical activity in terms of response rate, time to progression, and overall survival but also the improved tolerability. As a consequence these agents have the potential to make a major impact on the natural history of endocrine-sensitive breast cancer.

The purpose of this chapter is to define the current role for these novel potent oral aromatase inhibitors in the endocrine therapy of breast cancer, particularly in advanced disease. The potential role of aromatase inhibitors in earlier stages of the disease, including primary (neoadjuvant), adjuvant, or chemopreventive therapy, is discussed separately elsewhere in this book. The questions addressed will include when and how to use aromatase inhibitors most effectively, and more importantly whether clinical or biological factors have the potential to predict which patients will benefit most from this new generation of potent inhibitors.

III. ROLE IN ADVANCED BREAST CANCER FOLLOWING TAMOXIFEN

In the metastatic setting, recent Phase III trials have been conducted in over 2000 postmenopausal women comparing each of the third-generation aromatase inhibitors with megestrol acetate as second-line therapy following failure on tamoxifen (11–14). As discussed in Chapter 4, there were differences between each of the studies, which make indirect comparisons between the activity of each aromatase inhibitor difficult. In particular, major differences across the trials were seen in the megestrol acetate control arm both for response rate and median survival, which could be accounted for by differences in response criteria used or in the demographics of the patient populations studied (e.g., estrogen receptor [ER] status, disease sites). In spite of this each of the randomized trials demonstrated clinical superiority for the third-generation aromatase inhibitors over megestrol acetate. For anastrozole 1 mg daily, this was manifested as a significant improvement in overall survival (risk ratio 0.78, $P < .025$) (12), whereas for letrozole 2.5 mg daily, there was a significant improvement in response rate, response duration, and time to treatment failure (13). In the recently reported trial with exemestane, time to disease progression, time to treatment failure, and overall survival were all significantly better than megestrol acetate (14).

These improvements in clinical endpoints, together with the superior tolerability profile shown for each of the three drugs over megestrol acetate, have defined the role for these new third-generation aromatase inhibitors as the standard endocrine treatment following tamoxifen failure. Particularly impressive from some of these trials are the substantial improvements seen in response duration for those who benefited. For example, in the letrozole study, the median response duration was >33 months compared with 18 months for megestrol acetate (13). However, not all patients derived such clinical benefit (less than 50% in each of the three trials), and thus it remains important (as discussed below) to see whether predictive factors can be used to tailor treatments for specifically identified breast cancer patient subgroups.

IV. WHEN AND HOW TO USE AROMATASE INHIBITORS IN ADVANCED DISEASE

The current data have shown substantial improvements in clinical efficacy for the novel aromatase inhibitors when given as second-line therapy. It is obvious, therefore, to ask whether these drugs should challenge tamoxifen as the first-line agent of choice. Large randomized trials are in progress with anastrozole, letrozole, and exemestane, although tamoxifen's widespread use in the adjuvant setting has made recruiting tamoxifen-naïve patients challenging. At this stage, preliminary data are available only from the anastrozole trials (15), one of which shows no difference in primary efficacy endpoints to tamoxifen, whereas the other has reported a significant improvement in time to treatment failure in favor of anastrozole. The great potential of these studies is to see if complete estrogen blockade provides greater control of tumor growth than tamoxifen, thereby circumventing the problem of acquired tamoxifen resistance where a proportion of ER-positive tumor regenerates following an initial response to tamoxifen (16). Several groups, including our own, have established in experimental tumor models that estrogen deprivation provides longer lasting tumor control than tamoxifen (17,18). However, to date randomized clinical trials of tamoxifen versus the early-generation aromatase inhibitors such as aminoglutethimide (19–21), fadrozole (22,23), and formestane (24) have all failed to show improved time to treatment failure/disease progression or prolonged response duration. Thus, it remains to be seen whether the greater potency of new-generation inhibitors will translate into substantial gains in clinical efficacy over tamoxifen as first-line therapy for advanced disease.

An alternative strategy in the past has been to see if endocrine therapy for advanced disease could be improved when aromatase inhibitors were combined with tamoxifen. There are five randomized trials (with a total of 445 patients) which have evaluated tamoxifen alone or in combination with aminoglutethimide (25–29). There was no indication of a therapeutic advantage for the combination, and subsequent studies showed a pharmacokinetic interaction whereby aminoglutethimide significantly lowered tamoxifen concentrations by 70% through induction of hepatic metabolism (30). As a consequence, it has been important to establish whether any similar interaction occurs with the third-generation aromatase inhibitors. Neither anastrozole (31) nor letrozole (32) had any impact on tamoxifen concentrations, although tamoxifen resulted in a 30% reduction in letrozole levels (33). Thus, in advanced disease there is little evidence to support the combined use of aromatase inhibitors with tamoxifen, although it should be noted that in the adjuvant setting such a design remains as one arm in the ATAC trial.

In contrast, there may be some logic to the sequential combination of aromatase inhibitors, which results in stepwise estrogen suppression and the possibility of further prolonged benefit for patients (34). This was first demonstrated in a study of premenopausal women with ER-positive advanced breast cancer who were given formestane in addition to goserelin, a luteinizing hormone–releasing hormone (LHRH), following relapse after an initial response or disease stabilization to goserelin alone (35). Estradiol levels fell further, and the majority

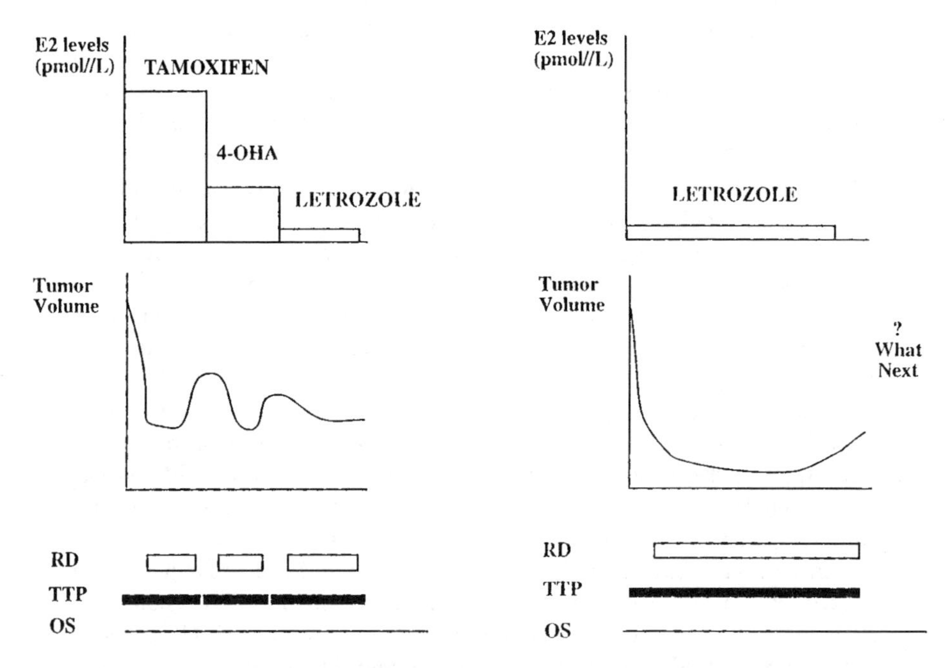

FIGURE 2 A biological rationale for a randomized trial of stepwise estrogen suppression following first-line tamoxifen versus up-front complete estrogen suppression in postmenopausal women with advanced breast cancer. The primary question is whether the degree of tumor shrinkage with complete estrogen suppression would result in longer response duration and time to progression, or alternatively whether stepwise suppression following tamoxifen with a response on each occasion would ultimately provide longer overall control in the natural history of the disease (i.e., survival). RD, response duration; TTP, time to progression; OS, overall survival.

of patients obtained a further remission. In a separate study of 11 patients who progressed on formestane, three achieved a further response when aminoglutethimide was added (36); and it was demonstrated that the degree of aromatase inhibition and estrogen suppression was greater in combination than either agent alone (37). The availability of aromatase inhibitors of differing potency now gives us the opportunity to induce stepwise estrogen suppression. The clinical question remains, however, as to whether overall disease control (i.e., time to treatment failure and survival) would be better achieved by stepwise estrogen suppression with the introduction of novel and more potent aromatase inhibitors at progression or alternatively full and complete estrogen suppression from the start. Now that we have an abundance of inhibitors with differing potency, a randomized trial should be established to investigate whether this is a more effective strategy for the use of aromatase inhibitors in advanced breast cancer (Fig. 2).

V. WHICH DRUG IN ADVANCED DISEASE: STEROIDAL VERSUS NONSTEROIDAL AROMATASE INHIBITOR?

There are emerging data suggesting a lack of cross resistance between the two classes of aromatase inhibitors (steroidal vs. nonsteroidal), giving the option for their use sequentially in advanced breast cancer at the time of progression. Two studies treated patients relapsing on the nonsteroidal inhibitor aminoglutethimide, with formestane 250 mg intramuscularly every 2 weeks, and observed objective response rates of 21 (38) and 10% (39). There have been two recent studies of the potent third-generation steroidal inhibitor exemestane as third-line therapy following failure of a nonsteroidal aromatase inhibitor. The first was a European multicenter study of 200 mg exemestane in 78 postmenopausal women with advanced breast cancer who had progressed on aminoglutethimide, having previously also received tamoxifen (40). The objective response rate was 26%, with a further 13% achieving disease stabilization for >6 months. The median duration of objective response was 59 weeks. In those patients who were resistant to previous treatment with aminoglutethimide (n = 33), 12% had an objective response with exemestane and 15% showed disease stabilization. In the larger multicenter study, patients had also failed on either aminoglutethimide (n = 136) or one of the potent third-generation nonsteroidal inhibitors such as anastrozole, letrozole, or vorozole (n = 105) in addition to tamoxifen (41). The objective response rate was 7%, with an overall clinical benefit rate (CR + PR + NC > 6 months) of 25%. For those patients who failed aminoglutethimide, the objective response and clinical benefit rates were 8 and 27%, respectively, whereas for those failing third-generation nonsteroidal inhibitors, the rates were 5 and 21%, respectively. Responses were more frequent in soft tissue and bone sites of disease, and the median response duration was 58 weeks.

In contrast, a recent study examined the effect of the nonsteroidal inhibitor anastrozole following progression on the steroidal compound formestane, and reported that 5 of the 12 patients who had relapsed on formestane had disease stabilization for >6 months with anastrozole (34). The results of some of these trials could be interpreted as stepwise estrogen suppression due to the greater potency of the second aromatase inhibitor (i.e., anastrozole > formestane > aminoglutethimide). However, in the study of Lønning et al. (41), the potency of the steroidal compound exemestane is similar to nonsteroidal inhibitors anastrozole and letrozole, implying that further clinical activity observed must result from non–cross resistance possibly related to differences in mechanism of action. These data suggest that exemestane in particular may have useful third-line activity and show some lack of cross resistance with nonsteroidal inhibitors.

VI. WHO BENEFITS FROM AROMATASE INHIBITORS—CLINICAL FACTORS

Clinical predictors of response to second-line endocrine therapy in advanced disease have traditionally included factors such as soft tissue versus visceral sites of disease, long disease-free interval, and prior response to tamoxifen. The enhanced efficacy of the more potent aromatase inhibitors, including evidence for activity in groups of patients previously deemed to have a low chance of endocrine response, means that some of these traditional clinical factors may be less of a discriminator. Subset analyses of the recent Phase III trials with third-generation aromatase inhibitors (11–14) have demonstrated significant efficacy in sites of visceral disease for aromatase inhibitors compared with progestins. For example, the clinical benefit (CR, PR, and stabilization of disease [SD]) seen in liver metastases was much greater with anastrozole than megestrol acetate (26 vs 17%), with a corresponding longer median duration of benefit of 545 days and 302 days (42). Equally, the objective response rate (CR and PR) in visceral metastases was greater for exemestane versus megestrol acetate (14 and 11%, respectively) (43); and for letrozole against megestrol acetate, the values were 16 and 8%, respectively (44). As a consequence, the presence of asymptomatic visceral metastases in those with hormone-sensitive breast cancer should no longer be the sole reason to favor chemotherapy over effective endocrine therapy.

Prior sensitivity to tamoxifen in advanced disease has been a clinical factor often cited as a predictor for the likelihood of response to a further endocrine agent (16). Patients who have received tamoxifen for advanced disease may be categorized as responders if they show an objective response (CR or PR) or have stabilization of disease for at least 6 months (SD). When these patients subsequently relapse, they are often described as having developed acquired (or secondary) resistance. Nonresponders typically progress straight through tamoxifen and are deemed to have de novo (or primary) resistance. Those who relapse on

TABLE 1 Response to Aromatase Inhibitors in Phase III Trials and Influence of Prior Response to Tamoxifen

Phase III Trial	Tamoxifen responder (%)	Tamoxifen nonresponder (%)	Tamoxifen adjuvant failure (%)
Anastrozole	16	5	12
Megestrol acetate	11	9	12
Letrozole	33	29	14
Megestrol acetate	10	15	16
Exemestane	17	12	13
Megestrol acetate	15	17	8

Source: From Refs. 11, 13, 14.

adjuvant tamoxifen cannot be categorized in the same way, as no prior response to tamoxifen can be ascertained; although it is generally accepted that prolonged adjuvant therapy (>2–5 yrs) and subsequent relapse is biologically similar to acquired resistance, whereas relapse after only 6–12 months of tamoxifen is more likely to represent de novo resistance. It is well recognized that patients who were previously sensitive to tamoxifen and then developed acquired resistance are more likely to respond to further endocrine therapy. We previously reported from our own historical series of studies with various aromatase inhibitors that 70% of tamoxifen responders had an objective response (CR/PR) or SD to second-line aromatase inhibitors compared with less than 15% who had shown de novo resistance to tamoxifen (45). However, in the recent randomized trials, especially with letrozole and exemestane, significant response rates have also been seen in so-called tamoxifen nonresponders (Table 1), although the definition of nonresponse is not always clear. The enhanced efficacy of these agents may mean that a benefit can be seen in patients deemed to have shown no initial sensitivity to tamoxifen. If this is true, one might expect significantly higher response rates for aromatase inhibitors in the randomized first-line studies versus tamoxifen, although no evidence has been seen for this to date (15).

VII. ER STATUS—PREDICTOR OF ENDOCRINE RESPONSE TO AROMATASE INHIBITORS

The presence of a functional ER in breast cancer implies a hormone-sensitive tumor which is dependent on estrogen for growth. Thus, ER-positive tumors should be highly sensitive to estrogen-deprivation therapy in comparison with tumors which are completely ER negative. The ER status of the primary tumor is now known to be the strongest predictor of response to adjuvant tamoxifen in the latest world overview (46), with patients in whom the ER status is negative deriving very little benefit. In the primary medical setting where tamoxifen has been used as initial therapy in elderly patients, ER expression has been proved to be the best predictor of response (47,48). In the Edinburgh series, the quantitative value of ER became a significant discriminator for response in such patients as determined by ultrasound after 3 months of tamoxifen (49). In particular, all tumors with an ER $>$ 200 fmol/mg protein responded, whereas tumors with ER values below this were equally likely to respond or not. As discussed later, other factors which might determine the function of an ER-positive tumor are required to improve the discrimination of moderate ER values in predicting response to tamoxifen.

In terms of similar data for predicting response to aromatase inhibitors, there have been recent studies both with anastrozole and letrozole in the neoadjuvant setting (49,50), which are described in more detail in Chapter 6. In these studies, the response rates and median reduction in tumor volume by ultrasound of primary ER-positive cancers appear much higher than previous patients (non-randomized) treated with tamoxifen. Patients were selected on the basis of their tumor being ER positive (biochemical score $>$20 fmol/mg protein or immunohistochemical H-score $>$80) (50). It remains to be seen whether the quantitative ER score correlates with the degree of tumor shrinkage in these studies, and can thus be a more accurate predictor of response to aromatase inhibitors than tamoxifen.

It is clear that ER status (i.e., ER positive or negative), rather than quantitative ER value, remains the most logical and reliable predictor of response to endocrine therapy both in the primary (neoadjuvant) and adjuvant settings. In contrast, when patients develop recurrent metastatic disease, it has been a concern as to whether hormone sensitivity due to the presence of ER in the original primary tumor is lost over time. This could account for lower response rates seen with tamoxifen in advanced metastatic breast cancer compared with recent studies when given as primary medical therapy, as it is recognized that 30–40% of metastases from original ER-positive primary breast cancer fail to respond to endocrine therapy. However, several studies have confirmed that a significant number of patients retain ER expression when they develop recurrent metastatic disease (51–53). Knowledge of the change in ER status may be more helpful in predicting response to endocrine therapy, in the metastatic setting, than relying on the ER status of the previous primary tumor. A Finnish study compared the ER status of recurrent tumors (both at locoregional and distant sites of metastases) with the primary breast cancer in 50 patients who had not received intervening adjuvant therapy (51). The median time between primary tumor diagnosis and recurrence was 25 months. Estrogen receptor status was concordant in 65% of those developing breast or nodal recurrences and in 57% of those developing distant metastases (sites sampled included skin nodules, liver, lung, and bone secondar-

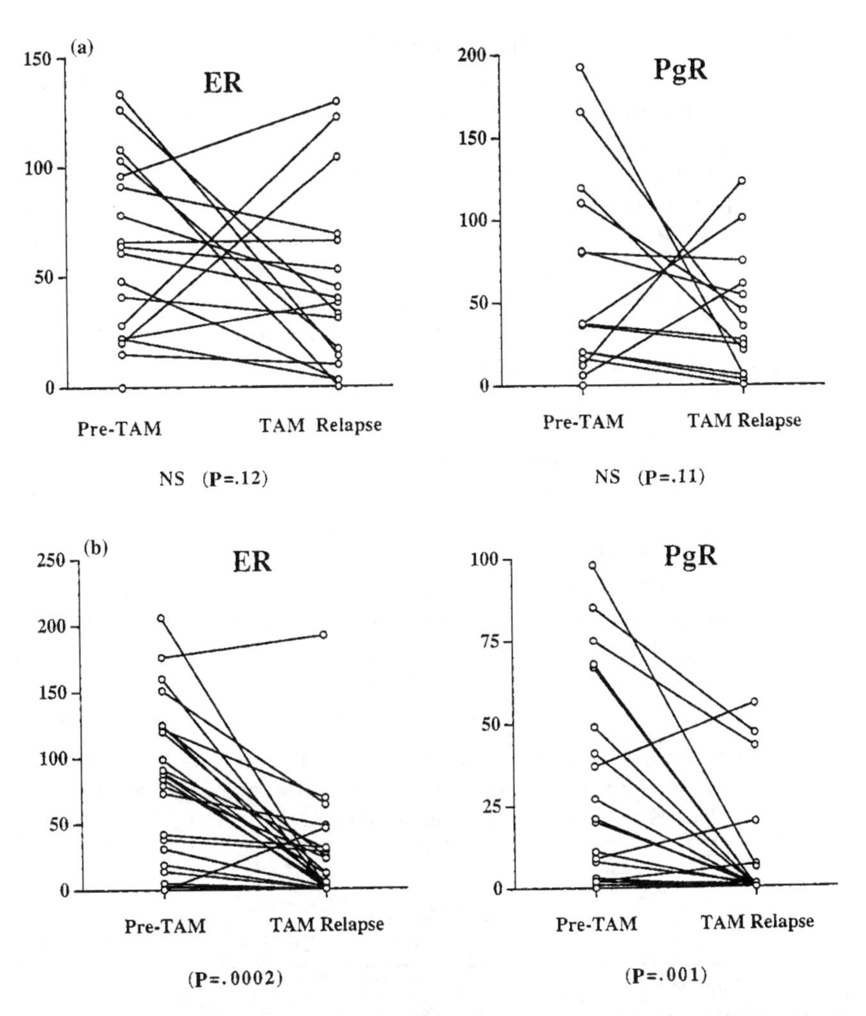

FIGURE 3 Change in ER/PgR expression following failure of tamoxifen given in either the adjuvant setting (n = 34) or in patients treated for advanced disease who initially responded and subsequently relapsed with acquired resistance (n = 18). Tumor samples included pretamoxifen core-cut biopsies (advanced disease) or primary tumor (adjuvant therapy), whereas tamoxifen relapse samples included repeat core-cut biopsies, locoregional recurrences which were excised, or metastatic skin nodules. ER and PgR were assessed by validated immunohistochemical assays and reported as H-scores (0–300). (From Ref. 53.)

ies). Moreover, this proved to be an effective predictor of response to first-line tamoxifen, which was then given for advanced disease; of those who remained ER positive at relapse, 74% responded to tamoxifen compared with only 13% in whom ER expression was lost. Thus, loss of ER with the development of advanced disease was a highly significant predictor ($P = .008$) of poor response to subsequent endocrine therapy (51).

Although ER status may be retained in many tumors during progression to advanced disease, it was envisaged that tamoxifen given in the adjuvant or first-line metastatic setting may significantly modify ER expression, with clonal selection of ER-negative tumors at relapse (54). The known response rates to aromatase inhibitors following tamoxifen failure would argue against that. We examined the role of intervening tamoxifen given in the adjuvant or first-line setting on preservation of ER status from the primary tumor to the relapsed sample in 72 patients who developed tamoxifen resistance (53). Overall we confirmed that ER expression was often retained, although quantitative values were reduced most frequently owing to a reduction in the percentage of cells staining positive for ER. The trend for reduction in ER score was most obvious in those relapsing on adjuvant tamoxifen after a median of 25 months, whereas those who initially responded to tamoxifen for advanced disease and then developed acquired resistance often retained high ER scores (Fig. 3).

TABLE 2 Second-Line Response to Either Third-Generation Nonsteroidal Aromatase Inhibitor (Anastrozole, Letrozole, or Vorozole) or Progestins in a Series of 29 Postmenopausal Women with Advanced Breast Cancer*

		ER Status	
	2nd line response	ER/PgR^+	ER/PgR^-
Aromatase inhibitor			
Tamoxifen responder	CR/PgR	5	—
	PD	—	2
Tamoxifen nonresponder	PD	—	2
Progestin			
Tamoxifen responder	CR/PR	3	2
	PD	2	1
Tamoxifen nonresponder	PgR	—	1
	PD	—	1

* All had previously received tamoxifen as first-line therapy for advanced disease, and second-line endocrine response is expressed in relation to prior tamoxifen response and change in ER/PgR status as assessed in a tumor biopsy taken at relapse on tamoxifen (i.e., just prior to commencement of second-line therapy).

Our subsequent data suggest that preservation of ER at relapse on tamoxifen may prove the best predictor of response to second-line endocrine therapy, especially to aromatase inhibitors (55). Of the 72 patients in the series above, there were 29 who had received tamoxifen as first-line therapy for advanced disease and then went on to receive second-line endocrine therapy with either third-generation aromatase inhibitors (anastrozole, letrozole, vorozole) or progestins (medroxyprogesterone acetate). Estrogen receptor status at relapse on tamoxifen was a significantly better predictor of response than prior response to tamoxifen (Table 2). In particular, tumors which were ER or PgR positive at progression on tamoxifen all responded to aromatase inhibitors, whereas in those which were completely ER or PgR negative, no responses were seen. In contrast, ER expression at relapse was less discriminatory for second-line response to progestins. These data are entirely consistent with the mechanism of action for aromatase inhibitors, and they suggest that, if possible (accepting that many times it is not feasible), an analysis of ER expression in recurrent tumor samples would provide a very accurate predictor of response to second-line aromatase inhibitors in advanced disease.

VIII. PREDICTING RESPONSE TO ENDOCRINE THERAPY—BIOLOGICAL STUDIES

It is clear that appropriate patient selection will maximize the chance of response to the new aromatase inhibitors. To date, clinical factors together with the ER status have been the only information available to clinicians, and, as discussed above, there are some limitations in their usefulness. Additional tumor factors, which might characterize hormone-sensitive versus hormone resistant tumors, might improve on the predictive power of ER status alone for response to endocrine therapy. In particular, the ER phenotype manifested by coexpression of estrogen-dependent proteins such as PgR, pS2, and bcl-2 is thought to represent a functional receptor pathway, but in the Edinburgh series of primary endocrine therapy, these provided no significant added value in predicting response to tamoxifen (49). Other independent tumor biomarkers such as p53, c-erbB2, and epidermal growth factor receptor (EGFR) are known to be associated with resistance to tamoxifen (56,57), although this in itself may relate to the fact that the majority of such tumors are known to be ER negative. Other assays of ER function which might improve its predictive power have included DNA binding assays, which in our experience remained positive in tumors with acquired tamoxifen resistance (58). To date, none of these additional biological factors significantly improves on the predictive power of tumor ER status alone.

A change in tumor biomarker expression following the introduction of systemic therapy has become an active area of research, and it may provide a more accurate predictor of tumor response to both chemotherapy and endocrine therapy. In particular, much recent interest has focused on changes in tumor cell proliferation and induction of apoptosis (59). Such measurements can be made in serial fine-needle aspirate cytology (FNAC) samples or repeat core-cut biopsies, and their potential to predict for response has been uniquely exploited in clinical trials of primary (neoadjuvant) therapy. In relation to endocrine therapy with tamoxifen or aromatase inhibitors, our group at the Royal Marsden Hospital has explored two different clinical scenarios. These were either short-term (2 weeks) exposure prior to surgery or sequential sampling during a 3-month preoperative assessment enabling tumor response by ultrasound also to be assessed (59).

Our early studies with tamoxifen clearly demonstrated that a fall in tumor cell proliferation was significantly associated with response to tamoxifen. In a series of 19 patients, those responding showed a significant fall in Ki-67 during 1–3 months (mean Ki-67 score 28.0 ± 8.4% pretreatment falling to 9.2 ± 4.8% posttreatment), whereas there was no consistent change in nonresponders (32.8 ± 15.8% pretreatment to 20.0 ± 10.4% posttreatment (60). Likewise, the Edinburgh group reported on a series of elderly patients receiving primary tamoxifen for 3 months in whom the majority of responding tumors demonstrated a fall in the proliferation marker Ki-S1, whereas nonresponding tumors displayed either no change or an increase in staining (61). More recently, our group has studied changes in proliferation markers after just 14 days of treatment in patients who then continued to receive tamoxifen therapy for up to 3 months in whom response could be assessed (62). Using sequential FNAC with proliferation assessed on cytospin preparations, the early change in these biomarker data from 22 patients (11 responders and 11 nonresponders) clearly demonstrate the ability strongly to predict subsequent response to tamoxifen (Fig. 4).

Similar biomarker studies are now being undertaken with aromatase inhibitors. In a small series of patients treated for 14 days with the second-generation steroidal inhibitor 4-hydroxyandrostenedione (formestane) prior to surgery, we have demonstrated in ER-positive tumors that there was a significant fall in proliferation which was not observed in ER-negative tumors (Fig. 5). Response could not be correlated in this study, but, at the Royal Marsden Hospital, ongoing randomized trials of 3 months of therapy have been undertaken in 53 postmenopausal women looking at the third-generation aromatase inhibitor vorozole versus tamoxifen (63). In addition, a large 400-patient study of anastrozole versus tamoxifen versus combined tamoxifen + anastrozole (IMPACT trial) is also ongoing. In these two trials, early and sequential biomarker studies should allow us to determine with greater confidence the ability to predict for response to aromatase inhibitor therapy through a fall in cell proliferation, induction of apoptosis, or downregulation of ER-dependent protein expression. Likewise, in the Edinburgh study of primary letrozole, a marked reduction in proliferation (Ki-67 staining) was noted in all 24 cases (96% of patients showed a > 25% reduction in tumor

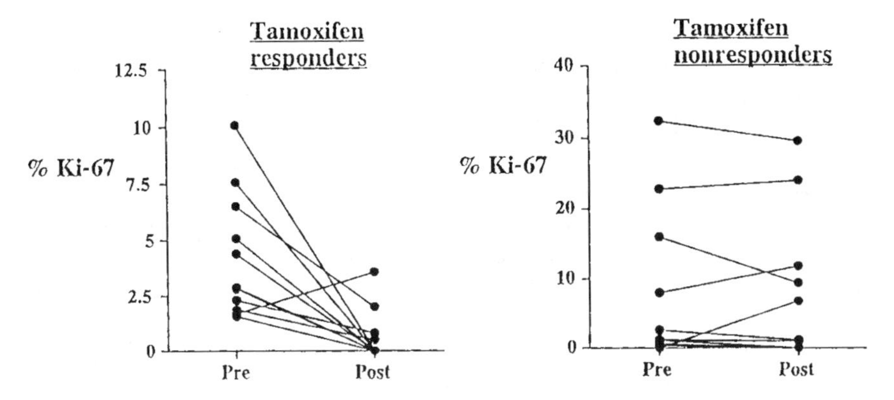

FIGURE 4 Change in cell proliferation as determined by Ki-67 (%) score in postmenopausal women with primary breast cancer after 14 days of treatment with tamoxifen in 11 women who subsequently responded over 3 months and in 11 who were nonresponders. Immunocytochemical measurements were made on cytospin preparations made from paired pretreatment and posttreatment fine-needle aspirates (62). In all but one patient, who then responded to tamoxifen, there was a significant fall in Ki-67 within 14 days, whereas, in contrast, there was no fall observed in those who were subsequent nonresponders.

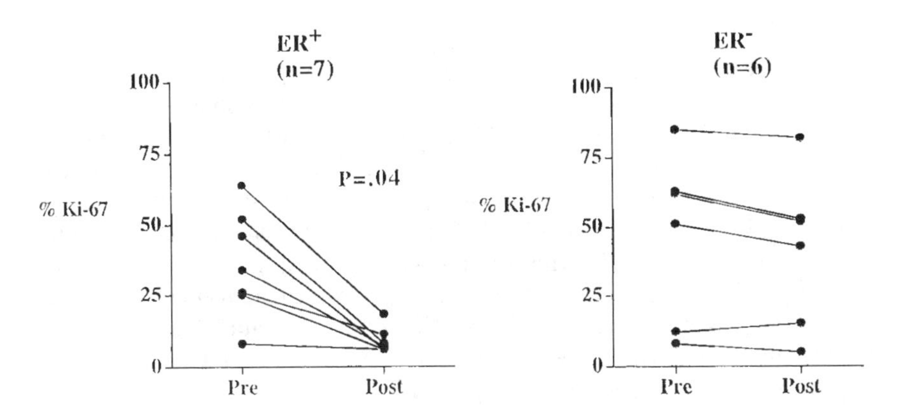

FIGURE 5 Change in cell proliferation as determined by Ki-67 (%) score in 11 postmenopausal women with primary breast cancer treated preoperatively with the second-generation steroidal aromatase inhibitor 4-hydroxyandrostenedione (formestane). Immunocytochemical measurements were made on cytospin preparations made from paired pretreatment and posttreatment fine-needle aspirates (62). In those with ER-positive tumors, there was a significant reduction in Ki-67 after 14 days, whereas no change was seen in those with ER-negative tumors.

volume). This contrasts with less consistent changes in proliferation seen in the historical cases treated with tamoxifen where the response rate and median tumor volume reduction was less (49). It remains to be seen whether, in the context of the current randomized trials of aromatase inhibitors versus tamoxifen, early changes in proliferation provide a more consistent predictor of response to aromatase inhibitors than tamoxifen.

At present, these research studies remain in the primary tumor setting where they can be conducted in a controlled and rigorous manner. If early (i.e., 2 weeks) changes can be reliably detected (i.e., sensitivity), and they have the ability accurately to predict response (i.e., specificity), then sequential FNAC samples could become of value for patients with accessible locally advanced/metastatic disease, or indeed in the preoperative setting, in determining the most appropriate adjuvant endocrine therapy. Advances in technology, especially the arrival of microarrays which can determine gene expression patterns in sequential samples, may significantly improve this approach. At present, our research has chosen fairly crude biomarkers (i.e., ER, PgR, Ki-67, apoptosis), and it is possible that their sensitivity and specificity may be limited. By analyzing changes in gene expression and detecting patterns (i.e., genes which are inactivated or switched on in response to endocrine therapy), our hope is to identify more reliable markers, whereas enhancing our understanding of the mechanisms of tumor response/resistance to current potent endocrine agents.

IX. CONCLUSIONS

The third-generation aromatase inhibitors have resulted in a significant advance in endocrine therapy of breast cancer. Their role is clearly established as second-line therapy following tamoxifen failure, with the consistent finding being that they significantly enhance tumor response duration, time to progression, and overall survival. If this efficacy is transferable to first-line therapy, then there is great optimism that the current trials in first-line therapy and the adjuvant setting against tamoxifen may yield further substantial gains in outcome for patients. This will be significantly helped if we can develop accurate methods to predict positive response to aromatase inhibitors. To this end, the current interest in biomarker studies is providing early promising data that may allow us to optimize the use of this new generation of drugs now reaching the clinic.

REFERENCES

1. Beatson GT. On the treatment of inoperable cases of carcinoma of the mamma: suggestions for a new method of treatment with illustrative cases. Lancet 1896; 2: 104–107.
2. Santen RJ, Samojlik E, Lipton A, Harvey R, Ruby EB, Wells SA, Kendall J. Kinetic, hormonal and clinical studies with aminoglutethimide in breast cancer. Cancer 1977; 39:2948–2958.

3. Harris AL, Dowsett M, Cantwell BM, Sainsbury JR, Needham G, Farndon J, Wilson R. Endocrine effects of low-dose aminoglutethimide with hydrocortisone—an optimal hormonal suppressive regimen. Breast Cancer Res Treat 1986; 7:69–72.
4. Dowsett M, Stein RC, Mehta A, Coombes RC. Potency and selectivity of the nonsteroidal aromatase inhibitor CGS 16949A in postmenopausal breast cancer patients. Clin Endocrinol 1990; 32:623–624.
5. Coombes RC, Goss P, Dowsett M, Gazet JC, Brodie A. 4-hydroxyandrostenedione in treatment of postmenopausal women with advanced breast cancer. Lancet 1984; 2:1237–1239.
6. Buzdar AU, Smith R, Vogel C, Bonomi P, Keller AM, Favis G, Mulagha M, Cooper J. Fadrozole HCL (CGS-16949A) versus megestrol acetate in treatment of postmenopausal patients with metastatic breast cancer. Cancer 1996; 77:2503–2513.
7. Thurlimann B, Castiglione M, Hsu-Schmitz SF, Cavalli F, Bonnefoi H, Fey MF, et al. Formestane versus megestrol acetate in postmenopausal breast cancer patients after failure of tamoxifen; a phase III prospective randomized cross-over trial of second-line hormonal treatment (SAKK 20/90). Eur J Cancer 1997; 33(7):1017–1024.
8. Plourde PV, Dyroff M, Dukes M. Arimidex; a potent and selective fourth generation aromatase inhibitor. Breast Cancer Res Treat 1994; 30:103–111.
9. Iveson TJ, Smith IE, Ahern J, Smithers DA, Trunet PF, Dowsett M. Phase I study of the oral nonsteroidal aromatase inhibitor CGS 20267 in postmenopausal patients with advanced breast cancer. Cancer Res 1993; 53:266–270.
10. Evans TRJ, Di Salle E, Ornati G, Lassus M, Benedetti MS, Pianezzola E, Coombes RC. Phase I endocrine study of exemestane (FCE 24304), a new aromatase inhibitor, in postmenopausal women. Cancer Res 1992; 52:5933–5939.
11. Buzdar A, Jonat W, Howell A, Jones SE, Blomqvist C, Vogel CL, Eiermann W, Wolter M, Azab M, Webster A, Plourde PV. Anastrozole, a potent and selective aromatase inhibitor, versus megestrol acetate in postmenopausal women with advanced breast cancer: results of overview analysis of two phase III trials. J Clin Oncol 1996; 14:2000–2011.
12. Buzdar A, Jonat W, Howell A, Jones SE, Blomqvist C, Vogel CL, Eiermann W, Wolter M, Steinberg M, Webster A, Lee D. Anastrozole versus megestrol acetate in the treatment of postmenopausal women with advanced breast carcinoma: results of a survival update based on a combined analysis of data from two mature phase III trials. Cancer 1998; 83:1142–1152.
13. Dombernowsky P, Smith IE, Falkson G, Leonard R, Panasci L, Bellmunt J, Bezwoda W, Gardin G, Gudgeon A, Morgan M, Fornasiero A, Hoffmann W, Michel J, Hatschek T, Tjabbes T, Chaudri HA, Hornberger U, Trunet P. Letrozole, a new oral aromatase inhibitor for advanced breast cancer: double-blind randomized trial showing a dose-effect and improved efficacy and tolerability compared with megestrol acetate. J Clin Oncol 1998; 16:453–461.
14. Kaufmann M, Bajetta E, Dirix LY, Fein LE, Jones SE, Cervek J, et al. Survival advantage of exemestane over megestrol acetate in postmenopausal women with advanced breast cancer refractory to tamoxifen; results of a phase III randomized double-blind study. Proc Am Soc Clin Oncol 1999; 18:109 (A412).
15. Thuerlimann BJK, Nabholtz JM, Bonneterre J, Buzdar AW, Robertson JFR, Webster A, et al. The preliminary results of two comparative multi-center clinical trials com-

paring the efficacy and tolerability of Arimidex (anasrozole) and tamoxifen in postmenopausal women with advanced breast cancer. Breast 1999; 8:214 (A004).

16. Johnston SRD. Acquired tamoxifen resistance; potential mechanisms and clinical implications. Anti-Cancer Drugs 1997; 8:991–930.
17. Osborne CK, Coronado EB, Robinson JP. Human breast cancer in the athymic nude mouse: cytostatic effects of long-term anti-estrogen therapy. Eur J Cancer Clin Oncol 1987; 23(8):1189–1196.
18. Johnston SRD, Boeddinghaus IM, Riddler S, Haynes BP, Grimshaw R, Hardcastle I, Jarman M, Dowsett M. Idoxifene antagonizes estradiol-dependent MCF-7 breast cancer xenograft growth by sustained induction of apoptosis. Cancer Res 1999; 59: 3332–3339.
19. Gale KE, Anderson JW, Tormey JC, Mansor EG, Davis TE, Horton J, Wolter JM, Smith TJ, Cummings FJ. Hormonal treatments for metastatic breast cancer, an Eastern Cooperative Oncology Group Phase III trial comparing aminoglutethimide to tamoxifen. Cancer 1994; 73:354–361.
20. Lipton A, Harvey HA, Santen RJ, Boucher A, White D, Bernath A, Dixon R, Richards G, Shafik A. Randomized trial of aminoglutethimide versus tamoxifen in metastatic breast cancer. Cancer Res 1982; 42:3434–3436.
21. Smith IE, Harris AL, Morgan M. Tamoxifen versus aminoglutethimide versus combined tamoxifen and aminoglutethimide in the treatment of advanced breast carcinoma. Cancer Res 1982; 42:3430–3433.
22. Falkson CI, Falkson HC. A randomized study of GCS 16949A (fadrozole) versus tamoxifen in previously untreated postmenopausal patients with metastatic breast cancer. Ann Oncol 1996; 7:465–469.
23. Thuerlimann B, Beretta K, Bacchi M, Castiglioni-Gertsch M, Goldhirsch A, Jungi WF, Cavalli F, Senn HJ, Fey M, Lohnert T. First-line fadrozole HCL (CGS 16949A) versus tamoxifen in postmenopausal women with advanced breast cancer. Ann Oncol 1996; 7:471–479.
24. Perrez-Carrion R, Alberola Candel V, Calabresi F. Comparison of the selective aromatase inhibitor formestane with tamoxifen as first-line hormonal therapy in postmenopausal women with advanced breast cancer. Ann Oncol 1994; 5(Suppl 7):19–24.
25. Rose C, Kamby C, Mouridsen HT, Bastholt L, Brincker H, Skovgaard-Poulsen H, Andersen AP, Loft H, Dombernowsky P, Andersen KW. Combined endocrine treatment of postmenopausal patients with advanced breast cancer. Breast Cancer Res Treat 1986; 7:45–50.
26. Ingle JN, Green SJ, Ahmann DL, Long HJ, Edmonson JH, Rubin J, Chang MN, Creagan ET. Randomised trial of tamoxifen alone or combined with aminoglutethimide and hydrocortisone in women with metastatic breast cancer. J Clin Oncol 1986; 4:958–964.
27. Alonso-Munoz MC, Ojeda-Gonzalez MB, Beltran-Fabregat M, Dorca-Ribugent J, Lopez-Lopez L. Randomized trial of tamoxifen versus aminoglutethimide and versus combined tamoxifen and aminoglutethimide in advanced postmenopausal breast cancer. Oncology 1988; 45:350–353.
28. Milsted R, Habeshaw T, Kaye S, Sangster G, Macbeth F, Campbell-Ferguson J, Smith D, Calman K. A randomized trial of tamoxifen versus tamoxifen with amino-

glutethimide in postmenopausal women with advanced breast cancer. Cancer Chemother Pharmacol 1985; 14:272–273.
29. Corkery J, Leonard RCF, Henderson IC, Gelman RS, Hourihan J, Ascoli DM, Salhanick HA. Tamoxifen and aminoglutethimide in advanced breast cancer. Cancer Res 1982; 42:3409–3414.
30. Lien EA, Anker G, Lønning PE, Solheim E, Ueland PM. Decreased serum concentrations of tamoxifen and its metabolites induced by aminoglutethimide. Cancer Res 1990; 50:5851–5857.
31. Dowsett M, Tobias J, Howell A, Blackman GM, Welch H, King N, Ponzone R, Von Euler M, Baum M. The effects of anastrozole on the pharmacokinetics of tamoxifen in postmenopausal women with early breast cancer. Br J Cancer 1999; 79:311–315.
32. Ingle JN, Suman VJ, Jordan VC, Dowsett M. Combination hormonal therapy involving aromatase inhibitors in the management of women with breast cancer. Endocr Rel Cancer 1999; 6:265–269.
33. Dowsett M, Pfister CU, Johnston SRD, Houston SJ, Miles DW, Verbeek JA, Smith IE. Pharmacokinetic interaction between letrozole and tamoxifen in postmenopausal patients with advanced breast cancer. Breast 1997; 6:245.
34. Coombes RC, Harper-Wynne C, Dowsett M. Aromatase inhibitors and their use in sequential setting. Endocr Rel Cancer 1999; 6:259–263.
35. Stein RC, Dowsett M, Hedley A, Coombes RC. The clinical and endocrine effects of 4-hydroxyandrostenedione alone and in combination with goserelin in premenopausal women with advanced breast cancer. Br J Cancer 1990; 62:679–683.
36. Geisler J, Johannessen DC, Anker G, Lønning PE. Treatment with formestane alone and in combination with aminoglutethimide in heavily pre-treated cancer patients: clinical and endocrine effects. Eur J Cancer 1996; 32A:789–792.
37. MacNeil FA, Jones AL, Jacobs S, Lønning PE, Powles TJ, Dowsett M. Combined treatment with 4-hydroxyandrostenedione and aminoglutethimide; effects on aromatase inhibition and oestrogen suppression. Br J Cancer 1994; 69:1171–1175.
38. Murray R, Pitt P. Aromatase inhibition with 4-hydroxyandrostenedione after prior aromatase inhibition with aminoglutethimide in women with advanced breast cancer. Breast Cancer Res Treat 1995; 35:249–253.
39. Lønning PE, Dowsett M, Jones A, Ekse D, Jacobs S MacNeil, F., Johannessen DC, Powles TJ. Influence of aminoglutethimide on plasma estrogen levels in breast cancer patients on 4-hydroxyandrostenedione treatment. Breast Cancer Res Treat 1992; 23:57–62.
40. Thurlimann B, Paridaens R, Senn D, Bonneterre J, Roche H, Murray R, Consonni A, Lanzalone S, Polli A, Arkhipov A, Piscitelli G. Third-line hormonal treatment with exemestane in postmenopausal patients with advanced breast cancer progressing on aminoglutethimide; a phase II multi-center multinational study. Eur J Cancer 1997; 33:1767–1773.
41. Lønning PE, Bajetta E, Murray R, Tubiana M, Eisenberg PD, Mickiewicz E, Fowst C, Arkhipov A, Polli A, Di Salle E, Massimini GA. Phase II study of exemestane in metastatic breast cancer failing non-steroidal aromatase inhibitors. Breast Cancer Res Treat 1998; 50:304 (A435).
42. Howell A, Buzdar A, Jonat W. Arimidex (anastrozole)—effective in advanced

breast cancer patients with visceral and liver metastases. Breast Cancer Res Treat 1998; 50:304 (A434).

43. Tedeschi M, Kvinnslan S, Jones SE, Kaufmann M, Polli A, Fowst C, et al. Hormonal therapy in breast cancer and predominant visceral disease: effectiveness of the new oral aromatase inhibitor, Aromasin (exemestane) in advanced breast cancer patients having progressed on anti-oestrogens. Eur J Cancer ECCO 10 Proc 1999(A1270): (A1270).
44. Smith IE, et al. Efficacy of letrozole and other endocrine therapies in advanced breast cancer with visceral metastases (manuscript in preparation).
45. Dowsett M, Johnston SRD, Iveson TJ, Smith IE. Response to pure anti-estrogen (ICI 182, 780) in tamoxifen-resistant breast cancer. Lancet 1995; 345:525.
46. Early Breast Cancer Trialists Group. Tamoxifen for early breast cancer; an overview of the randomised trials. Lancet 1998; 351:1451–1467.
47. Low SC, Dixon AR, Bell J. Tumor oestrogen receptor content allows selection of elderly patients with breast cancer for conservative tamoxifen treatment. Br J Surg 1992; 79:1314–1316.
48. Gaskell DJ, Hawkins RA, Sangster K, Chetty U, Forrest APM. Relation between immunoctyochemical estimation of estrogen receptor in elderly patients with primary breast cancer and response to tamoxifen. Lancet 1992; 1:1044–1046.
49. Miller WR, Anderson TJ, Hawkins RA, Keen J, Dixon JM. Neoadjuvant endocrine treatment; the Edinburgh experience. In: Howell A, Dowsett M, eds. European School of Oncology Update Vol 4—Primary Medical Therapy for Breast Cancer. Amsterdam: Elsevier, 1991:89–99.
50. Dixon JM, Love CDB, Renshaw L, Bellamy C, Cameron DA, Miller WR, Leonard RCF. Lessons from the use of aromatase inhibitors in the neoadjuvant setting. Endocr Rel Cancer 1999; 6:227–230.
51. Kuukasjarvi T, Kononen J, Helin H, Holli K, Isola J. Loss of estrogen receptor in recurrent breast cancer is associated with poor response to endocrine therapy. J Clin Oncol 1996; 14:2584–2589.
52. Encarnacion CA, Ciocca DR, McGuire WL, Clark GM, Fuqua SAW, Osborne CK. Measurement of steroid hormone receptors in breast cancer patients on tamoxifen. Breast Cancer Res Treat 1993; 26:237–246.
53. Johnston SRD, Saccani-Jotti G, Smith IE, Salter J, Newby J, Coppen M, Ebbs SR, Dowsett M. Changes in estrogen receptor, progesterone receptor and pS2 expression in tamoxifen-resistant human breast cancer. Cancer Res 1995; 55:3331–3338.
54. Howell A, DeFriend D, Anderson E. Mechanisms of response and resistance to endocrine therapy for breast cancer and the development of new treatments. Rev Endocr Rel Cancer 1993; 43:5–21.
55. Johnston SRD, Saccani-Joti G, Ebbs SR, Smith IE, Dowsett M. ER expression at relapse predicts for response to second-line endocrine therapy. Br J Cancer 1997; 7 5(S1):16 (A6.3).
56. Wright C, Nicholson S, Angus B, Sainsbury JRC, Farndon JR, Cairns J, Harris AL, Horne CHW. Relationship between c-erb-B2 protein expression and response to endocrine therapy in advanced breast cancer. Br J Cancer 1992; 65:118–121.
57. Sainsbury JRC, Farndon JR, Sherbet G, Harris AL. Epidermal growth factor receptors and oestrogen receptors in human breast cancers. Lancet 1985; 1:364–366.

58. Johnston SRD, Lu B, Dowsett M, Liang X, Kaufmann M, Scott GK, Osborne CK, Benz CC. Comparison of estrogen receptor DNA-binding in untreated and acquired anti-estrogen resistant human breast tumors. Cancer Res 1997; 57:3723–3727.
59. Dowsett M, Smith IE, Powles TJ, Salter J, Ellis PA, Johnston SRD, Makris A, Mainwaring P, Gregory RK, Archer C, Chang J, Asserhon L. Biological studies in primary medical therapy of breast cancer; the Royal Marsden Hospital Experience. In: Howell A, Dowsett M, eds. European School of Oncology Vol 4—Primary Medical Therapy of Breast Cancer. Amsterdam: Elsevier, 1999:113–125.
60. Dowsett M, Johnston SRD, Detre S. Cytological evaluation of biological variables in breast cancer patients undergoing primary medical treatment. In: Motta M, Serio M, eds. Sex Hormones and Anti-Hormones in Endocrine Dependent Pathology. Amsterdam: Elsevier, 1994:329–336.
61. Keen JC, Dixon JM, Miller WR. Expression of Ki-S1 and Bcl-2 and the response to primary tamoxifen therapy in elderly patients with breast cancer. Breast Cancer Res Treat 1997; 44:123–134.
62. Makris A, Powles TJ, Allred DC, Dowsett M. Changes in hormone receptors and proliferation markers in tamoxifen treated breast cancer patients and the relationship with response. Breast Cancer Res Treat 1998; 48:11–20.
63. Harper-Wynne C, Shenton K, Dowset M, MacNeil F, Sauven P, Laidlaw I, et al. A randomized multi-center study of vorozole compared to tamoxifen as primary therapy in postmenopausal breast cancer. Proc Am Soc Clin Oncol 1999; 18:72 (A272).

4

Clinical Overview of Aromatase Inhibitors

Ian E. Smith
The Royal Marsden Hospital
London, England

I. INTRODUCTION

Aminoglutethimide, the first clinically useful aromatase inhibitor, was shown in the late 1970s to have clinical efficacy in patients with advanced breast cancer (1,2). Progress in the development of aromatase inhibitors since then has been slow and steady rather than dramatic.

Aminoglutethimide itself was shown to be at least as effective as tamoxifen (3–5), and it was also shown to have potential as adjuvant therapy (6). Aminoglutethimide never established itself as a major front-line treatment for advanced breast cancer, principally because of side effects, including rash and dose-related drowsiness or lethargy.

Gradually so-called ''second-generation'' aromatase inhibitors emerged, including formestane and fadrozole. Formestane, a highly selective, steroidal aromatase inhibitor, is similar in structure to androstenedione, the naturally occurring substrate for estrogen synthesis. A series of Phase II noncomparative studies established that formestane has good antitumor efficacy (7) combined with a good side effect profile. Fadrozole is a nonsteroidal imidazole aromatase inhibitor many times more potent than aminoglutethimide (8). Its clinical development was interrupted in many countries by the demonstration that serum aldosterone levels were significantly suppressed by doses of 1 mg twice daily and greater (9).

These in turn have now been replaced by the ''third-generation'' aromatase inhibitors, including letrozole, anastrozole, vorozole, and exemestane. Letrozole, anastrozole, and vorozole are highly potent nonsteroidal triazole derivatives with minimal side effects. Fadrazole and anastrozole are 1000 times more potent, and letrozole is said to be in excess of 10,000 times more potent than aminoglutethimide (10). Letrozole and anastrozole have established roles in the treatment of advanced breast cancer and are under investigation as adjuvant therapy; vorozole, despite also being highly active, has been withdrawn from development by the manufacturer. Exemestane is a steroidal irreversible aromatase inhibitor similar to formestane but with the advantage of oral administration. It is under study both for advanced disease and as adjuvant therapy.

Each generation of aromatase inhibitor has achieved a greater degree of inhibition than its predecessor, such that 98–99% aromatase inhibition can now be achieved in patients using third-generation agents (11). In addition side effects have been minimized, such that these agents are now among the least toxic used in cancer medicine.

In this chapter, the key clinical trials involving aromatase inhibitors in advanced breast cancer will be reviewed. Most of these have compared modern aromatase inhibitors with megestrol acetate or aminoglutethimide as second-line endocrine therapy after tamoxifen failure. Some trials have also compared aromatase inhibitors as first-line endocrine therapy against tamoxifen.

II. TRIALS OF AROMATASE INHIBITORS AS SECOND-LINE THERAPY

Randomized Phase III trials of aromatase inhibitors as second-line endocrine therapy against megestrol acetate or aminoglutethimide are listed in Table 1. These provide by far the largest body of data on second-line endocrine therapy in advanced breast cancer currently available. Despite this there are limitations with these trials which prevent meaningful inter-trial comparisons.

First, the selection of the primary endpoint varied between trials. These include response rate, time to progression (an endpoint of limited value in endocrine therapy given that more than 50% of patients do not respond and rapidly progress), clinical benefit (a better endpoint based on complete response, partial response, and no change for >6 months; however, this endpoint has not been accepted in any published response evaluation criteria), response duration, and survival. Second, only the trials of letrozole versus megestrol acetate were double blinded. Third, follow-up varied between trials, with differences between the frequency and nature of imaging techniques. Finally, entry criteria varied between

TABLE 1 Randomized Phase III Trials of Aromatase Inhibitors as Second-Line Therapy

Trials versus megestrol acetate (40 mg × 4 daily)			
Fadrozole	1.0 mg orally bd	2 trials	(9)
Anastrozole	1.0 mg orally daily		
	10.0 mg orally daily	2 trials	(10)
Letrozole	0.5 mg orally daily		
	2.5 mg orally daily		(11)
Vorozole	2.5 mg orally daily		(12)
Exemestane	25 mg orally daily		(13)
Trials versus aminoglutethimide (250 mg orally bd)			
Letrozole	0.5 mg orally daily		(14)
	2.5 mg orally daily		
Vorozole	2.5 mg orally daily		(15)

trials, allowing the potential for populations of different prognostic outlook to emerge, in particular variable endocrine sensitivity, which is reflected in a comparison of control arm outcomes (Table 2).

A. Second-Line Aromatase Inhibitors Versus Megestrol Acetate

1. Fadrozole Versus Megestrol Acetate

The second-generation imidazole aromatase inhibitor fadrozole, dosed at 1 mg twice daily, has been compared with megestrol acetate 40 mg four times daily in two randomized double-blind multi-institutional trials reported by Buzdar in-

TABLE 2 Differences in Control Arm Outcomes in Phase III Aromatase Inhibitor Trials

	Response rate (%)	Survival (months)
Anastrozole vs. megestrol acetate	8	23 (10)
Letrozole vs. megestrol acetate	16	22 (11)
Vorozole vs. megestrol acetate	8	29 (12)
Exemestane vs. megestrol acetate	12	28 (13)
Letrozole vs. aminoglutethimide	12	20 (14)
Vorozole vs. aminoglutethimide	18	22 (15)

volving 672 patients (9). Objective response rates in the two trials (fadrozole data first) were 11 versus 16% and 13 versus 11%; no change was seen in 25 versus 20% and 24 versus 30%. None of these differences was significant. Likewise, there was no significant difference in time to treatment progression, response duration, or survival.

2. Anastrozole Versus Megestrol Acetate

Anastrozole (Arimidex; Zeneca) is a third-generation nonsteroidal triazole-selective aromatase inhibitor which has been compared at two doses (1 and 10 mg once daily) with megestrol acetate (40 mg four times daily) in two large trials in the treatment of postmenopausal women with advanced breast cancer whose disease had progressed after treatment with tamoxifen. These two multicenter trials were identical in design, double blind for dose of anastrozole, and open label for megestrol acetate. Their results have been presented as a combined analysis involving a total of 764 patients (13). Median follow-up for survival was 31 months.

Objective response rate was 10% for anastrozole 1 mg, 9% for anastrozole 10 mg, and 8% for Megace. Anastrozole 1 mg daily showed a significant survival advantage over megestrol acetate with a risk ratio of 0.78 ($P < .025$). Anastrozole 10 mg also showed a trend toward a survival benefit over megestrol acetate with a risk ratio of 0.83 ($P = .09$). Anastrozole was associated with less weight gain than megestrol acetate. The trial, therefore, demonstrated a significant survival advantage for anastrozole over megestrol acetate, but it did not show any dose-response effect between the two anastrozole doses.

3. Letrozole Versus Megestrol Acetate

A large Phase III multinational multicenter randomized trial has been carried out comparing letrozole 0.5 mg against letrozole 2.5 mg against megestrol acetate 40 mg four times daily (14).

In this trial involving 551 patients with locally advanced metastatic breast cancer in postmenopausal patients previously treated with antiestrogens, results were as follows. The objective response rate for letrozole 2.5 mg of 23.6% was significantly higher than letrozole 0.5 mg at 12.8% ($P = .04$) and megestrol acetate at 16.4% ($P = .02$). The median duration of objective response was not reached for letrozole 2.5 mg compared with 18 months for letrozole 0.5 mg and 18 months for megestrol acetate ($P = .02$).

Clinical benefit (including both patients with objective response and stable disease for ≥6 months) was 35% for letrozole 2.5 mg, 27% for letrozole 0.5 mg, and 32% for megestrol acetate (no significant difference). Median duration of clinical benefit was 23.5 months for letrozole 2.5 mg, 18.0 months for letrozole 0.5 mg, and 14.5 months for megestrol acetate.

Median survival was 25 months for letrozole 2.5 mg compared with 21.5 months for letrozole 0.5 mg, which was statistically significant ($P = .03$). Median survival for megestrol acetate was also 21.5 months.

4. Vorozole Versus Megestrol Acetate

Vorozole, a nonsteroidal triazole aromatase inhibitor whose development has been discontinued (see earlier), has been compared with megestrol acetate in a Phase III randomized multicenter trial whose results have only been presented in abstract form (12). A total of 452 postmenopausal patients previously treated with tamoxifen were randomized to vorozole 2.5 mg orally daily or megestrol acetate 40 mg four times daily. Response rate was 9.7% for vorozole compared with 6.8% for megestrol acetate, and respective durations of response were 18.2 and 12.5 months. Despite the trend toward improved duration of response, none of these differences was significant, and vorozole had no demonstrated advantage over megestrol acetate in terms of efficacy or overall survival. Nevertheless, it was better tolerated, specifically in terms of weight gain (15).

5. Exemestane Versus Megestrol Acetate

Exemestane, a steroidal irreversible aromatase inhibitor, has been compared in a dose of 25 mg/day with megestrol acetate 40 mg orally four times daily in a Phase III randomized trial of 769 postmenopausal patients with advanced breast cancer refractory to tamoxifen (13). Objective response rate was 15% for exemestane versus 12% for megestrol acetate and respective overall clinical benefit 37% versus 25%. Median duration of response was 17.5 versus 16.3 months, respectively, and median duration of overall benefit 13.8 versus 11.3 months ($P = .025$). Median survival has been reported as being significantly prolonged for exemestane (not reached) versus megestrol acetate (28.4 months) ($P = .039$).

B. Newer Aromatase Inhibitors Versus Aminoglutethimide

1. Letrozole Versus Aminoglutethimide

Letrozole in two doses (0.5 g or 2.5 mg once daily) was compared with aminoglutethimide 250 mg twice daily with corticosteroid support in a Phase III multicenter trial in 555 postmenopausal women with advanced breast cancer previously treated with antiestrogens (17). Objective response rates were 19.5% for letrozole 2.5 mg and 16.7% for 0.5 mg and 12.4% for aminoglutethimide. These differences were not quite significant. The respective median durations of response were 24 months versus 21 months versus 15 months.

Median duration of clinical benefit (response + stable disease for at least 6 months) was 21 months for letrozole 2.5 mg, 18 months for letrozole 0.5 mg, and 14 months for aminoglutethimide. Overall survival was 28 months for letrozole 2.5 mg versus 21 months for 0.5 mg, and this difference was statistically significant ($P = .04$). Letrozole 2.5 mg was likewise superior to aminoglutethimide with a median survival of 20 months ($P = .002$).

2. Vorozole Versus Aminoglutethimide

Vorozole is the only other third-generation aromatase inhibitor to be assessed directly against aminoglutethimide as second-line endocrine therapy in a multicenter Phase III trial involving 556 patients. Response rate for vorozole 2.5 mg was 23% compared with 18% for aminoglutethimide ($P = .07$). Respective response durations were 21 months versus 20 months (not significant) and overall survivals 26 months versus 22 months (not significant). Quality of life was reported as better on vorozole ($P = .014$) and drug-related side effects were significantly less (31% vs 53%; $P < .001$). In particular, this involved a lower incidence of rash and somnolence. These data have so far only been presented in abstract form (18).

III. FIRST-LINE ENDOCRINE THERAPY: AROMATASE INHIBITORS VERSUS TAMOXIFEN

A. Aminoglutethimide Versus Tamoxifen

In 1981, the first randomized trial, of which we are aware, of an aromatase inhibitor against another endocrine agent was published; 117 patients with advanced breast cancer were randomized to receive aminoglutethimide orally four times daily with hydrocortisone 25 mg twice daily or tamoxifen 20 mg orally daily (3). The great majority of patients had received no previous endocrine therapy, but 13 patients had undergone oophorectomy. Objective responses were seen in 30 patients in each group, and median response duration was likewise 15 months for each treatment. Tamoxifen had fewer side effects. In a subsequent analysis no significant survival differences emerged. In two further randomized studies similar results were obtained confirming the suggested equivalence between tamoxifen and aminoglutethimide (Table 3) (4,5).

B. Fadrozole Versus Tamoxifen

Fadrozole has been compared with tamoxifen as first-line treatment of advanced or metastatic breast cancer in postmenopausal patients in two randomized Phase III trials.

In the first, carried out in South Africa, 80 patients were randomized to fadrozole 1 mg twice daily or tamoxifen 20 mg orally daily (19). Response rates were 48% for fadrozole and 43% for tamoxifen. Median response duration was 343 days for fadrozole and was greater than this with the median not being

TABLE 3 Phase III Trials of Aromatase Inhibitors Versus Tamoxifen as First-Line Therapy for Advanced Disease

Aminoglutethimide	250 mg orally 4 × daily with HC	(3)
Fadrozole	1 mg orally bd	(16)
	1 mg orally bd	(17)
Formestane	250 mg im every 2 weeks	(18)
Anastrozole (Trial 0030)	1 mg orally daily	(19)
Anastrozole (Trial 0027)	1 mg orally daily	(19)
Exemestane	25 mg orally daily	EORTC[a]

[a] In progress.

reached for tamoxifen (P = .009). Survival was likewise significantly increased for tamoxifen (34 months) compared with fadrozole (26 months) (P = .046).

In the second trial, carried out in Switzerland, 105 patients were randomized to receive fadrozole 1 mg orally twice daily compared with 107 patients to tamoxifen 20 mg orally twice daily (20). Response rate was 20% for fadrozole compared with 27% for tamoxifen. Response rate was 15 months versus 20 months, respectively (no significant difference), and no significant difference was seen in survival. There was almost a significant benefit in time to treatment failure for tamoxifen, 8.5 months, compared with fadrozole, 6 months (P = .05).

C. Formestane Versus Tamoxifen

A Phase III randomized trial of formestane 250 mg every 2 weeks by intramuscular injection versus tamoxifen 30 mg orally daily was carried out in 409 postmenopausal patients with advanced breast cancer, none of whom had received previous endocrine therapy for advanced disease (21). Only 348 were evaluable for response; 33% had an objective response to formestane compared with 37% to tamoxifen (P = .48). Median response was 15 months for formestane compared with 20 months for tamoxifen (P = .17) and median survival was 35 and 38 months, respectively (P = .64). Time to disease progression was, however, significantly longer for tamoxifen (294 days) than for formestane (213 days) (P = .01). Tolerance was excellent for both treatments with no significant differences.

D. Anastrozole Versus Tamoxifen

Two randomized Phase III trials have recently been carried out comparing anastrozole to tamoxifen in postmenopausal women with advanced breast cancer who had received no previous endocrine therapy for metastatic disease and had stopped adjuvant tamoxifen treatment for at least 12 months (22).

The first, carried out in the United States and Canada (Trial 0030), involved 353 patients randomized to anastrozole 1 mg versus tamoxifen 20 mg. Of these, 21% had an objective response to anastrozole compared with 17% to tamoxifen. In the anastrozole group, 59% achieved a clinical benefit (objective response + stable disease for at least 6 months) compared with 46% for tamoxifen. Median time to disease progression was 11 months for anastrozole compared with 5.6 months for tamoxifen ($P = .005$), with a risk ratio of 1.44. There was a higher incidence of vaginal bleeding and thromboembolic disease with tamoxifen than with anastrozole, but both drugs were otherwise well tolerated.

In the second trial, carried out worldwide (Trial 0027), 33% achieved an objective response to both anastrozole and to tamoxifen, and clinical benefit was achieved in 56% in both groups. Median time to disease progression was identical for both groups at 8.3 months. Thromboembolic events occurred in 7.3% patients on tamoxifen compared with 4.8% with anastrozole; treatment was otherwise well tolerated and no other significant differences emerged.

The reason for the apparent discrepancy between the two trials in which one but not the other showed a significant improvement in time to disease progression with anastrozole compared with tamoxifen have been considered by the trialists. Approximately around 90% of patients in Trial 0030 were confirmed as estrogen receptor (ER)–positive and/or progesterone receptor (PgR)–positive compared with only 45% in Trial 0027. This implies the likelihood of there being a higher proportion of patients with unknown receptor status who would in fact be ER negative and therefore nonresponders to endocrine therapy in Trial 0027. A subgroup analysis of Trial 0027 in patients confirmed as receptor positive again showed a clear separation of time to progression curves in favor of anastrozole, although the benefit did not reach that observed in Trial 0030. There is, therefore, a hypothesis generated by this trial that true ER-positive patients achieve greater clinical benefit with anastrozole than with tamoxifen.

Another significant result to come out of the two studies is the apparent reduction of thromboembolic events in patients treated with anastrozole. This is important for two reasons. First, in the metastatic setting, hormone treatments are preferable in the frail, elderly in whom thromboembolic events may be a particular problem due to other predisposing factors such as reduced mobility. Second, there is increasing interest in the use of anastrozole in the adjuvant setting; obviously a lower incidence of this potentially fatal complication is of particular desirability in patients who have potentially received curative treatment.

E. Exemestane Versus Tamoxifen

A Phase II randomized trial of exemestane versus tamoxifen as first-line hormonal therapy in postmenopausal ER- or PR-positive patients with locally recurrent or metastatic breast cancer is planned by the EORTC with a projected accrual of 100 patients; no data are available so far.

IV. OTHER ISSUES

A. Letrozole in the Treatment of Visceral Metastases

It is generally assumed that visceral metastases in advanced breast cancer are better treated with chemotherapy than with endocrine therapy. In the two Phase III letrozole trials against megestrol acetate and aminoglutethimide, the opportunity arose to assess efficacy against visceral metastases. In the trial against megestrol acetate, 73 patients with predominantly visceral metastases were randomized to letrozole 2.5 mg. Sixteen percent of the letrozole group achieved a response compared with 8% in 77 patients randomized to megestrol acetate. Median response durations were 33 versus 15 months, and median clinical benefit durations were 26 versus 11 months ($P = .02$), both in favor of letrozole.

In the parallel trial against aminoglutethimide, 17% of 90 patients with predominantly visceral metastases randomized to receive letrozole 2.5 mg achieved response compared with only 3% of 71 patients randomized to receive aminoglutethimide ($P = .01$). Response duration was 38 months versus 9 and 24 months for the two responding aminoglutethimide patients. Respective response rates for clinical benefit were 31 versus 30%, and median duration of clinical benefit was 24 versus 14 months ($P = .009$).

These limited data indicate that letrozole, and perhaps other aromatase inhibitors, can achieve useful clinical benefit of more than 2 years median in approximately 25% of patients with visceral metastases. The data also provide further evidence that letrozole is superior to the much less potent aromatase inhibitor aminoglutethimide.

B. Is There a Dose-Response Effect for Letrozole?

An unexpected finding in both these trials was the small but significant increase in clinical efficacy with letrozole 2.5 mg over 0.5 mg. This is despite the fact that no further plasma estrogen suppression or in vivo aromatase inhibition has been demonstrated with the higher dose (11,23). If the observation is real, then how can this result be explained?

It has been postulated that the letrozole data may indicate the role of intratumoral aromatase inhibition as a more important and sensitive parameter than plasma inhibition or estrogen suppression. If this is the case, then why is a further dose-response effect not seen with other aromatase inhibitors? The answer to this may lie in the structure/function relationship between the drug and enzyme. Letrozole is certainly a more powerful aromatase inhibitor than aminoglutethimide or fadrozole, and noncomparative data suggest that it is probably biochemically more active than anastrozole. Likewise, there is a similar suggestion from noncomparative clinical data that the response rates in the two Phase III letrozole trials were significantly higher than those for anastrozole, which may not simply be based on patients selection. The hypothesis would, therefore, be that letrozole, because of a better "fit" with the aromatase enzyme complex, can achieve higher intratumoral inhibition than anastrozole, which continues below detectable further plasma biochemical changes. Noncomparative data suggest that, in this respect, vorozole is more similar to letrozole than to anastrozole, and it is a pity that similar dose-response studies have not been carried out with vorozole. Further data are required to determine whether this unpredictable observation of a dose-response effect can be confirmed.

V. CONCLUSIONS

The following conclusions can be drawn from these clinical trials:

1. Aromatase inhibitors are active agents in the endocrine treatment of patients with advanced breast cancer and, with the exception of aminoglutethimide, they all have a very low toxicity profile.
2. The currently available third-generation agents, letrozole, anastrozole, and exemestane, all show relatively small but statistically significant clinical benefit over megestrol acetate as second-line endocrine therapy for advanced disease, although the specific clinical benefit parameters vary for the different trials. In addition, they all have better side effect profiles and are to be recommended in preference to megestrol acetate.
3. Where tested, third-generation aromatase inhibitors (letrozole and vorozole) have been shown to have significantly greater clinical benefit than the much less potent parent aromatase inhibitor aminoglutethimide.
4. In the context of first-line endocrine therapy, a small early trial suggested that aminoglutethimide had equivalent, but not superior, clinical activity to tamoxifen. Trials of the second-generation aromatase inhibitors, fadrozole and formestane, suggest a slight clinical superiority for tamoxifen. Trials of the third-generation aromatase inhibitor anastrozole provocatively suggest that this might be clinically more active than tamoxifen and further data are required here. This issue is of interest in the context of adjuvant trials now running comparing third-generation aromatase inhibitors with tamoxifen.
5. The letrozole trials emphasize that a small, but clinically important, group of patients with visceral metastases achieve clinical benefit with letrozole (~25%). The clinical importance of this lies principally in the median duration of clinical benefit which is around 2 years and well in excess of that normally achieved with chemotherapy.

REFERENCES

1. Santen RJ, Samojlik E, Lipton A, Harvey H, Ruby EB, Wells SA, Kendall J. Cancer 1977; 39:2948.
2. Smith IE, Fitzharris BM, McKinna JA, Fahmy DR, Nash AG, Neville AM, Gazet

J-C, Ford HT, Powles TJ. Aminoglutethimide in treatment of metastatic breast carcinoma. Lancet 1978; 23:646–649.
3. Smith IE, Harris AL, Morgan M, Ford HT, Gazet J-C, Harmer CL, White H, Parsons CA, Villardo A, Walsh G, McKinna JA. Tamoxifen versus aminoglutethimide in advanced breast carcinoma: a randomized cross-over trial. BMJ 1981; 283:1432–1434.
4. Lipton A, Harvey HA, Santen RJ, Boucher A, White D, Bernath A, Dixon R, Richards G, Shafik A. A randomised trial of aminoglutethimide versus tamoxifen in metastatic breast cancer. Cancer. 1982; 50(11):2265–2268.
5. Gale KE, Andersen JW, Tormey DC, Mansour EG, Davis TE, Horton J, Wolter JM, Smith TJ, Cummings FJ. Hormonal treatment for metastatic breast cancer. An Eastern Cooperative Oncology Group Phase III trial comparing to aminoglutethamide Cancer. 1994; 73(2):354–361.
6. Coombes RC, Powles TJ, Easton D, Chilvers C, Ford HT, Smith IE, McKinna A, White H, Bradbeer J, Yarnold J, Nash A, Bettelheim R, Dowsett M, Gazet J-C. Adjuvant aminoglutethimide therapy for postmenopausal patients with primary breast cancer. Cancer Res 1987; 47:2496–2499.
7. Stein RC, Dowsett M, Hedley A, et al. Treatment of advanced breast cancer in postmenopausal women with 4-hydroxyandrostenedione. Cancer Chemother Pharmacol 1990; 26:75–78.
8. Steele RF, Mellor LB, Sawyer WK, Wasvary JM, Browne LJ. In vitro and in vivo studies demonstrating potent and selective estrogen inhibition with the nonsteroidal aromatase inhibitor CGS 16949A. Steroids 1987; 50:147–161.
9. Dowsett M, Stein RC, Mehta A, Coombes RC. Potency and selectivity of the nonsteroidal aromatase inhibitor CGS 16949A in postmenopausal breast cancer patients. Clin Endocrinol 1990; 32:623–624.
10. Bhatnagar AS, Hausler A and Schieweck K. Inhibition of aromatase in vitro and in vivo by aromatase inhibitors. J Enzyme Inhib 1990; 4:179–186.
11. Iveson TJ, Smith IE, Ahern J, Smithers DA, Trunet PF, Dowsett M. Phase I study of the oral nonsteroidal aromatase inhibitor CGS 20267 in postmenopausal patients with advanced breast cancer. Cancer Res 1993; 53:266–270.
12. Buzdar AU, Smith R, Vogel C, Bonomi P, Keller AM, Favis G, Mulagha M, Cooper J. Fadrozole HCl (CSG-16949A) versus megestrol acetate treatment of postmenopausal patients with metastatic breast carcinoma: results of two randomized double blind controlled multi-institutional trials. Cancer 1996; 77:2503–2513.
13. Buzdar A, Jonat W, Howell A, Jones SE, Blomqvist CP, Vogel CL, Eiermann W, Wolter JM, Steinberg M, Webster A, Lee D. Anastrozole versus megestrol acetate in the treatment of postmenopausal women with advanced breast carcinoma. Cancer 1998; 83:1142–1152.
14. Dombernowsky P, Smith I, Falkson G, Leonard R, Panasci L, Bellmunt J, Bezwoda W, Gardin G, Gudgeon A, Morgan M, Fornasiero A, Hoffmann W, Michell J, Hatscchek T, Tjabbes T, Chaudri HA, Hornberger U, Trunet PF. Letrozole, a new oral aromatase inhibitor for advanced breast cancer: double-blind randomized trial showing a dose effect and improved efficacy and tolerability compared with megestrol acetate. J Clin Oncol 1998; 16:453–461.
15. Goss P, Wine E, Tannock I, Schwartz IH, Kermer AB. Vorozole vs Megace® in

postmenopausal patients with metastatic breast carcinoma who had relapsed following tamoxifen (abstr). Proc Am Soc Clin Oncol 1997; 16:155a.

16. Kaufmann M, Bajetta E, Dirix LY, Fein LE, Jones SE, Cervek J, Fowst C, Polli A, di Salle E, Arkhipov A, Piscitelli G, Massimini G. Survival advantage of exemestane (EXE, Aromasin®) over megestrol acetate (MA) in postmenopausal women with advanced breast cancer (ABC) refractory to tamoxifen (TAM): results of a phase III randomized double-blind study (abstr). Proc Am Soc Clin Oncol 1999; 18:109.
17. Gershanovich M, Chaudri HA, Campos D, Lurie H, Bonaventura A, Jeffrey M, Buzzi F, Bodrogi I, Ludwig H, Reichardt P, O'Higgins N, Romieu G, Friederich P, Lassus M, for the Letrozole International Trial Group. Letrozole, a new oral aromatase inhibitor: randomized trial comparing 2.5 mg daily, 0.5 mg daily and aminoglutethimide in postmenopausal women with advanced breast cancer. Ann Oncol 1998; 9:639–645.
18. Bergh J, Bonneterre J, Illiger HJ, Murray R, Nortier J, Paridaens R, Rubens RD, Samonigg H, Van Zyl J, for the Vorozole Study Group. Vorozole (Rivizor™) versus aminoglutethimide in the treatment of postmenopausal breast cancer relapsing after tamoxifen (abstr). Proc Am Soc Clin Oncol 1997; 16:155a.
19. Falkson CI, Falkson HC. A randomized study of CGS 16949A (fadrozole) versus tamoxifen in previously untreated postmenopausal patients with metastatic breast cancer. Ann Oncol 1996; 7:465–469.
20. Thurlimann B, Beretta K, Bacchi M, Castiglione-Gertsch M, Goldhirsch A, Jungi WF, Cavalli F, Senn HJ, Fey M, Lohnert T. First-line fadrozole HCl (CGS 16949A) versus tamoxifen in postmenopausal women with advanced breast cancer. Prospective randomized trial of the Swiss Group for Clinical Cancer Research SAKK 20/88. Ann Oncol 1996; 7:471–479.
21. Perez Carrion R, Alberola Candel V, Calabresi F, Michel RT, Santos R, Delozier T, Goss P, Mauriac L, Feuilhade F, Freue M, Pannuti F, van Belle S, Martinez J, Wehrle E, Royce CM. Comparison of the selective aromatase inhibitor formestane with tamoxifen as first-line hormonal therapy in postmenopausal women with advanced breast cancer. Ann Oncol 1994; 5(suppl 7):S19–S24.
22. Thuerlimann BJK, Nabholtz JM, Bonneterre J, Buzdar AU, Robertson JFR, Webster A, Steinberg M, von Euler M, on behalf of the 'Arimidex' Study Group. Preliminary results of two comparative multicentre clinical trials comparing the efficacy and tolerability of Arimidex™ (anastrozole) and tamoxifen (TAM) in postmenopausal women with advanced breast cancer (ABC) (abstr). The Breast 1999; 8:214.
23. Dowsett M, Jones A, Johnston SRD, Jacobs S, Trunet P, Smith IE. In vivo measurement of aromatase inhibition by letrozole (CGS 20267) in postmenopausal patients with breast cancer. Clin Cancer Res 1995; 1:1511–1515.

5

Who Benefits Most from Second-Line Treatment with Letrozole?

Manfred B. Wischnewsky
University of Bremen
Bremen, Germany

Peter Schmid and Kurt Possinger
Humboldt University Berlin
Berlin, Germany

Rainer Böhm and Hillary A. Chaudry
Novartis Pharma AG
Basel, Switzerland

I. ABSTRACT

The individual prognosis of patients with metastatic breast cancer depends on a complex interaction of biological factors. Knowledge discovery in databases (KDD) can describe these multiple interactions and generate specific decision structures on the basis of decision rules. Using KDD, we reanalyzed the data of a key trial comparing letrozole with megestrol acetate (ARBC2) to characterize patients who benefit most from second-line hormonal treatment with respect to response, time to progression, time to treatment failure, and overall survival.

A total of 552 patients with metastatic breast cancer were randomly assigned to receive letrozole 2.5 mg (n = 174), letrozole 0.5 mg (n = 188), or megestrol acetate 160.0 mg (n = 190) once daily until progression of disease.

Multivariate analysis by KDD delivered a different set of seven to nine predictive parameters for each endpoint. In contrast, conventional multivariate analysis revealed only two to five parameters. Each of the factors delivered by KDD has an influence on the outcome which varies between the subgroups. Several of the predictive parameters do not reach significance in conventional analysis for the entire population but may be highly significant within subgroups.

KDD describe common parameters like the dominant site of metastases. However, they also reveal new prognostic factors, which are not detected in conventional data analysis, as they are only relevant for subsets of patients but not for the whole population; for example, the body mass index (BMI) had no predictive value in univariate and multivariate analyses for the entire population. However, KDD showed that patients with visceral and osseous metastases and a BMI of less than 30 kg/m^2 responded significantly better than patients with a BMI of 30 kg/m^2 or more. Furthermore, KDD can be used to classify patients according to their risk with respect to a certain endpoint; for example, KDD divided the population automatically into two groups with low and high risk to die within 2 years. Mean time to die (TTD) was 416 days for low-risk (95% CI 365–467) and 822 days for high-risk patients (95% CI 725–920).

KDD can be used to predict the clinical outcome for an individual patient. In contrast to conventional methods, the level of confidence for the predictions reaches 90% and more. Thus, therapeutic strategies might be adjusted to the individual risk; for example, high-risk patients can be identified before initiation of hormonal therapy and might subsequently be considered for a more intensive treatment.

For visualization, prediction, clustering, and modeling of the (individual) outcome, oncological maps, which are optimal representations of high-dimensional data have been used. The generated model, based on the ARBC2 data, have been tested on 100 new randomly selected cases from another clinical trial. The error rate between predicted and actual objective response is 4%.

II. INTRODUCTION

Approximately one-third of human breast cancers are estrogen dependent and exhibit regression after estrogen deprivation. In postmenopausal women, the synthesis of estrogens occurs mainly in peripheral tissues, which convert androgen into estrogen using the aromatase enzyme. Peripheral aromatase is predominantly located in fat, liver, and muscle tissue. However, aromatase activity is also found in about two-thirds of breast tumors; apparently providing a local source of estrogen within the tumor itself. Complete inhibition of aromatase could accomplish effective estrogen blockade. Twenty to thirty percent of the patients who fail antiestrogen treatment respond to aromatase inhibitor treatment.

Letrozole is a new, orally active, highly selective, nonsteroidal competitive inhibitor of the enzyme aromatase (1). It has shown substantial activity in postmenopausal patients with metastatic breast cancer, where objective tumor response rates of up to 25% after failure of previous therapy have been reported. Two large randomized studies comparing letrozole with other antiestrogen therapies have been published to date.

In a first multinational clinical trial two different doses of letrozole, 0.5 and 2.5 mg orally once daily, were compared with 160 mg megestrol acetate (MA) (BC2 Trial) (2). The purpose of this chapter is to characterize patients within the ARBC2 trial, which benefit most from second-line hormonal treatment in terms of response, time to progression (TTP), time to treatment failure (TTF), and overall survival (TTD) using techniques from knowledge discovery in databases (KDD). For this data reanalysis we used Cox, Kaplan-Meier, logistic regression, chi-squared interaction detector, CART, entropy-based decision tree algorithms, and self-organizing maps (SOMs).

III. PATIENTS AND METHODS

A. Patients

A total of 552 postmenopausal women with metastatic breast cancer, positive or unknown estrogen (ER) or progesterone (PgR) receptor status were randomly assigned to receive either letrozole 0.5 mg, letrozole 2.5 mg, or MA once daily in a double-blind, peer-reviewed, multinational trial. Patient characteristics are listed in Table 1. The treatment groups were well balanced across baseline covariates.

Postmenopausal status was defined by no spontaneous menses for at least 5 years, amenorrhea for at least 12 months, luteinizing hormone (LH) and follicle-stimulating hormone (FSH) within postmenopausal range of the laboratory involved, or amenorrhea for at least 3 months following bilateral oophorectomy or radiation castration, respectively. Patients were regarded as ER or PgR positive if any assay of the primary or secondary tumor tissue was positive. Previous hormonal therapy other than antiestrogens, oophorectomy, or radiation castration was not allowed. Patients had either relapsed on adjuvant hormonal therapy or within 12 months of stopping treatment or had progressed on first-line antiestrogen treatment for metastatic disease. Adjuvant and one chemotherapy regimen for advanced disease were allowed, but most patients had not received prior chemotherapy.

Patients were ineligible if they had central nervous system (CNS) involvement, diffuse lymphangitis carcinomatosa of the lungs, inflammatory breast can-

TABLE 1 Patient Baseline Characteristics

	Evaluable patients		Letrozole 0.5		Letrozole 2.5		Megestrol acetate	
	N	(%)	N	(%)	N	(%)	N	(%)
Dominant site of metastases	541	(100)	186	(100)	169	(100)	186	(100)
Soft tissue	150	(27.7)	57	(30.6)	44	(26.0)	49	(27.7)
Bone	171	(31.6)	59	(31.7)	52	(30.8)	60	(31.6)
Viscera	220	(40.7)	70	(37.6)	73	(43.2)	77	(40.7)
Number of anatomical sites of metastatic disease	541	(100)	186	(100)	169	(100)	186	(100)
1	119	(58.8)	119	(64.0)	101	(59.8)	98	(52.7)
2	55	(32.9)	55	(29.6)	53	(31.4)	70	(37.6)
3	12	(8.3)	12	(6.5)	15	(8.9)	18	(9.7)
Disease-free interval	552	(100)	188	(100)	174	(100)	190	(100)
Stage IV at presentation	56	(10.1)	21	(11.2)	13	(7.5)	22	(11.6)
<24 months	160	(29.0)	48	(25.5)	55	(31.6)	57	(30.0)
≥24 months	336	(60.9)	119	(63.3)	106	(60.9)	111	(58.4)
Hormone receptor status	552	(100)	188	(100)	174	(100)	190	(100)
Both receptors unknown	236	(42.8)	84	(44.7)	74	(42.5)	78	(41.1)
ER^+ or PgR^+	119	(21.6)	35	(18.6)	43	(24.7)	41	(21.6)
ER^+ and PgR^+	197	(35.7)	69	(36.7)	57	(32.8)	71	(37.4)
Performance status	552	(100)	188	(100)	174	(100)	190	(100)
WHO grade 0	270	(48.9)	94	(50.0)	89	(51.1)	87	(45.8)
WHO grade 1	28	(39.5)	72	(38.3)	60	(34.5)	86	(45.3)
WHO grade 2	63	(11.4)	22	(11.7)	24	(13.8)	17	(8.9)
WHO grade 3	1	(0.2)	0	(0)	1	(0.6)	0	(0)
Body mass index	544	(100)	84	(100)	172	(100)	188	(100)
<30 kg/m^2	421	(77.4)	148	(80.4)	128	(74.4)	125	(77.1)
≥30 kg/m^2	123	(22.6)	36	(19.6)	44	(25.6)	54	(22.9)
Previous chemotherapy	552	(100)	188	(100)	174	(100)	190	(100)
None	347	(62.9)	114	(60.6)	120	(69.0)	113	(59.5)
Adjuvant only	118	(21.4)	41	(21.8)	36	(20.7)	41	(21.6)
Therapeutic only	69	(12.5)	29	(15.4)	13	(7.5)	27	(14.2)
Adjuvant and therapeutic	18	(3.3)	4	(2.1)	5	(2.9)	9	(4.7)
Previous antiestrogen therapy	552	(100)	188	(100)	174	(100)	190	(100)
Adjuvant only	183	(33.2)	65	(34.6)	57	(32.8)	61	(32.1)
Therapeutic only	306	(55.4)	108	(57.4)	93	(53.4)	105	(55.3)
Adjuvant and therapeutic	63	(11.4)	15	(8.0)	24	(13.8)	24	(12.6)
Response to prior antiestrogens	552	(100)	188	(100)	174	(100)	190	(100)
Objective response (CR or PR)	113	(20.5)	40	(21.3)	33	(19.0)	40	(21.1)
NC or response unknown but antiestrogen given ≥ 6 months	178	(32.2)	60	(31.9)	61	(35.1)	57	(30.0)
PD or response unknown but antiestrogen given < 6 months	63	(11.4)	15	(8.0)	21	(12.1)	27	(14.2)
Not applicable (e.g., adjuvant therapy)	198	(35.9)	73	(38.8)	59	(33.9)	66	(34.7)

cer, extensive hepatic metastases involving more than one-third of the liver, or disease limited to pleural effusion or ascites. Further exclusion criteria included a history of prior malignancy other than contralateral breast cancer, in situ carcinoma of the cervix or adequately treated basal or squamous cell carcinoma of the skin, uncontrolled cardiac disease or diabetes mellitus, adrenal disease, porphyria, or confirmed peptic ulceration. Written, informed consent was required.

Patients were staged at baseline and assessed at three monthly intervals until disease progression and thereafter followed up for survival. The primary endpoint was overall objective tumor response assessed using UICC criteria. Objective remissions and stable disease had to be confirmed on two occasions at least 4 weeks apart. Secondary endpoints included time to progression, time to treatment failure, and time to death.

B. Methods

The methods used are exploratory data analysis, knowledge discovery, and machine learning techniques.

Knowledge discovery is the nontrivial extraction of implicit, previously unknown, and potentially useful knowledge from data. *Exploratory data analysis* and data mining are used for KDD. The emphasis in exploratory data analysis is on the whole interactive process of knowledge discovery; that is, the discovery of novel patterns or structures in the data.

One important application of knowledge discovery is the construction of a classification procedure from a set of data for which the true classes (i.e., the individual outcomes) are known. This type of procedure can be termed pattern recognition, discrimination, or supervised learning. The major strands in current classification practice and research (3) are: extensions to linear discrimination, decision tree and rule-based methods, and density estimates.

In this chapter, we use techniques from these three areas, such as Cox regression, entropy-based inductive and chi-square detection algorithms (i.e., decision trees), boosting algorithms, and Kohonen's self-organizing maps.

Decision trees are defined as follows: The root of the tree is the top node, and examples are passed down the tree, with decisions being made at each node until a terminal node or leaf is reached. Each nonterminal node contains a question (e.g., previous chemotherapy: adjuvant, therapeutic, adjuvant and therapeutic, or none) on which a split is based. Each leaf contains the label of a classification (e.g., low risk or high risk or mean time to death).

To determine the variability of the results and, therefore, the accuracy of the classifier *on unknown cases*, the decision trees are *cross validated* (3). A possible alternative would be bootstrapping (4). This represents a nonparametric procedure for estimating parameters in general and error rates in particular and to reduce variability in small data sets.

A classifier is usually evaluated in terms of its error rate on new cases. However, misclassification errors of one kind are more serious than those of another kind. The consequence of misclassifying a high-risk patient as one with low risk is more serious and "costs" more than the converse. Therefore, a *cost function* is introduced to calculate a decision tree for which the total cost of misclassification is minimized. Furthermore, *boosting*, another innovative technique (5), is applied. This means that a number of classifiers are constructed. When a case is classified, all these classifiers are consulted before a decision is made. Boosting gives higher predictive accuracy at the expense of increased classifier construction time. As the first step, a single decision tree or rule-set is constructed as before from the training data. This classifier will usually make mistakes on some cases in the data, and the first decision tree possibly gives the wrong class for some cases. When the second classifier is constructed, more attention is paid to these cases. As a consequence, the second classifier will generally be different from the first. It also will make errors on some cases, and these will be focused on when constructing the third classifier. This process continues for a predetermined number of iterations.

Self-organizing maps (SOMs), special types of neural nets, allow:

1. Clear and intuitive analysis by track breaking visualization technique
2. Direct identification of system attributes
3. Identification of nonlinear dependencies between parameters
4. Easy to handle preprocessing and postprocessing capabilities
5. Built-in support for numerous applications, such as clustering, filtering, and prediction

SOMs are imparticularly designed to learn from new data and predict individual outcome. This approach, developed by Kohonen in 1982, can be used to produce visual displays or maps of the similarities and dissimilarities in the data (i.e., oncological maps); a concept used for the first time in oncology, as far as the authors know. Each point in an oncological map represents a patient.

The SOM is realized by a two-dimensional hexagonal grid. Starting from a set of numerical, multivariate, high-dimensional data, the "nodes" on the grid gradually adapt themselves to the intrinsic shape of the data distribution. Since the order on the grid reflects the neighborhood within the data, attributes and features of the data distribution can be read off from the emerging "landscape" on the grid. A map (one for each *parameter*) represents the (local average) parameter value at each node. Some variables carry gently over the map windows; hence they may be assumed to be dominant variables. Together they define a complete order of the data space. In contrast, other parameters are distributed nonuniformly. It does not play a major role in the overall distribution.

IV. RESULTS

A. Response

Overall 192 patients (34.8%) responded to the hormonal treatment with 35 (6.3%) complete and 78 (14.1%) partial remissions, accounting for an objective response rate of 20.4%. Although there was a trend favoring letrozole 2.5 mg, no significant difference in response rates was obtained between the three different treatments in total ($P = .1036$) (Table 2). However, within subsets of patients, letrozole 2.5 mg induced significantly higher response rates then letrozole 0.5 mg or MA. For example, patients with predominantly soft tissue metastases and treatment with letrozole 2.5 mg responded significantly better (RR 69.4%) than patients on letrozole 0.5 mg or MA (RR 45.5%; $P = .0211$).

Multivariate analysis by KDD (C4.5, an entropy-based decision tree algorithm combined with CHAID, a chi-squared interaction detector) delivered nine parameters predictive for response: dominant site of metastases, disease-free interval, objective response (CR or PR) to prior antiestrogen therapy, body mass index, World Health Organization (WHO) performance status, age, receptor status, previous antiestrogen therapy (therapeutic is more favorable than only adjuvant) as well as the type of treatment (Table 3). In contrast, forward conditional logistic regression revealed only three significant parameters (dominant site of metastases, performance status, and disease-free interval) (error rate 32.7%). The logistic model just consisting of dominant site of metastases has an error rate of 32.9%; that is, there is no gain by adding additional parameters.

TABLE 2 Overall Tumor Response

	Evaluable patients							
	Total		Letrozole 0.5		Letrozole 2.5		Megestrol acetate	
Response	N 552	(%) (100)	N 188	(%) (100)	N 174	(%) (100)	N 190	(%) (100)
Response[a] (CR/PR/NC)	192	(34.8)	57	(30.3)	71	(40.8)	64	(33.7)
CR	35	(6.3)	8	(4.3)	18	(10.3)	9	(4.7)
PR	78	(14.1)	21	(11.2)	30	(17.2)	27	(14.2)
NC	79	(14.3)	28	(14.9)	23	(13.2)	28	(14.7)
No response[a]	360	(65.2)	131	(69.7)	103	(59.2)	126	(66.3)
PD	307	(55.6)	106	(55.9)	96	(55.2)	106	(55.8)
NA/NE	53	(9.6)	26	(4.0)	7	(4.0)	20	(10.5)

[a] UICC criteria.

TABLE 3 Univariate Analysis of Clinical Benefit

	Clinical benefit			(=CR + PR + SD + TTP > 6 months)
	Megestrol acetate	Letrozole 0.5	Letrozole 2.5	P-value (Letrozole 2.5 mg)
Overall clinical benefit	32.1	28.7	39.1	
Age				
36–56 yrs	20.5	26.1	18.9	0.0046
57–93 yrs	35.1	29.6	44.5	
Performance status				
WHO grade 0	42.5	36.2	47.2	0.0419
WHO grade 1	25.6	23.6	33.3	
WHO grade 2	11.8	13.6	20.8	
Dominant site of metastases				<0.0001
Bone	30.0	30.5	25.0	
Viscera	22.1	18.6	28.85	
Soft tissue	51.0	40.4	68.2	
Number of metastatic sites				0.0033
1	39.8	32.8	47.5	
2	25.7	25.5	22.6	
3	16.7	8.3	26.7	
Disease-free interval				0.1688
Stage IV at presentation	31.8	19.0	15.4	
<24 months	26.3	12.5	43.6	
≥24 months	35.1	37.0	39.6	
Hormone receptor status				0.2513
ER^+ or PgR^+	43.9	8.6	32.6	
ER^+ and PgR^+	26.8	29.0	43.9	
Body mass index				0.4612
<30 kg/m^2	30.3	29.7	40.6	
≥30 kg/m^2	34.9	27.8	36.4	
Previous chemotherapy				0.5758
None	34.5	28.9	38.3	
Adjuvant and therapeutic	22.2	—	20.0	
Previous antiestrogen therapy				0.0160
Adjuvant only	41	20.0	26.3	
Therapeutic only	31.4	33.3	44.1	
Adjuvant and therapeutic	25	33.3	50.0	
Response to prior antiestrogens				0.0603
CR or PR	20	37.5	51.5	
NC	38.6	31.7	34.4	
PD	22.2	13.3	57.1	

The decision structure generated by KDD shows that each of the factors listed above had an influence on the outcome which varies between the subgroups (Fig. 1). The most important prognostic parameter for response was the dominant site of metastases. Although 52.8% of the patients with predominantly soft tissue metastases showed a partial or complete response or remained stable for at least 6 months, only 30.4 and 25.0% of the patients with bone and visceral metastases, respectively, responded to study treatment ($P < .0001$). Response rates of patients with predominantly soft tissue, visceral, or bone metastases were influenced by different parameters.

In patients with visceral metastases receiving letrozole 2.5 mg, the body mass index (BMI) has a strong influence on the response. Patients with a BMI of less than 30 kg/m^2 responded significantly better than patients with a BMI of

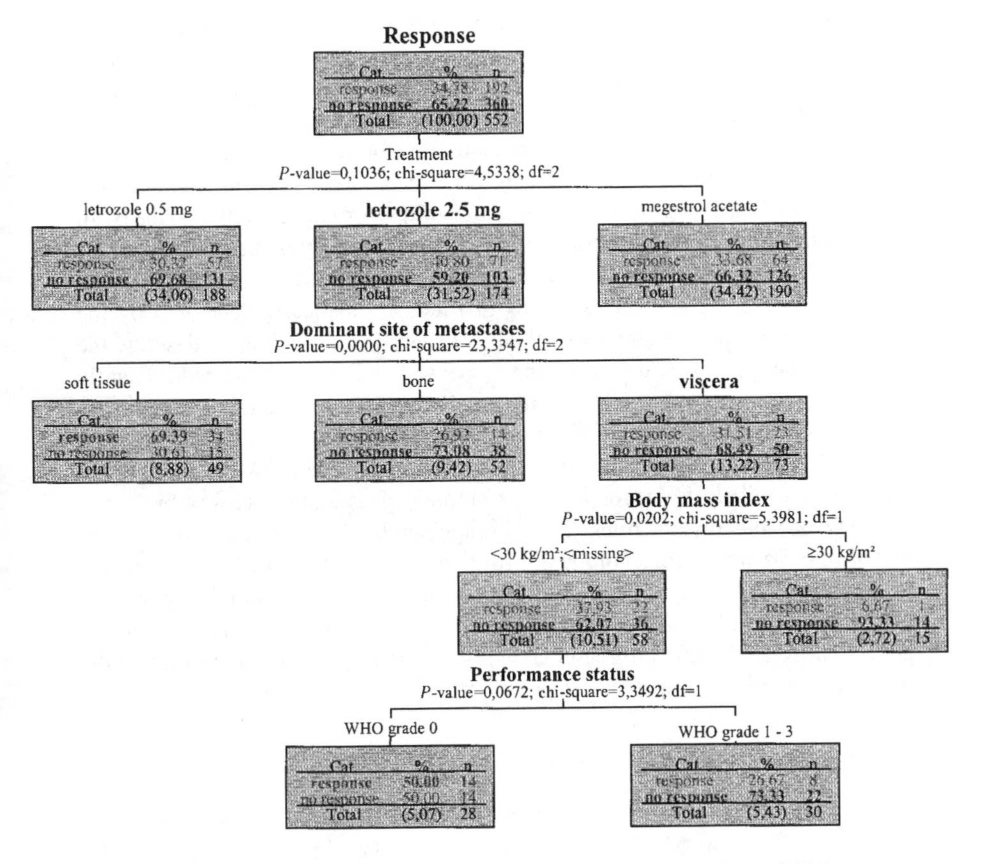

FIGURE 1 Decision structure with respect to response (CR, PR, and SD).

30 kg/m^2 or more. Only 6.7% of the patients with visceral metastases and a BMI $\geq$30 kg/m^2 remained stable under treatment with letrozole 2.5 mg and no patient achieved an objective response. Similarly, in patients with bone metastases, a strong trend favoring patients with a lower BMI was obtained, although the differences in response rates were not significant. However, significance is reached in several subsets defined by at least one additional parameter. The influence of the BMI on the response might indicate that the dose of letrozole is inadequate in patients with a high BMI and visceral or osseous metastases. It therefore can be hypothesized that the results might be improved if the dose were adapted to the BMI or the weight.

In contrast, BMI was of less prognostic value in patients with soft tissue metastases. This could be due to the fact that response rates were significantly higher in this subset compared to patients with visceral or osseous metastases, with the consequence that other parameters might have become more dominant, whereas the BMI was only relevant for smaller subsets. For patients with predominantly soft tissue metastases, the most important parameter was the hormone receptor status. Patients with both estrogen and progesterone receptor positive had a high chance of responding to treatment, whereas patients with only one receptor positive were only likely to respond if prior hormonal therapy was used in the adjuvant setting. In patients with both receptors unknown, response was mainly dependent on the disease-free interval, the performance status, and the BMI.

KDD converts the decision structure into a set of rules. Each rule consists of a condition, a predicted class, and the level of confidence of the given prediction. Table 5 shows several examples of rules automatically generated by the computer. The rules are directly evaluable. Rules 1–3, for example, describe the impact of the type of treatment on the response for patients with predominantly soft tissue metastases. According to these rules, patients within the subset receiving letrozole 2.5 mg are more likely to respond than patients with letrozole 0.5 mg or MA. Rules 4 and 5 demonstrate that patients with a BMI below 30 kg/cm^2 are more likely to respond. As the number of generated rules is high, and not only one but several rules might be applicable for each individual patient and have to be weighted, a computer program has been generated to analyze and aggregate automatically all rules for an individual patient to simplify the process of reaching a verdict. Based on the individual data and the decision structure generated by KDD, this program can apply all the information and estimate the outcome for individual patients.

B. Time to Treatment Failure

The median times to treatment failure (MTTF) with respect to response are shown in Table 6. Whereas MTTF was similar in patients with stable disease under letrozole 2.5 mg or MA, MTTF was 7 months and almost 1 year longer, respec-

TABLE 4 Clinical Benefit Response with Respect to Dominant Site of Metastases and Body Mass Index for Patients Treated with Letrozole 2.5 mg

	Total	BMI $<$ 30 mg/m^2	BMI $\geq$ 30 mg/m^2	*P*-value
Clinical benefit (response) with respect to dominant site of metastases (%)				
Viscera	28.8 (31.5)	37.5 (37.5)	6.7 (6.7)	0.0286 (0.0341)
Bone	25.0 (26.9)	27.5 (30.0)	16.7 (16.7)	0.4472 (0.36)
Viscera and bone	27.2 (29.6)	32.3 (34.7)	11.1 (11.1)	0.0629 (0.0175)
Soft tissue	68.2 (69.4)	64.5 (65.5)	76.9 (76.4)	0.4202 (0.43)
Objective response rate with respect to dominant site of metastases (%)				
Viscera	19.2	25.0	0.0	0.0307
Bone	19.2	20	16.7	0.7972
Viscera and bone	19.2	22.9	7.4	0.0724
Soft tissue	54.6	48.4	69.3	0.2052

TABLE 5 Examples for Decision Rules Automatically Generated by KDD with Respect to Response

Rule 1: if (dominant site of metastases = soft) and treatment = letrozole 0.5 mg, then prediction = response (probability = 0.407)
Rule 2: if (dominant site of metastases = soft) and treatment = letrozole 2.5 mg, then prediction = response (probability = 0.694)
Rule 3: if (dominant site of metastases = soft) and treatment = megestrol acetate, then prediction = response (probability = 0.509)
Rule 4: if (dominant site of metastases = bone or dominant site of metastases = viscera) and performance status = WHO grade 0 and (previous antiestrogen therapy = therapeutic only or previous antiestrogen therapy = both adjuvant and therapeutic) and body mass index = < 30 kg/m^2, then prediction = response (probability = 0.427)
Rule 5: if (dominant site of metastases = bone or dominant site of metastases = viscera) and performance status = WHO grade 0 and (previous antiestrogen therapy = therapeutic only or previous antiestrogen therapy = both adjuvant and therapeutic) and body mass index = ≥ 30 kg/m^2, then prediction = response (probability = 0.273)
Rule 6: if (dominant site of metastases = viscera) and performance status = WHO grade 0 and body mass index = <30 kg/m^2 and treatment = letrozole 2.5 mg, then prediction = response (probability = 0.500)

tively, in patients with partial or complete responses under letrozole 2.5 mg compared to MA (Fig. 2). However, there were no significant differences in terms of survival between letrozole and MA indicating that a short TTF does not necessarily have a negative impact on survival.

MTTF for patients with osseous, visceral, and soft tissue metastases were 263 (104) days, 231 (91) days, and 439 (257) days, respectively; showing that patients with osseous and visceral metastases had similar mean and median TTF, but MTTF in both groups were markedly lower compared to patients with soft tissue metastases. The mean TTF for patients with clinical benefit was 604 days on an average. For MA it was 473 days, for letrozole 0.5 mg 622 days, and for letrozole 2.5 709 days (P-value = .0002).

Cox regression (forward stepwise) calculated the following variables for TTF: dominant site of metastases, number of different anatomical sites of metastases, disease-free interval, performance status, previous chemotherapy, and treatment.

C. Time to Progression

The mean time to progression (MTTP) was 342.5 days, with 335.3 days on letrozole 0.5 mg, 414.1 days on letrozole 2.5 mg, and 268.9 days on MA (see Table 6; 20.2% were censored).

TABLE 6 Time to Treatment Failure, Time to Progression, and Overall Survival with Respect to Treatment for All Patients and Patients with Objective Response

	Evaluable patients								
	Letrozole 0.5 mg			Letrozole 2.5 mg			Megestrol acetate		
All	Mean (days)	(95% CI)	Median (days)	Mean (days)	(95% CI)	Median (days)	Mean (days)	(95% CI)	Median (days)
TTF	289.5	(231.4; 347.7)	98.0	388.6	(321.1; 456.2)	155.0	243.4	(204.8; 282.0)	118.0
TTP	335.3	(267.4; 403.5)	104.0	414.1	(341.4; 486.7)	169.0	268.9	(225.6; 312.2)	168.0
TTD	735.1	(665.5; 804.9)	654.0	807.9	(737.3; 878.5)	767.0	724.2	(657.3; 791.2)	655.0
Objective response									
TTF	800.6	(648.9; 952.4)	735.0	881.6	(768.9; 994.4)	796.0	560.8	(469.7; 652.0)	513.0
TTP	829.5	(652.9; 1006.0)	851.0	925.0	(792.8; 1057.1)	1002	594.0	(497.8; 690.1)	548.0
TTD	1200.1	(1084.4; 1315.9)	—	1182.1	(1082.0; 1282.3)	—	1045.0	(916.4; 173.6)	1144.0

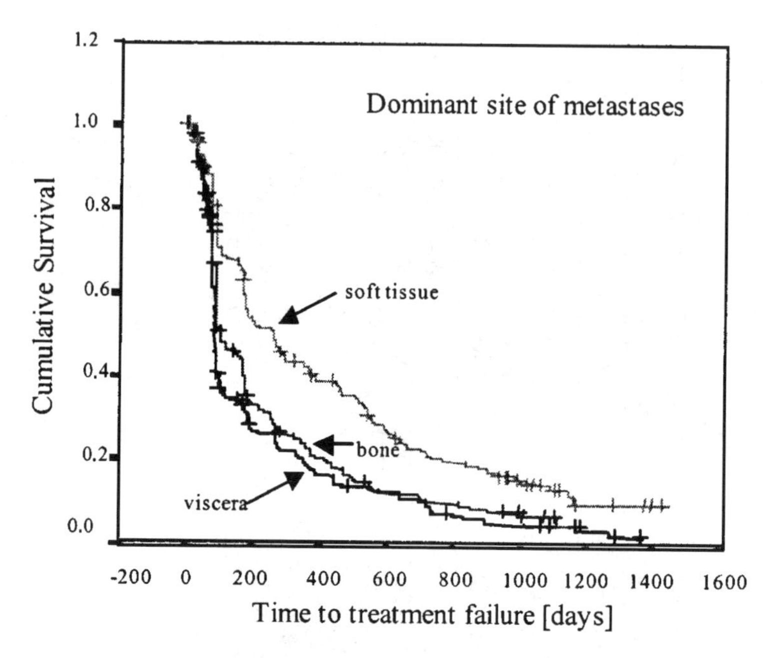

FIGURE 2 Kaplan-Meier function compares time to treatment failure with respect to dominant site of metastases.

This indicates a significant dose effect in favor of letrozole 2.5 mg compared to letrozole 0.5 mg, and superiority of letrozole 2.5 mg compared to MA ($P = .001$) (Fig. 3). The differences in MTTP with respect to the type of treatment were even higher within several subgroups of patients (Fig. 4).

KDD delivered several parameters predictive for TTP: severity of pain, age, dominant site of metastases, body mass index, disease-free interval, response to previous antiestrogen therapy, and type of treatment. Cox regression (forward stepwise) delivers age, number of different metastases, disease-free interval, performance status, and previous antiestrogen therapy.

D. Overall Survival

The mean time to death (MTTD) was 735 days for letrozole 0.5 mg, 808 days for letrozole 2.5 mg, and 724 days for MA. Although these results were not significant, there was a trend favoring letrozole 2.5 mg over MA and a dose effect in favor of letrozole 2.5 mg. KDD delivered eight parameters predictive for survival: previous chemotherapy, receptor status, age, number of anatomical sites of metastatic disease, prior bisphosphonates, WHO performance status, body

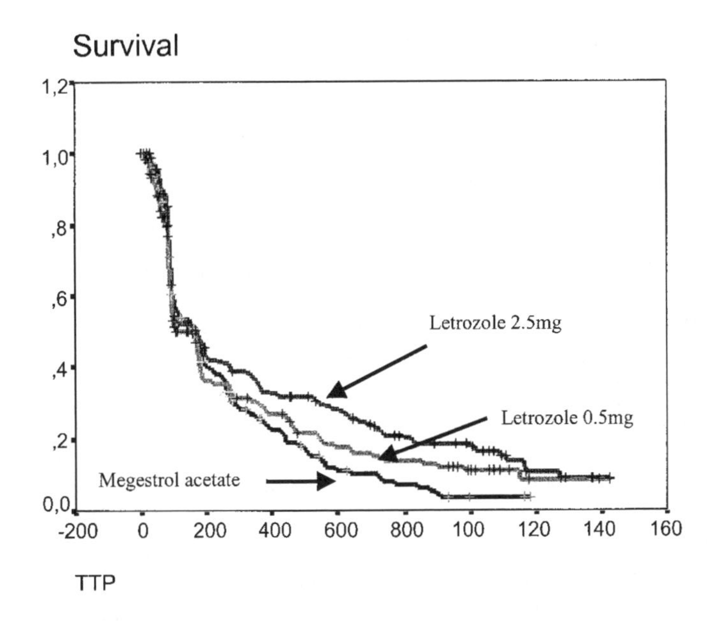

FIGURE 3 Kaplan-Meier function compares time to progression with respect to type of treatment.

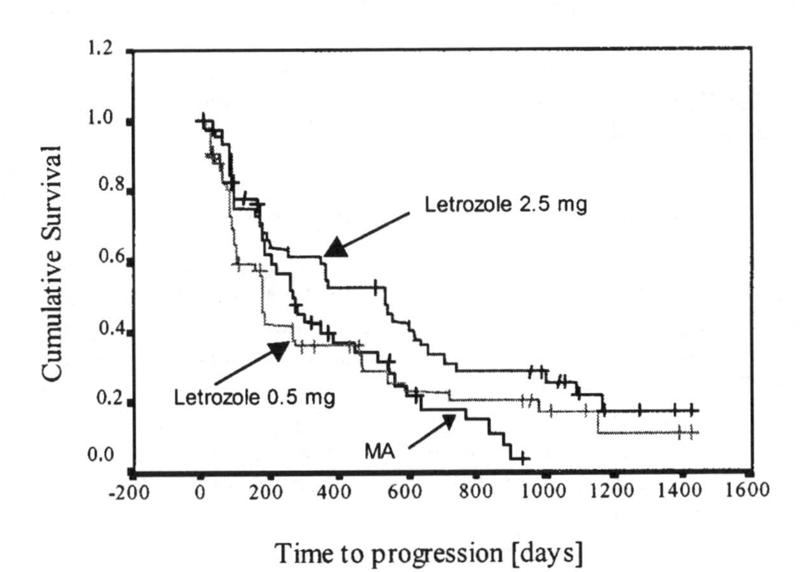

FIGURE 4 Kaplan-Meier function compares time to progression with respect to type of treatment in the subset of patients with soft tissue metastases.

mass index, and dominant site of metastases. In contrast, forward conditional Cox analysis (forward stepwise) delivered only the performance status and the number of metastatic sites.

The decision structure generated by KDD shows that each of the factors listed above had an influence on the survival which varies between the subgroups (Fig. 5). Most important parameters for overall survival were the number of metastatic sites and the dominant site of metastases. Whereas 56% of patients with only one metastatic area lived for at least 24 months, 68.6% of patients with two or three sites involved died within 2 years ($P < .0001$). Patients with two or three metastatic sites and predominantly bone metastases survived significantly longer than those with visceral involvement ($P = .0063$).

KDD defined three groups with respect to age. They differed significantly in terms of survival within several subsets. Patients aged between 56 and 72 years had a significantly better clinical outcome compared to older patients and especially younger patients. Whereas, for example, within patients with two or three metastatic sites, predominantly visceral metastases, and therapeutic prior hormonal treatment, 45.0% of women aged between 56 and 72 years survived longer than 2 years and 91.2% of younger women died within 24 months.

KDD classified patients automatically in two groups with low and high risk to die within 2 years. The cut-off point of 24 months was selected, as it is close to the median survival. MTTD was 416.5 days for high-risk patients (95% CI 365–467) and 822 days for low-risk patients (725–920), respectively. Owing to boost evaluation techniques, the misclassification rate for letrozole 2.5 mg was 2.9%, with a sensitivity and specificity of 95.1% and 90.7%, respectively. The corresponding Kaplan-Meier survival functions of these two subgroups automatically generated by the algorithm are shown in Fig. 6.

E. Oncological Maps and Prediction on New Cases

Oncological maps (self-organizing maps) support, as mentioned above, the following technical tasks: unsupervised clustering of data, association and prediction, pattern recognition, nonlinear regression, identification of nonlinear dependencies, visualization, and animated monitoring. Oncological maps have also the capability to generalize. This means that an oncological map can recognize or characterize patients it has never encountered before. A new patient is assimilated with a patient in the map which has the smallest distance. Furthermore, patients even with missing data can be used to look up or forecast the values of the missing data based on a trained oncological map. In the following we will use oncological maps to predict the individual outcome of patients treated with letrozole 2.5 mg. As a model we use the set of oncological maps (Oncological Atlas) generated by the previously described 174 (letrozole 2.5 mg) patients of the BC2 study. This oncological atlas was tested with 100 patients from another study on the

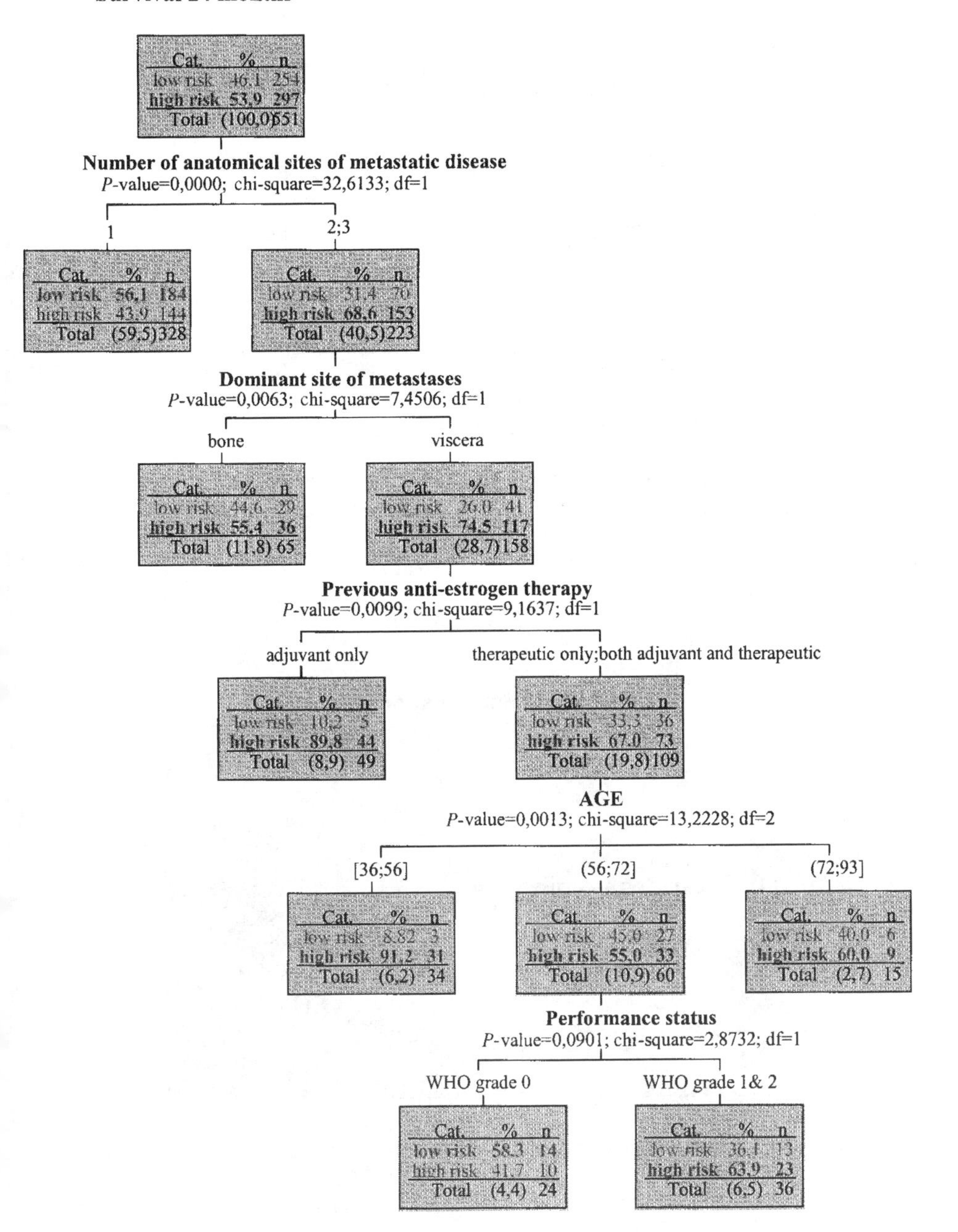

FIGURE 5 Decision structure with respect to survival for more than 24 months.

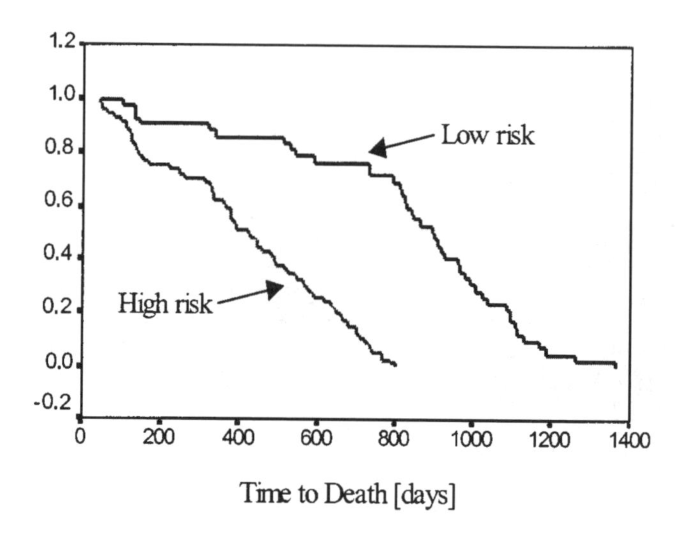

FIGURE 6 Kaplan-Meier survival function generated by the KDD algorithm for patients with high and low risk, respectively, to die within 24 months.

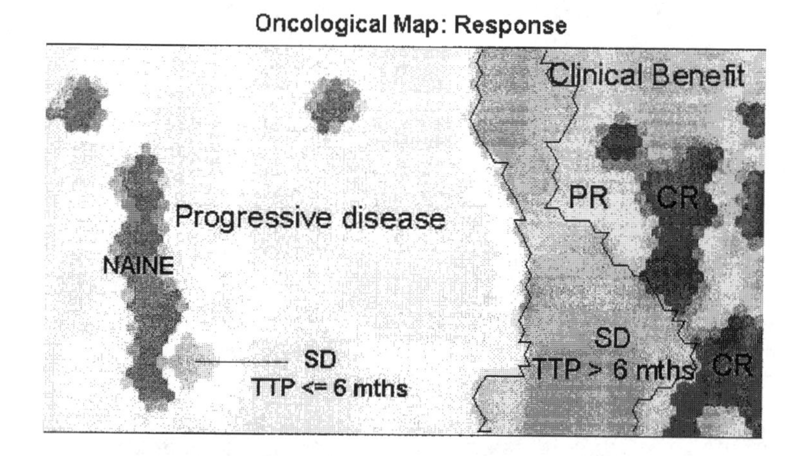

FIGURE 7 The oncological map for the parameter response. The response is nicely separated in areas with patients with CR, PR, SD, PD and NA/NE. Patients with SD $\leq$ 6 months are put into the area for patients with PD by the algorithm.

TABLE 7 Confusion Matrix: The Actual and Predicted Individual Response Values of the 100 Test Patients

	Actual					
Prediction	CR	PR	SD	PD	NA/NE	Prediction totals
CR	2	0	0	0	0	2
PR	3	12	1	0	0	16
SD	1	1	16	0	0	18
PD	0	0	6	46	9	61
NA/NE	0	1	0	2	0	3
Actual totals	6	14	23	48	9	

basis of all documented parameters except the corresponding clinical outcome. Then the calculated and the actual outcomes were compared.

The response is nicely separated in areas with patients with CR, PR, SD, PD, and NA/NE (Fig. 7). This means that the response can be reasonably deduced from the other values. If response values were randomly distributed over the whole map, we would have to conclude that there was no relationship between the response and the other values.

The error rate for objective response was 4%, for clinical benefit 8%, for response (CR + PR + SD) 7%, and for survival 24 months 7%. In Table 7, we represent the so-called confusion matrix (a generalization of the notion sensitivity) for the actual and calculated response of the individual patients. Of 18 patients predicted an objective response, only one had a documented SD. Furthermore, there is no significant difference between the actual and calculated values for mean and median TTP and TTF (Table 8). In contrast to this, the calculated median TTD is 10.8% higher than the actual value. One reason for this might be the fact that exactly 50% of these patients were censored with respect to TTD.

V. DISCUSSION

The current progress of computer technologies has opened new applications with algorithms that integrate the knowledge of a scientific or technical discipline (e.g., advanced breast cancer) and use artificial intelligence to provide decision solutions independent from human intelligence. Our analysis demonstrated that there are different sets of prognostic factors for predicting response, TTF, TTP, and TTD. The relevance of the different parameters varies within subsets of patients. KDD can describe these interactions between the prognostic factors and generate decision structures and sets of decision rules. They can be used to predict the

TABLE 8 Actual and Predicted Mean and Median Time to Treatment Failure, Time to Progression, and Time to Death (Kaplan-Meier survival analysis of the results derived from the oncological maps)

		Actual			Calculated		
		Survival time	Standard error	95% Confidence	Survival time	Standard error	95% Confidence
TTF	Mean	274.3	37.0	201.7;346.9	277.0	29.8	218.6;335.4
	Median	95.0	11.7	72.1;117.9	126.0	22.0	82.9;169.1
	Mean	312.5	42.8	228.7;396.4	308.7	33.7	242.7; 374.8
TTP	Median	108	30.0	49.1;166.9	143.0	25.5	93.1; 192.9
	Mean	879.5	51.7	778.3;980.8	872.0	46.5	781.7; 963.9
TTD	Median	899	69.7	762.4;1035.6	996.0	109.1	782.1;1209.8

clinical outcome for an individual patient. In contrast to conventional methods, the level of confidence for the predictions reaches 90% and more.

It is hardly possible to demonstrate the entire structure for all of the endpoints mentioned above, as the interactions between the different parameters are to complex. This chapter provides instead only several examples for the use of KDD in the analysis of patients receiving second-line hormonal therapy. KDD describe well-known parameters like the dominant site of metastases. However, they also reveal new prognostic factors, which are not detected in conventional data analysis, as they are only relevant for subsets of patients and not for the whole population. The body mass index may serve as example. Although, in the past, many clinicians have suspected that the BMI influences the outcome of endocrine therapy, it had been impossible to demonstrate this, as the BMI had no predictive value in univariate and multivariate analyses for the entire population. With the use of KDD and analysis of the decision structure, it becomes apparent that the BMI is highly predictive but only within subsets of patients. Patients especially with visceral and osseous metastases are much more likely to respond if the BMI is below 30 kg/m^2.

Another example is the age as a prognostic factor for survival. It is generally accepted that younger women have a poorer prognosis. However, KDD shows that within certain subsets patients older than 70 years also have a worse outcome compared to patients between 56 and 70 years old. The reasons behind this observation are not yet understood, but the first step is done by describing the phenomenon. It has to be stressed that the range was automatically defined by the algorithm. It is therefore important to avoid categorization of parameters in data analysis, as much information may get lost.

Since the decision structures are extremely complex, it is important to simplify the information for the clinical use. A possible way is to program a computer with the decision rules generated by KDD. This computer can subsequently analyze and aggregate automatically all rules for an individual patient and can provide an estimation for the individual outcome which is based on all the underlying information.

The oncological maps presented in this chapter are a new concept in data analysis in oncology. The corresponding background are improvements and extensions of Kohonen's self-organizing maps. The oncological maps capture the topology and probability distribution of input data. Self-organization in this context is a process of "unsupervised" learning whereby significant patterns or features in the input data are discovered. This type of data analysis provides valuable insight into any kind of numerical data in oncology in a distinct visual form, which allows the user to discover intuitively, analyze, and interpret relationships within the data.

Finally, it remains up to the clinicians how to interpret the predictions. The predictions are based on the clinical courses of patients treated according to a more or less defined strategy. Therefore, it is impossible to predict the outcome if the subsequent treatment strategy changes. If, for example, the survival of an individual patient is estimated to be less than 24 months, it is not clear whether the prognosis might be changed by using another therapeutic strategy. It can be hypothesized that this patient might, for example, benefit from more a aggressive treatment. However, it might also be possible that the survival is more or less independent from the type of treatment and the patient rather undergo a symptomatic treatment than intensive therapy. KDD provide a means to analyze biological structures, which have been thought to be too complex to understand, and to help to individualize treatment with respect to the individual risk. Of course, their task will not be to replace but to assist the clinician in therapeutic decisions in order to optimize treatment strategies.

REFERENCES

1. Bhatnagar AS, Häusler A, Schieweck, K, Lang M, Bowman R. Highly selective inhibition of estrogen biosynthesis by CGS 20267, a new non-steroidal aromatase inhibitor. J Steroid Biochem Mol Biol 1990; 37:1021–1027.
2. Dombernowsky P, Smith I, Falkson G, Leonard R, Panasci L, Bellmunt J, Bezwoda W, Gardin G, Gudgeon A, Morgan M, Fornasiero A, Hoffmann W, Michel J, Hatschek T, Tjabbes T, Chaudri HA, Hornberger U, Trunet PF. Letrozole, a new oral aromatase inhibitor for advanced breast cancer: double-blind randomized trial showing a dose effect and improved efficacy and tolerability compared with megestrol acetate. J Clin Oncol 1998; 16(2):453–461.
3. Ripley BD. Statistical aspects of neural networks. In: Barndorff-Nielsen OE, et al., eds. Chaos and networks—statistical and prohabilistic aspects. London: Chapman & Hall, 1993.
4. Efron B. Estimating the error rate of a prediction rule: improvements of crossvalidation. J Am Stat Assoc 1983; 78:316–331.
5. Freund Y, Schapire RE. Experiments with a new boosting algorithm. In: Machine Learning: Thirteenth International Conference, 1996.

Panel Discussion 1

Advanced Breast Cancer November 12, 1999

List of Participants

Ajay Bhatnagar Basel, Switzerland
Mitch Dowsett London, England
Harold Harvey Hershey, Pennsylvania
Stephen Johnston London, England
William Miller Edinburgh, Scotland
Henning Mouridsen Copenhagen, Denmark
Nam-Sun Paik Seoul, Korea
Carsten Rose Odense, Denmark
Ian Smith London, England
G. Stathopoulos Athens, Greece
Arnold Verbeek Basel, Switzerland
Manfred Wischnewsky Bremen, Germany

G. Stathopoulos: I have two questions for Ian Smith or Stephen Johnston. The first is how long should one wait to see a response? And, the second is about the visceral metastases and their relation to age? Is there a difference between a woman of 55 and 75.

Ian Smith: I will answer the first question on how long before one sees a response. I do not think that is clinically such a critical question. The key thing for clinical management is whether the patient is progressing or not. But, providing the disease remains stable, then we know that is of useful clinical benefit. The question of visceral metastases in relationship to age I will pass over to Stephen.

Stephen Johnston: I think the answer is that I am not aware of a breakdown in this sort of subgroups I have shown you of age. I would put myself in a position in the clinic where age is one factor. However, I think you also want to weigh up the other things we do in terms of making a decision on endocrine therapy or chemotherapy. This is, namely, the extent of the visceral involvement; whether or not they have got symptoms; and what you judge to be the rate of progression of that disease. For example, if you take a patient who clearly has got liver metastases as the dominant site, she has an enlarged liver with right upper quadrant pain and tenderness, is losing weight, is nauseous, and has a slightly elevated gamma-glutamyl transpeptidase (GT). That patient will benefit from chemotherapy far faster and easier than from endocrine therapy. Now, if you take a patient who has got liver as the dominant site, but has two to three lesions of about 2–3 cm in size, with no symptoms, steady weight, and you judge that the rate of change has been relatively slow with a normal liver function. If that patient has got other features that suggest she is suitable for endocrine therapy, namely, a tumor that is originally estrogen receptor (ER) positive, and a long disease-free interval of several years before relapse, then I think it is very suitable that such patients have endocrine therapy first. We have just demonstrated that in trials a third of them can respond. And, I have many clinical examples and slides of patients with large liver lesions shrinking away using endocrine therapy alone. So, I do not think that age is the sole factor. I think it is a clinical judgment that you make in the consulting room.

Carsten Rose: Ian, could you try to refresh my memory? You showed us these interesting data comparing anastrozole with tamoxifen. Isn't it true that there were no significant differences within the two trials concerning time to progression (TTP), and that it was only when you added up the figures that you came up with a significant benefit?

Ian Smith: Thank you very much for bringing up the topic, because Mitch, who constantly keeps me right, pointed out that I actually gave the data the wrong way around. I didn't realize that at the time I was doing it. It is the American trial that is mainly ER positive, which shows the difference in time to disease progression. Ninety percent of these patients were ER positive. The European trial is the trial that does not show any difference. As you might anticipate, the

European trial had more patients where ER was unknown—I suppose that is because there are British patients in amongst them—there are only about 45% where the ER was positive. So the difference at present in TTP is in the American trial, not the European, but if you take the patients that are ER unknown out of the European trial, you see a similar trend in the TTP. I still have not answered your question about the survival differences. I am actually not aware of up-to-date details on whether there are survival differences. Maybe somebody else on the panel—Mitch do you know?

Mitch Dowsett: Could I just make a couple of points about these data? Firstly, although they are obviously very exciting data, I do not think we should spend a lot of time with them, largely because they are very immature and very early data. There is one particular point I have taken from it—if you look at the time to progression in the American trial, it is only between 5 and 6 months. The median time to progression for tamoxifen in the European study which has more ER negative in it, is in fact between 8 and 9 months, which is rather surprising. I know they are two different trials—which may explain the difference. I think we need a lot more time for these trials to become mature before we rely on them very much.

Nam-Sun Paik: I have a question for Dr. Dowsett. Tamoxifen has a bone-preserving effect, but aromatase inhibition does not have a bone-preservation effect. Is there any necessity to preserve the bone?

Mitch Dowsett: There really are not any substantive data as yet on bone-preserving effects. We have recently done some work from a primary medical treatment study of vorozole versus tamoxifen, and bone biomarker analyses within that. We can see quite distinct differences in the bone biomarker analyses after 3 months, i.e., with tamoxifen we see a suppression of the cross-laps, which is an index of bone resorption, and that is therefore consistent with the preservation of bone mineral density that we see with tamoxifen. We only had a very small number of patients who were on the aromatase inhibitor. They clearly did not show a suppression of the index; neither did they show a very marked increase. So, whilst those data would indicate, as we would expect, that you are not getting bone preservation with the aromatase inhibitor, there does not seem to be a marked exacerbation in bone loss. These are relatively early studies—we will get a lot more data of that type from the ongoing trials.

Nam-Sun Paik: To Dr. Harvey. In the case of a survey, positive Her-2/neu expression cannot be expected to be a good effect. Even in the negative survey would you recommend a cyclophosphamide, methotrexate, 5-fluorouracil (CMF) regimen?

Harold Harvey: The prediction of Her-2/neu expression of resistance to CMF is suggested in the literature. My own review, however, suggests that there really is not enough evidence absolutely to conclude that in patients who have Her-2/neu expression, they should be denied CMF. I think the evidence that they should be given an anthracyclin is somewhat stronger. That would be the interpretation I would use.

Nam-Sun Paik: I have a question for Stephen Johnston. For good effects, what kind of drug do you recommend in case of a bone metastasis?

Stephen Johnston: Traditionally, endocrine therapy has been the mainstay in the management of patients with bone metastases, and I think it continues to remain so. Nothing has emerged from the data with the aromatase inhibitor trials to change that precept. However, what has emerged is increasing data on the role of bisphosphonates in addition to endocrine therapy. There are at least two or three randomized trials now where either chemotherapy versus chemotherapy with bisphosphonates or endocrine therapy versus endocrine therapy with bisphosphonates have shown the benefits of adding a bone-preserving agent. I think that the algorithm that Dr. Harvey showed is useful in helping to delineate that, particularly in patients who were ER negative, progesterone receptor (PgR) negative, and failed endocrine therapy should certainly move on to the bisphosphonates. Whether you should be combining them right up front in a patient who has got just bone metastases, who otherwise is suitable for endocrine therapy, I think that in the terms of UK management, this is an area where practice varies. I do not know what it is like in the United States. Traditionally, we have still gone with our second-line endocrine agent of choice, and we still get very long time to progression, response duration, and control with aromatase inhibitors in that setting. However, I think there are no randomized trials with the new potent inhibitors versus those with bisphosphonates. I do not know what practice in the United States is like.

Harold Harvey: I agree with what you said, Dr. Johnston. The tendency, however, is to use bisphosphonates earlier rather than later, and then they are only adjunctive. Except in the studies we did very early on to assess bisphosphonates, where patients did not receive any antitumor therapy, all patients with bone metastases are in fact receiving either chemotherapy or endocrine therapy, and the bisphosphanate is clearly adjunctive—better used early rather than late. And, better used until such time as the patient is no longer at risk, in your view, from some skeletal morbidity.

Henning Mouridsen: Just referring to the presentation by Ian Smith. You mentioned the first-line studies—I do not understand why you call them first-line studies, because the majority of patients have had tamoxifen already in the adjuvant situation. Isn't that right? And in the future, the majority of patients who relapse will have had adjuvant tamoxifen already. So, this gives rise to the question, which treatment therapy (antiestrogen or an aromatase inhibitor) do you choose after a treatment-free period of say 6 months or 2 years?

Ian Smith: Well, certainly first-line means in general first-line for metastatic diseases. Of course, you are right, a lot of patients have had adjuvant tamoxifen. In terms of the specific question you ask, these trials are difficult to run, nowadays, for all the obvious reasons. Steve has had some practice in that. I am not aware of the answer to your question of what is the duration of treatment, whereby you can say it is short—tamoxifen will not work—or it is a long time and, therefore, tamoxifen will work.

Stephen Johnston: You have raised a valid point. I think the entry criteria in a lot of these current studies dictate that you have to have at least a 12-month free interval at the end of adjuvant to tamoxifen before rechallenging. I am aware of one previous publication by Hyman Muss several years ago, where he actually looked at the interval. At that stage, they were rechallenging with tamoxifen. And, if I can recall correctly, I think, provided that the patient had at least 18 months or more, then the chances of responding to tamoxifen again was still high. I think the generation of trials has therefore taken an arbitrarily 12-month cut-off. But, I do not think there is any clear proper data on the best interval. I certainly think that, if the patients are relapsing within 6 months, we will not rechallenge with tamoxifen, but we would go straight to aromatase inhibitor. But, in the studies of tamoxifen versus the selective estrogen receptor modulators (SERMS), one of which I have been involved in running, which is difficult actually to do, they have chosen at least a year, if not longer, to maximize the chance of response. But, your point is well taken. There are not necessarily very good data to tell us what that interval should be.

Ian Smith: I think there is an additional issue here, Henning, which is increasing. We are in an area where we treat patients with tamoxifen for 5 years, whereas a lot of patients in these trials may have been on tamoxifen for considerably less than that. So, even if we have an answer for what the duration should be, if you have been on for 1 or 2 years, it may be a very different story than if you have been on tamoxifen for 5 years, and it is beginning to act as an agonist.

Mitch Dowsett: Can I just extend that for a second: Henning, are you really saying that these trails are unfair to tamoxifen?

Henning Mouridsen: Yes, that was what I meant. I think they are unfair to tamoxifen, and I would be very surprised if the aromatase inhibitors were not superior to tamoxifen in these trials. But, let us see the data.

Mitch Dowsett: So, presumably you would be looking with keenness to the neoadjuvant studies, where patients will be randomized to first-line treatment.

Ian Smith: The only thing, Henning, is that the tamoxifen versus the aminoglutethimide trial certainly was a front-line. Nobody had had adjuvant tamoxifen then. Now the numbers are very small. But, the results seemed to be pretty identical. We have data that third-generation aromatase inhibitors are clinically more effective than aminoglutethimide. So you could extrapolate that; at least it is plausible through these two observations that third-generation inhibitors might be more effective than tamoxifen. I agree with you about overinterpreting the data, but it is a plausible hypothesis.

Mitch Dowsett: I would just like to finish that off, because I think that in the trials we have just alluded to on anastrozole, about 40% of the patients had had prior adjuvant therapy, and this was not necessarily endocrine therapy. So, it does leave us 60% of the population that were receiving treatment for the first time. I think in many of those cases, these patients were presenting with locally advanced disease. So, they may be powerful enough to give a valid comparison of tamoxifen and the inhibitor. Hopefully, when we get the letrozole study data, this will be the case there as well.

William Miller: I wonder whether I could ask two separate questions. The first question really relates to the correlation between efficacy of estrogen suppression and clinical response in advanced disease. I guess the whole panel would back the theory that, as we have progressed from first- to third-generation inhibitors, we have seen an increase in efficacy of estrogen suppression, and that is related in general with the improved clinical efficacy. But there was a mention, especially in Ian Smith's talk as to efficacy differences between doses of individual aromatase inhibitors. Certainly, with letrozole there is the impression that the higher dose of 2.5 mg seems to be more efficacious clinically than the 0.5-mg dose. I wonder whether the panel could confirm how strong the data are, if indeed this is the case, is this a reflection of an increased efficacy of estrogen suppression, or do Mitch Dowsett's data actually show that there is very little difference between the two doses?

Mitch Dowsett: In terms of the plasma estrogen suppression, I think the effects the two letrozole dosages are entirely indistinguishable. But, the main reason for that is we are below the detection limit of even the most sensitive assays. Once we go on to the aromatization data, there is just a suggestion—we did not have enough patients—that the 2.5 mg was marginally more effective than the 0.5 mg, but again we get into the very limits of the detectability of that assay. So, there are two potential explanations. If we accept that 2.5 mg is clinically more effective than 0.5 mg, one explanation is that this tiny residual mass of estrogen is important. As I showed you, the estrogen-deprivation curves indicate that even 1 fmol/L of estradiol can stimulate proliferation of estrogen-deprived cells. So there may be very tiny, but significant, differences between the estrogen levels in these patients which we cannot detect. The second possibility is that we are affecting other signaling pathways. To date, they have not been identified. We need to be cautious that there may be something there that the higher dose is affecting which the lower dose is not.

Ian Smith: As far as the clinical data are concerned, the evidence that the 2.5 mg of letrozole is better than 0.5 mg is every bit as compelling as the evidence that letrozole is better than megestrol. So we accept that letrozole is better than megestrol, so I think in that sense we have to accept that dose-response data. The fact that there is a second trial against aminoglutethimide that also shows the effect gives more credence. Conversely in the anastrozole trials, there is absolutely no suggestion of dose-response effect between 1 and 10 mg. So it is very interesting, but I think one must not create too much in a way of a hypothesis to exchange this without further confirmatory data, perhaps from other studies or trials as they emerge, if indeed such data are around. Because there is this conflict in the literature between the two agents, and I think there is still a question mark over whether there really is a dose-response effect.

William Miller: Can I ask my second question, and that is really, I guess to Steve Johnston, because he presented some quite interesting data which suggested that response to letrozole may not be too different in patients who have or have not had a previous response to tamoxifen. I guess my basic question is: Can you give us some explanation for the mechanism of these beneficial effects of an aromatase inhibitor in patients who evidently have not responded initially to tamoxifen? There is also a supplementary question. In the same slide, unless I misread your data for megestrol acetate, although the response rate was lower compared with letrozole, again there did not seem to be a difference between patients previously responding to tamoxifen and those who did not. In fact, if I remember rightly, there was a 15% response rate in patients who have not previously responded to tamoxifen and only 10% in those who have. What is the explanation for it?

Stephen Johnston: That particular slide is basically taken from the three published trials, all of which have attempted to do subgroup analysis looking at prior response to tamoxifen, and putting them on one slide to see what the overall message is. The more I look at this slide—and I was looking at it last night, because I knew that would be an area that people would pick up on—the more I get confused about it. This is because clearly in the letrozole study the response rate is around 30% irrespective of prior response to tamoxifen. However, that does not appear to be a persistent trend in the trials, and as you rightly point out, the megestrol arms do not necessarily appear to show the differences either. So, I think we have to be relatively cautious in overinterpreting those data. The only reason that I highlighted it was to take the point further in terms of mechanisms and to suggest that we know that the response rate to tamoxifen in advanced disease may not be greater than 60%—even if you select ER positive—and there are a lot of reported mechanisms as to why that may be the case. Furthermore, you may have patients with ER-positive disease, who for various reasons specific to tamoxifen, do not show a response. However, when you give them an aromatase inhibitor, if they are truly dependent on estrogen through an ER pathway, then universally you would expect them logically to respond to the aromatase inhibitor. Again, I think the only way of actually demonstrating that in a clearer way will be in the neoadjuvant setting. Mike Dixon's data, again nonrandomized, compare the response rates, but the tamoxifen response rate is also only about 60%. This is, in terms of percentage reduction, this is not the overall response rate; that is, how much tumor shrinkage you get. It is actually greater with the aromatase inhibitors. But, I think in the front-line trials versus tamoxifen or even in the neoadjuvant you will be able to see if there is really a true difference in terms of nonresponders. What you ideally would like to do is a cross-over design in that setting. You would like to take the nonresponders to tamoxifen in the first-line study and cross them over to an aromatase inhibitor. I think Ian did that originally with the aminoglutethimide trial, and found evidence that it worked the one way but not the other way. Is that right?

Ian Smith: The crossover supported Steve's data, that there was more cross-over response to aminoglutethimide than other agent. But, the numbers were very small.

Harold Harvey: Mr. Chairman, I would like to return to the question of dose with letrozole. I think Ian Smith is correct that we have to be careful of not forming too many strong hypotheses on a fairly small body of data. I was also very surprised, however, by your response Dr. Dowsett. You said it could be explained by different pathways. Well surely, an equally tenable hypothesis is that the difference could be explained on not only plasma suppression, but on an effect on the intratumoral target of aromatase. But, I would like to know whether either Dr. Bhatnagar or Dr. Brodie or you yourself has data suggesting that there is a dose-response effect that can be related to the amount of estrogen production suppression by the tumor, not simply measurement of the plasma levels?

Mitch Dowsett: It is quite a demanding question to answer. The hypothesis I am drawing about other pathways is really based on a completely different type of drug which was developed for prostatic cancer. It was found that this drug which was expected to inhibit androgen synthesis did that, but it was found to be as effective in androgen receptor–negative prostatic tumors as it was in androgen receptor positive. Further exploration led to the discovery that this was actually a retinoic acid metabolism inhibitor. Also, I think that the lesson I went through at some length at the start of my talk today about aminoglutethimide is relevant—that was used as an antiepileptic, but was found to be an adrenal suppressant. And then, finally, we think now that it is aromatase inhibitor. So, I think especially when we use these drugs at relatively high dosages, they may have an impact on other pathways we do not expect. Nonetheless, my preferred hypothesis would be that it is actually a dose effect on estrogen.

Ajay Bhatnagar: I would like to look into the future a little bit with the panel. I think it has been very clear from all the talks given today that the aromatase inhibitors are established as second-line agents in advanced breast cancer. As the trials go along, they will move along up the scale to first-line, and then into early breast cancer. I think that in the short-term you will see a great deal of sequential treatment, but always after tamoxifen. And I have a question for Mitch and to the rest of the panel. The drug-interaction studies that we did with letrozole and with anastrozole are obviously relevant to a combination situation where you competently administer both drugs at the same time. Do you think that the data from those sorts of drug-interaction trials are relevant to the sequential situation, where you, for example, will stop tamoxifen treatment and start treatment with one of the new agents? Or will we have to do new types of studies to show that there is an interaction between the waning levels of tamoxifen with the increasing levels of the new aromatase inhibitors that are going to come after them?

Mitch Dowsett: I do think they are relevant. The clearance of tamoxifen from the tissues as patients come off tamoxifen is a very prolonged event. Ernst Lien measured tamoxifen in metastases of patients after they had come off tamoxifen, and found that he could measure tamoxifen, I believe, in a brain lesion over a year after the patient had ceased tamoxifen. So, yes, the effects of tamoxifen can be prolonged. We had to cater for this in our endocrine studies: for example, catering for the clearance of a tamoxifen effect on the SHBG, and the effect on gonadotrophins, etc. So, relevant? Yes, but significant? I am not so sure. The difference we are talking about with letrozole, for example, is about 35%. As far as we are aware, it is an effect which is consistent across the patients. So, this essentially reduces the dose of letrozole from 2.5 mg to 1.5 mg per day, and therefore, as we have already been saying, we would expect still to get very profound estrogen suppression. So whilst it is relevant, it is questionable whether it is significant.

Ajay Bhatnagar: The reason I asked that question was because most of the trials that were reported on either by Ian or Steve have many many patients where treatment with aromatase inhibitor starts immediately after cessation of tamoxifen. And, we still get—as they all showed—very good efficacy in terms of response rates, duration of response, and everything else. So, the question I asked was relevant, because, despite the fact that we see these sorts of drug interactions, we see extremely good efficacy with these agents immediately after tamoxifen.

Ian Smith: To be provocative slightly, can I challenge the premises on which you imply that for a considerable time to come we are still going to be using tamoxifen as first-line treatment. The thing that strikes me about all these trials is that, if you look at response during all the older studies of tamoxifen and aminoglutethimide, you come out to a figure of round about 18 months median. And if you look at response duration in the third-generation of aromatase inhibitors, you come out with a figure of around 24–30 months. In one trial that has been done that I think is relevant so far, where you compare an old treatment with a new one, i.e., letrozole with aminoglutethimide, you get complete confirmation of that. You have a median duration of 24 months versus 15 months. The vorozole trial did not show such a big difference, so there is a bit of a question mark there. But, if it is true—and there is a lot of circumstantial evidence that it is true—then it is hard for me to believe that this is not going to be translated into adjuvant therapy. And, therefore, I think it is quite possible that the trials of adjuvant therapy may show an advantage here. The problem I predict will be with bone, as we will have to work out ways of protecting it.

Mitch Dowsett: Can I ask a question of Dr. Wischnewsky? One of the things that struck me today listening to various presentations here is the attempt to explain differences on the basis of heterogeneity of the population. Well, you took a sophisticated and new approach to that. I think there are at least three possible applications—one would be just in learning and interpreting the data we are getting, and I think that was the one you were focusing on. But, do you believe we will be able to use the data in improved design of trials? And secondly, do you think we will be able to use the data for enhancing the reliability of understanding in a way that is acceptable to regulatory authorities?

Manfred Wischnewsky: Yes, these are three questions.

1. Learning and interpreting data from clinical trials?

Machine learning (ML) or automatic knowledge discovery in databases (KDD) is generally taken to encompass automatic procedures based on logical or binary operations that learn a task from the data of clinical trials. The corresponding data mining tools can answer questions that traditionally were too time consuming to resolve. They scour databases for hidden patterns, finding predictive information that experts may miss, because it lies outside their expectations. New knowledge is recognized by such a discovery system via the autonomous use of evaluation criteria. From a practical point of view, KDD is:

a. A means of engineering rule-based systems (expert systems) from sample cases volunteered interactively.

b. A method of data analysis whereby rule-structured classifiers for predicting the classes of newly sampled cases are obtained from a training set of preclassified cases.

As a means of data analysis, KDD delivers an optimal set of predictor variables to answer a question in an ordering of decreasing importance. Furthermore, the automatically generated knowledge represented by decision trees or rules shows the various correlations of the predictor variables.

2. Are we able to use the data derived by these machine learning techniques in improved designer trials?

Randomized trials without additional stratification afford, in most cases, large numbers of patients in order to be representative and to derive a statistical model with a predefined accuracy. Using the results of the type of automatically generated decision trees presented in my talk, we are able to design new clinical trials in a more efficient and economical way, since the necessary information (minimal number of patients in subgroups, the proportions of the subgroups, the estimated risk rates, etc.) can be derived from decision trees (it is like looking at the landscape from the top of a mountain). This leads to a knowledge-based design of trials.

3. Are we able to predict the outcome of patients?

The oncological maps (visualizations of Kohonen nets; each map represents a parameter) allow not only to get a more or less complete overview of the patient's situation in connection with a therapy strategy (e.g., aromatase inhibitors as second-line therapy in advanced breast cancer) but allow at the same time to predict any of the parameter values (response, time to treatment failure, overall survival, etc.) from the set of given parameters. A set of given parameters characterizes a subgroup of patients. This subgroup is represented as an area in the oncological maps characterizing a certain area in each map. The corresponding possible values for the dependent variables (response, overall survival, etc.) can be read immediately from the corresponding maps for response.

Harold Harvey: The question to my colleague, Dr. Wischnewsky, has to do with the body mass index (BMI). Some years ago I was explaining to a group of medical students the biology of aromatase. I was asked why don't we change the dose for very overweight patients. And I said, ''Gee that is a good idea.'' I would like to know whether clinicians in the audience in general use a different dose of aromatase inhibitors based on body mass index. And I ask you specifically if indeed in your models that was truly an important finding?

Manfred Wischnewsky: The body mass index does not belong to the group of the most important predictor variables for clinical outcome like the variables dominant site of metastases or number of different anatomical sites for metastases. But when you look at subgroups, then the body mass index is clearly a significant parameter to discriminate responders from non responders. Let me give just some examples for the influence of the body mass index BMI (BMI < 30 kg/m^2 (= BMI-Normal) and BMI ≥ 30 kg/m^2 (= BMI-High) in connection with the FEMARA study AR/BC2: letrozole 2.5 mg: viscera (BMI-N: response rate (RR) = 37.9%; BMI-H: RR = 6.7%) or bone (BMI-N: RR = 30.0%; BMI-H: RR = 6.7%). For soft tissue, we find no difference (BMI-N: RR = 65.5%; BMI-H: RR = 76.4%; $P = .43$). Similar results are valid for megestrol acetate. For example, patients aged between 56 and 72 (RR = 40.5%) and soft tissue metastases (RR = 63.3%): BMI-N (RR = 78.9%), BMI-H (RR = 36.4%), or viscera: (RR = 26.9%): BMI-N (RR = 28.6%), BMI-H (RR = 20.0%); the mean time to progression in this study for responders was 604 days for patients with BMI-N and 511 days for patients with BMI-H.

Thus, the conclusion would be that certain subgroups of ABC patients, in particular those with visceral or bone metastases treated with aromatase inhibitors as second-line therapy, seem to show significantly higher response rates and longer time to progression if the body mass index is normal (BMI < 30 kg/m^2). This can imply that the dose for patients with a high body mass index has to be increased. An in-depth analysis with data from other clinical trials with aromatase inhibitors should be done.

Arnold Verbeek: A last question for Dr. Harvey. You have given a very nice presentation of chemotherapy and hormonal therapy. In the current metastatic setting, patients either receive hormone or chemotherapy or a sequence of either. Could you give some thoughts, if in the future, we ought to combine chemotherapy with hormones?

Also looking at very durable responses in patients with visceral disease, who normally would be treated with chemotherapy?

Harold Harvey: In the interest of time, I did not discuss that. I did try to look at the literature, but it is very confusing. I did not find anything to change the usual teaching that for most patients there is very little to be gained in combining chemotherapy and endocrine therapy in the setting of metastatic disease. You might see an initially slightly higher response rate, but there are no good studies to show that there is an ultimate benefit in using combined therapy compared to using these modalities in sequence. However, there were a couple of articles that talked about instances where you might combine therapies—and that was when you needed a quick response. An example would be a patient with bone marrow involvement in whom you could not give full-dose chemotherapy, but you had the idea she might have a hormone-dependent tumor, and you would not want to wait 12 weeks to be sure she responded. The problem is that the data in the literature are not good, and I think you end up relying on clinical experience. I seldom do it except for certain special circumstances.

Arnold Verbeek: Should we consider giving hormone therapy to patients who withdraw from chemotherapy because of tolerability problems?

Harold Harvey: Absolutely, in that little survey we did in our patients after we treated them. These patients expressed this often as a preferred therapy: "I wish you would have given me that little pill a little earlier, and I wouldn't have lost my hair, I would not have vomited, and perhaps I would not have been any worse off."

Part III

Early Breast Cancer/Chemoprevention

H. Mouridsen, Chair
W. Miller, Chair

6

Neoadjuvant Endocrine Therapy

J. M. Dixon
Western General Hospital
Edinburgh, Scotland

I. ABSTRACT

Few studies have evaluated neoadjuvant endocrine therapy. The first group studied were elderly patients who were given tamoxifen. Although 60–70% of patients with estrogen receptor (ER)–positive tumors responded, benefits in some patients were short-lived, and of those patients kept on tamoxifen for long periods, only 30–40% sustained response beyond 2 years. The high percentage of initial responses stimulated some groups to give tamoxifen for 3 months before surgery in selected postmenopausal large operable or locally advanced ER-positive breast cancer. Results of these studies have shown significant reductions in volume over 3 months, with over 70% of patients suitable for breast conservation at the end of the treatment period. More recent studies using aromatase inhibitors in neoadjuvant therapy have shown responses at least as good as those with tamoxifen, with even higher rates of conversion from mastectomy to breast conservation. Data from patients treated with letrozole suggest that less extensive surgery does not compromise local control providing it is followed by appropriate radiotherapy. Further studies of endocrine therapy are clearly warranted, and the results of a randomized study comparing letrozole with tamoxifen in large operable and locally advanced breast cancer are awaited with interest.

II. INTRODUCTION

In patients with large operable or locally advanced breast cancer, successful neoadjuvant therapy has the potential advantage of down-staging the primary tumor and permitting a more conservative and less extensive surgery (1). Although studies to date have concentrated principally on neoadjuvant chemotherapy, a few groups have evaluated neoadjuvant endocrine therapy in hormone-sensitive, large operable or locally advanced breast cancer (2–4). Substantial reductions in tumor volume over a 3-month period have been recorded in hormone-sensitive tumors with agents such as tamoxifen, aminoglutethimide, and 4-hydroxyandrostenedione (4-OHA). Although limited data are available to date on the use of aminoglutethimide or 4-OHA in the neoadjuvant setting (2,5) the newer synthetic aromatase inhibitors have now superseded these agents. One of these is letrozole (4,4′-[1H-1,2,4-triazol-1-ylmethylene] bis-benzonitrile), a synthetic achiral benzydrytriazole derivative. It is an orally active agent and a highly selective competitive inhibitor of the aromatase enzyme system (6). In Phase III studies in postmenopausal women who had either relapsed on adjuvant therapy or who had progressed during antiestrogen treatment for metastatic disease, letrozole at a dose of 2.5 mg produced a significantly better response rate in advanced breast cancer than megestrol acetate (2,7). In a second Phase III study, letrozole 2.5 mg resulted in a significant survival benefit in patients with advanced breast cancer compared with aminoglutethimide (8). Anastrozole is another well-tolerated oral aromatase inhibitor. A combined analysis of two large randomized trials in postmenopausal women with advanced breast cancer who had failed on tamoxifen has shown that 1 mg of anastrozole significantly increases survival time compared with megestrol acetate at a mean follow-up of 31 months (9). Another aromatase inhibitor currently undergoing study is exemestane, a steroidal aromatase inhibitor. There are few published data using any of these newer agents in the neoadjuvant setting.

Primary endocrine therapy has potential theoretical and practical advantages over chemotherapy. Inhibition of estrogen-stimulated enzyme release (e.g., plasminogen activator and collagenases) may reduce tumor cell shedding. In addition, there is strong evidence that estrogen withdrawal not only reduces growth factor synthesis, but also disrupts the function of a number of other growth factors (10,11) and hormone receptors and/or their second-messenger signaling pathways (12–14). Although there are only four randomized trials of preoperative endocrine therapy, there are least 12 small Phase II studies (15–29). This chapter looks at the results of these trials and presents new data on two series of patients treated with neoadjuvant letrozole and anastrozole.

III. RANDOMIZED TRIALS

There have been four randomized trials of primary endocrine therapy (11–13,30). In nearly all the trials, the major question asked was whether preoperative endocrine therapy could avoid the need for breast surgery and whether survival would be affected in relatively infirm and aged women. In two trials, tamoxifen therapy and no immediate surgery was compared with immediate surgery and no tamoxifen (13,30). In the other two trials, tamoxifen alone was compared with surgery and tamoxifen (11,12). Unsurprisingly, time to local relapse was significantly shorter in the tamoxifen alone arm in all trials. What is surprising is that in three of four trials, the number of patients with distant relapse was slightly lower in the group who had no immediate surgery. This contradicts findings from a more recent combined analysis showing a significant reduction in deaths from breast cancer in patients undergoing immediate surgery (31). Because these trials were designed to address the question of whether delayed surgery is detrimental, it is not possible to use these studies to compare the value of a fixed period of primary endocrine therapy followed by surgery.

IV. PROBLEMS WITH ASSESSMENT OF RESPONSE

One problem with endocrine studies has been evaluating response to agents such as tamoxifen. Response has generally been described using UICC criteria, but in some studies, more accurate attempts have been made to assess response either by measuring multiple diameters or assessing tumor volume using mammography or ultrasound (32). Various studies have reported complete response rates from 8 to 58%, partial response rates from 15 to 75%, stable disease from 7 to 50%, and progressive disease from 0 to 23% (15–29). The median duration of response to tamoxifen was approximately 2 years in studies where it was reported (15,32–34). However the duration in each category of response have been highly variable, often prolonged, and may relate to how rigorously response was assessed.

V. SELECTION OF PATIENTS

In most studies, ER status was not measured, and thus an unknown proportion of patients with ER-negative tumors have been treated. In the World Overview results, ER-negative tumor patients treated by tamoxifen show no significant advantage in time to relapse or survival, and it is unlikely they will benefit from hormonal agents given preoperatively. In ER-positive tumors, however, complete and partial response rates using tamoxifen have been much higher, ranging from 72 to 92% (2,5,16,35,36). These data emphasize the need to select patients with ER-positive tumors for entry into preoperative endocrine trials.

VI. USE OF BREAST CONSERVATION AFTER NEOADJUVANT ENDOCRINE THERAPY

The majority of preoperative endocrine trials have been performed in the elderly, and they have not been designed to assess breast conservation. The prolonged time to response seen in some studies would suggest that primary endocrine therapy is not suitable for this approach. However, as noted previously, some studies have shown impressive reductions in tumor size within 3 months of starting endocrine therapy (2,5,16,31).

VII. NEW DATA USING NEOADJUVANT AROMATASE INHIBITORS

In Edinburgh, we have performed a number of detailed studies using neoadjuvant endocrine therapies. Data from three of these groups are presented here. All patients were postmenopausal, with ER-positive breast cancers (>20 fmol of receptor/mg cytosol protein) or a histoscore of >80 (histoscore calculated by multiplying the percentage of cells staining by the intensity of staining graded from 0 to 3) (37). Only patients with large operable breast cancers or locally advanced breast cancer without evidence of metastases were included, $T_2 > 3$ cm, T_3, T_{4b}, N_{0-1}, and M_0. Patients were M_0 on the basis of normal biochemistry and no metastases on a chest radiograph or bone scan. Patients with inflammatory cancers, extensive peau d'orange, and satellite nodules were excluded.

A. Tamoxifen Patients

Sixty-five patients (aged 59–88 years) were treated as part of a standard protocol within the Edinburgh Breast Unit. All women took tamoxifen at a dose of 20 mg per day, with data on some of these patients being previously published (5).

B. Letrozole Patients

Twenty-four postmenopausal patients (aged 61–87 years) were treated in sequence: The first 12 received 2.5 mg letrozole and the second 12 were treated with 10 mg letrozole. Estrogen receptor levels varied from 40 to 890 fmol/mg cytosol protein, and histoscore levels ranged from 140 to 300 in all these patients.

C. Anastrozole Patients

Twenty-four postmenopausal patients were treated with anastrozole 1 or 10 mg per day in a randomized study. Histoscore levels ranged from 120 to 300 in these patients. Although it was known which patients were treated with 1- or 10-mg doses, the data are presented for the whole group. All the patients treated were ER positive. One patient withdrew because of side effects, and only 23 patients were therefore assessible.

D. Tumor Size Measurement

At the outset of the study, all tumors were measured clinically in four different directions, 45 degrees apart, and the tumor volume calculated using the formula (31):

$$V = \frac{D^3 \times \Pi}{6}$$

where V is the volume and D is the mean diameter.

Patients also had mammograms prior to any treatment, and by measuring the lesion on the oblique and craniocaudal mammogram views, four diameters were obtained. The mean diameter and tumor volume were calculated using the same formula as for clinical measurements.

Patients also had an ultrasound scan at the time of diagnosis. Four scans were performed 45 degrees apart across the tumor and its volume estimated according to the following formula:

$$V = \frac{D^2 \times d \times \Pi}{6}$$

where V is the volume, D is the mean diameter, and d is the mean thickness.

Patients were treated for a total of 3 months following an initial wedge biopsy to obtain tissue samples to confirm the tumor's ER type and enable further biological studies. At the end of the 3-month period, patients treated with tamoxifen were assessed and those deemed operable were treated surgically.

During the 3-month study period patients were monitored at monthly intervals and had clinical and ultrasound volumes calculated at 1, 2, and 3 months. Mammographic volumes were calculated on an initial and a second mammogram performed at the end of the study period; percentage change in volume was then used to assess response to therapy. Percentage change in volume was used in preference to UICC criteria, which requires a 50% reduction in the product of two diameters persisting for at least 1 month; an impractical measure in a 3-month trial.

Patients were only given letrozole or anastrozole if it was deemed that they would be operable after the 3-month treatment period. The extent of surgery was decided after neoadjuvant therapy based on whether breast conservation or mastectomy was needed to excise adequately the remaining tumor. All patients who received letrozole or anastrozole were surgically treated. Following surgery, the need for additional local or systemic treatment was dependent on the extent of the disease in the breast and any axillary node involvement.

VIII. RESULTS OF STUDIES WITH TAMOXIFEN AND AROMATASE INHIBITORS

The number of patients in the three groups who had a reduction in tumor volume by more than 50%, as assessed by ultrasound, is shown in Table 1. Only 2 patients had progression during treatment: 1 from the 65 tamoxifen patients and 1 from the 24 letrozole group. Thirty (46%) patients in the tamoxifen-treated group, 21

TABLE 1 Median Percentage Reductions in Tumor Volume as Assessed by Ultrasound

Drug	No. of patients	No. with >50% reduction	No. with <50% reduction or increase <25%	No. with >25% increase
Tamoxifen	65	30	34	1
Letrozole	24	21	2	1
Anastrozole	23	18	5	0

(88%) from the letrozole group, and 18 (78%) from the anastrozole group had greater than 50% reduction in tumor volume.

A. Changes in Tumor Volume

1. Letrozole

Individual letrozole-treated patients' changes in tumor volume, based on clinical, mammographic, and ultrasound assessments are shown in Figures 1 and 2. There was no significant difference between letrozole 2.5 mg and letrozole 10.0 mg. The median percentage reduction in volume, assessed by ultrasound, was 81% (95% CI 69–86).

2. Anastrozole

Individual patient data are not shown, but there was no significant difference between 1 and 10 mg of anastrozole. The median reduction in tumor volume on

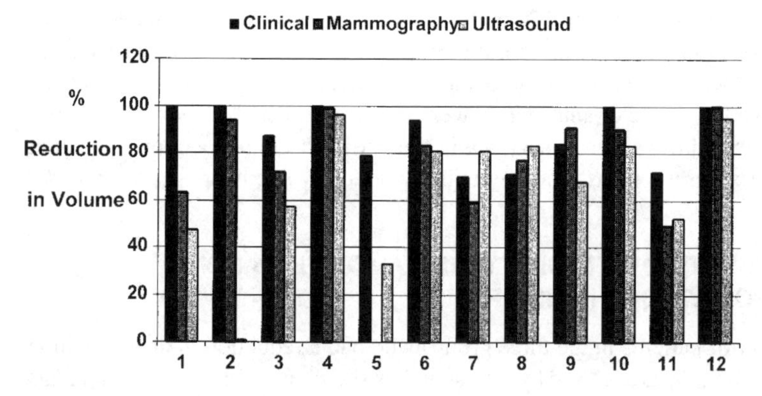

FIGURE 1 Percentage reduction in tumor volume as assessed by clinical examination, mammography, and ultrasound in patients receiving 2.5 mg letrozole daily.

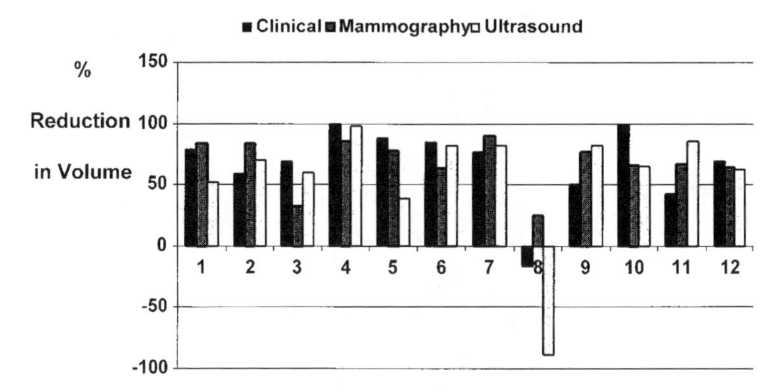

FIGURE 2 Percentage reduction in tumor volume as assessed by clinical examination, mammography, and ultrasound in patients receiving 10 mg letrozole daily.

ultrasound was 75.5% (95% CI 51–79). This is almost identical to that seen with letrozole.

3. Tamoxifen

The percentage median reduction in ultrasound volume with tamoxifen was 48%.

Data comparing the percentage reduction in tumor volume, as assessed by ultrasound, in the 65 tamoxifen patients, 24 letrozole patients, and 23 anastrozole assessible patients are shown in Figure 3.

B. Treatment Outcome of Patients

In the tamoxifen group, 41 of 65 patients were considered to require mastectomy prior to treatment. At the end of 3 months, 15 required mastectomy, 2 continued

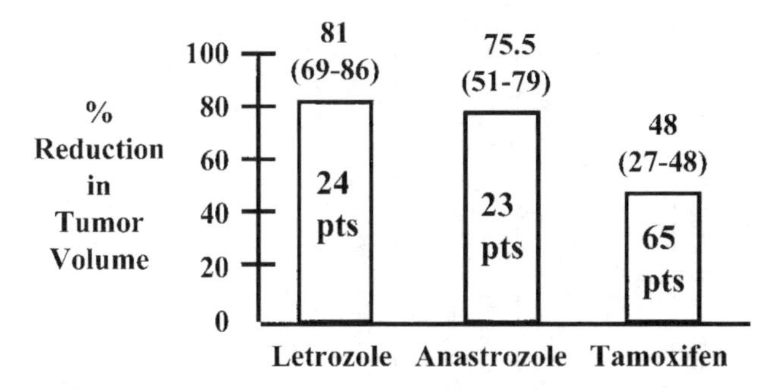

FIGURE 3 Reduction in tumor volume in patients receiving letrozole, anestrozole, and tamoxifen.

TABLE 2 Surgical Outcomes

Drug	No. of patients	No. requiring Mx prior to hormonal therapy	No. having Mx at the end of 3 months hormonal treatment	No. having breast conservation at the end of 3 months hormonal therapy	No. having no surgery
Tamoxifen	65	52	18	38	9
Letrozole	24	15	0	24	0
Anastrozole	23	15	2	21	0

Mx, mastectomy.

on tamoxifen, and 2 had radiotherapy (Table 2). The remaining 22 patients subsequently underwent breast conservation. All 15 patients (of 24) who would have required a mastectomy in the letrozole group underwent breast conservation after 3 months. In the anastrozole group, 15 patients were also considered to require mastectomy at the outset. Subsequently, two patients underwent mastectomy; of these, one had a very large tumor at presentation (the largest tumor in either the letrozole or the anastrozole study group), and the second had a tumor in which response was difficult to assess clinically. In this case, however, a later pathological examination identified a localized area of breast cancer that would, in retrospect, have been suitable for breast conservation.

C. Control Rates Following Breast Conservation

Patients who received letrozole 2.5 mg and 10.0 mg have median follow-up periods of 4 years and 3 years, respectively. All the patients from the letrozole group underwent breast conservation surgery. In the 2.5-mg group, seven patients had a wide local excision only, continuing on letrozole after surgery, and five patients had a wide local excision followed by radiotherapy and continuation on letrozole. Three of the letrozole 10.0-mg group had a wide excision and continuation of letrozole after surgery, and nine patients had a wide excision combined with radiotherapy. Two patients in the 10-mg group also had postoperative chemotherapy because of extensive nodal involvement.

Histology of the wide excision specimens demonstrated that all cancers were completely removed. Ten patients, five each from the 2.5-mg and 10.0-mg dosage groups, were node positive and 14 were node negative. There was one complete pathological response in a patient treated with letrozole 2.5 mg, and three patients had residual microscopic disease only; two in the 2.5-mg and one in the 10.0-mg group.

To date, there have been two local recurrences; both from the 2.5-mg group and in patients who did not have radiotherapy. Of these two patients, one had her therapy changed to tamoxifen; her disease remains under control. The other patient was treated by a reexcission followed by radiotherapy, and she had no local recurrence in the year following her second operation. Four of the 24 patients have died from metastatic disease, one at 15, one at 18, one at 27, and one at 33 months.

IX. DISCUSSION

Primary endocrine therapy is potentially superior to primary chemotherapy in patients with ER-positive tumors. Antiestrogens, most notably tamoxifen and the newer aromatase inhibitors letrozole and anastrozole, are major treatment options for postmenopausal women with large ER-positive tumors (35,38). In addition to their clinical efficacy, these drugs have comparatively few side effects and are well tolerated (39).

Tamoxifen has been frequently used to treat breast cancer in the elderly. Although 60–70% of patients with ER-positive tumors respond to agents such as tamoxifen, the benefits in some patients are short-lived, and it is estimated that only 30–40% of patients sustain a response beyond 2 years (40). However, those patients who do respond can have significant reductions in tumor volume within a 3-month period (5).

New data presented in this chapter have shown substantial reductions in tumor volume both by tamoxifen, letrozole, and anastrozole. Although there appears to be greater reductions in tumor volume with the newer aromatase inhibitors, there are a number of possible explanations for this. The first is patient selection; all patients treated by letrozole and anastrozole had very high levels of ER, and all had potentially operable disease even at the outset of treatment. This contrasts with some of the patients treated with tamoxifen, who had tumors that were not even operable after 3 months of treatment, and these patients may have had more biologically aggressive tumors. Second, tamoxifen takes up to 5 weeks to attain steady-state plasma levels and to have its maximum impact; in contrast to letrozole and anastrozole, which both build up plasma concentrations quickly.

Previous studies using neoadjuvant chemotherapy have demonstrated that significant numbers of patients, even those with inflammatory breast cancer, can have down-staging of their disease allowing breast conservation surgery (41,42). In one series of 226 patients with breast cancer tumors larger than 3 cm in greatest dimension who were candidates for mastectomy, 203 (90%) demonstrated a reduction in tumor size following chemotherapy sufficient to justify breast conservation surgery. Of these 203 patients, there were 12 local recurrences (5.9%) compared with five local recurrences in the 23 patients treated by mastectomy. (21.7%). The mean follow-up in this study was 36 months. This study suggests that breast conservation is safe following neoadjuvant chemotherapy. Results presented here have shown neoadjuvant endocrine therapy produces reductions in tumor volume such that breast conservation is possible in a high percentage of carefully selected patients.

As the majority of patients treated with letrozole and anastrozole were elderly and some were unfit for major surgery, postoperative radiotherapy was withheld following an apparently adequate local excision in patients who had negative nodes. In two of these cases, local recurrence developed. This contrasts to the absence of local recurrence in patients who were treated by wide local excision followed by radiotherapy. In the patients who developed recurrences, one patient was considered unfit for further operation and was treated by tamoxifen with a good response, and the other was able to undergo reexcision followed by radiotherapy, and her disease remains under control.

A potential advantage for elderly populations is that of neoadjuvant therapy reduces the need mastectomy (in appropriate ER- and node-positive subpopulation), which carries a significant mortality risk of approximately 1% (43). By avoiding mastectomy and limiting surgery, patients are kept in hospital for shorter periods of time, limiting the morbidity and mortality associated with operating on such elderly patients.

There are potential weaknesses with the new data presented from Edinburgh. All patients had a wedge biopsy prior to the start of drug treatment. Although the aim was always only to remove 1 g of tissue, this still reduced tumor volume. The biopsy may also have interfered with the tumor blood supply, which in itself could produce a reduction in subsequent tumor volume. Although surgery produces local swelling which can affect assessment of response, the swelling associated with wedge biopsy appeared to settle quickly and was certainly not a problem at the 3-month assessment of tumor size by mammography and ultrasound.

Both mammography and ultrasound proved particularly useful for assessing tumor volume in this group of elderly patients. Data from the anastrozole patients (not shown) comparing correlation between final volume assessed by the pathologist and the volume assessed by clinical examination, mammography, and ultrasound demonstrated ultrasound was the best predictor of final pathological volume (44).

X. CONCLUSIONS

This chapter has demonstrated the efficacy of neoadjuvant endocrine therapy in all studies conducted to date. Toxicity has been low, and in only one patient in the recent studies performed in Edinburgh has the toxicity lead to cessation of treatment. In the remaining patients, treatment was well tolerated. The reductions in tumor volume seen with neoadjuvant endocrine therapy are not that dissimilar to those seen with neoadjuvant chemotherapy (41). From a surgical perspective, the ability to perform less mutilating surgery is obviously an advantage. Furthermore, it does not appear that by performing less surgery there is any compromise in local control—particularly if radiotherapy is given following wide excision. Further studies of neoadjuvant endocrine therapy are clearly warranted, and the results of a randomized study comparing letrozole with tamoxifen in large locally advanced and operable breast cancer are awaited with interest.

REFERENCES

1. Miller WR, Anderson T, Hawkins RA, Keen J, Dixon JM. Neoadjuvant endocrine treatment: the Edinburgh experience. In: Howell A, Dowsett M, eds. ESO Scientific

Updates. Vol 4. Primary Medical Therapy for Breast Cancer. Amsterdam: Elsevier, 1999:1–11.
2. Anderson EDC, Forrest APM, Levack PA, Chetty U, Hawkins RA. Response to endocrine manipulation and oestrogen receptor concentration in large operable primary breast cancer. Br J Cancer 1989; 60:223–226.
3. Leal da Silva JM, Cardosa F, Oliveira F, Cunha H, Pinton Ferreira E. Neoadjuvant hormonal therapy in locally advanced breast cancer (abstr 52). Eur J Cancer 1998; 34(suppl 5):S15.
4. Valero V, Hoff PM, Singletary SE, Buzdar AU, Theriault RL, Strom E, Booser DJ, Asmar L, Frye D, McNeese MD, Hortobagyi GN. Combined modality treatment of locally advanced breast cancer (LABC) in elderly patients (PTS) using tamoxifen (TAM) as primary therapy (abstr 403). Proc Am Soc Clin Oncol 1989; 17:105a.
5. Keen JC, Dixon JM, Miller EP, Cameron DA, Chetty U, Hanby A, Bellamy C, Miller WR. The expression of Ki-S1 and BCL-2 and the response to primary tamoxifen therapy in elderly patients with breast cancer. Breast Cancer Res Treat 1997; 44:123–133.
6. Hamilton A, Piccart M. The third-generation non-steroidal aromatase inhibitors: a review of their clinical benefits in the second-line hormonal treatment of advanced breast cancer. Ann Oncol 1999; 10:377–384.
7. Dombernowsky P, Smith I, Falkson G, Leonard R, Panasci L, Bellmunt J, Bezwoda W, Gardin G, Gudgeon A, Morgan M, Fornasiero A, Hoffman W, Michel J, Hatschek T, Tjabbes T, Chaudri HA, Hornberger U, Trunet PF. Letrozole, a new oral aromatase inhibitor for advanced breast cancer: double-blind randomized trial showing a dose effect and improved efficacy and tolerability compared with megestrol acetate. J Clin Oncol 1998; 16:453–461.
8. Gershanovich M, Chaudri HA, Campos D, Lurie H, Bonaventura A, Jeffrey M, Buzzie F, Bodrogi I, Ludwig H, Reichardt P, O'Higgins N, Romieu G, Friederich P, Lassus M. Letrozole, a new oral aromatase inhibitor: randomized trial comparing 2.5 mg daily with 0.5 mg daily and aminoglutothemide in postmenopausal women with advanced breast cancer. Ann Oncol 1998; 9:639–645.
9. Buzdar AU, Jonat W, Howell A, Jones SE, Blomqvist CP, Vogel CL, Elermann W, Wolter JM, Steinberg M, Webster A, Lee D for the Arimidex Study Group. Anastrozole versus megestrol acetate in the treatment of postmenopausal women with advanced breast carcinoma: results of a survival update based on a combined analysis of data from two mature phase III trials. Cancer 1998; 83:1142–1152.
10. Gaskell DJ, Hawkins RA, de Carteret S, Chetty U, Sangster K, Forrest AP. Indications for primary tamoxifen therapy in elderly women with breast cancer. Br J Surg 1992; 79:1317–1320.
11. Mustacchi G, Milani S, Pluchinotta A, De Matteis A, Rubagotti A, Perrota A. Tamoxifen or surgery plus tamoxifen as primary treatment for elderly patients with operable breast cancer. The GRETA trial. Anticancer Res 1994; 14:2197–2200.
12. Gazet JC, Ford HT, Coombes RC, Bland JM, Sutcliffe R, Quilliam J, Lowndes S. Prospective randomized trial of tamoxifen vs surgery in elderly patients with breast cancer. Eur J Surg Oncol 1994; 20:207–214.
13. Van Dalsen AD, De Vries J. Treatment of breast cancer in elderly patients. J Surg Oncol 1995; 60:80–82.

14. Bergman L, van Dongen JA, van Ooijen B, van Leeuwen FE. Should tamoxifen be a primary treatment choice for elderly breast cancer patients with locoregional disease? Breast Cancer Res Treat 1995; 34:77–83.
15. Ciatto S. Tamoxifen as primary treatment of breast cancer in elderly patients. Neoplasma 1996; 43–45.
16. Dixon JM, Love CDB, Tucker S, Bellamy C, Cameron DA, Miller WR, Leonard RCF. Letrozole as primary medical therapy for locally advanced and large operable breast cancer. Breast Cancer Res Treat 1997; 46(suppl):54.
17. Clemons M, Leahy M, Valle J, Jayson G, Ranson M, Howell A. Review of recent trials of chemotherapy for advanced breast cancer: studies excluding taxanes. Eur J Cancer 1997; 13:2171–2182.
18. Fisher B, Mamounas EP. Preoperative chemotherapy: a model for studying the biology and therapy of primary breast cancer. J Clin Oncol 1995; 13:537–540.
19. Gunduz N, Fisher B, Saffer EA. Effect of surgical removal on the growth and kinetics of residual tumor. Cancer Res 1979; 39:3861–3865.
20. Fisher B, Gunduz N, Coyle J, Rudock C, Saffer E. Presence of a growth-stimulating factor in serum following primary tumor removal in mice. Cancer Res 1989; 49: 1996–2001.
21. Fisher B, Saffer E, Rudock C, Coyle J, Gunduz N. Effect of local or systemic treatment prior to primary tumor removal on the production and response to a serum growth-stimulating factor in mice. Cancer Res 1989; 49:2002–2004.
22. Goldie JH, Coldman AJ. A mathematical model for relating the drug sensitivity of tumors to their spontaneous mutation rate. Cancer Treat Res 1979; 63:1727–1733.
23. Skipper HE. Kinetics of mammary tumor cell growth and implications for therapy. Cancer 1971; 28:1479–1499.
24. Gregory H, Thomas CE, Willshire IR, Young JA, Anderson H, Baildam A, Howell A. Epidermal and transforming growth factor alpha in patients with breast tumors. Br J Cancer 1989; 59:605–609.
25. Noguchi S, Motomura K, Inaji H, Imaoka S, Koyama H. Down-regulation of transforming growth factor alpha by tamoxifen in human breast cancer. Cancer 1993; 72:131–136.
26. Vignon F, Bouton MM, Rochefort H. Anti-estrogens inhibit the mitogenic effect of growth factor on breast cancer cells in the total absence of estrogens. Biochem Biophy Res Commun 1987; 146:1502–1508.
27. Ignar-Trowbridge DM, Nelson KG, Bidwell MC, Cutis SW, Washburn TF, McLachlan JA, Korach KS. Coupling of dual signaling pathways: epidermal growth factor action involves the estrogen receptor. Proc Natl Acad Sci USA 1992; 89:4658–4662.
28. Katzenellenbogen BS, Montano MM, Ekena K, Herman ME, McInerney EM. Antiestrogens: mechanism of action and resistance in breast cancer. Breast Cancer Res Treat 1997; 44:23–48.
29. Early Breast Cancer Trialists' Collaborative Group. Tamoxifen for early breast cancer: an overview of the randomised trials. Lancet 1998; 351:1451–1467.
30. Bates T, Riley DL, Houghton J, Fallowfield L, Baum M. Breast cancer in elderly women: a Cancer Research Campaign trial comparing treatment with tamoxifen and optimal surgery with tamoxifen alone. Br J Surg 1991; 78:591–594.

31. Kenny FS, Robertson JFR, Ellis IO, Elston CW, Blamey RW. Primary tamoxifen versus mastectomy and adjuvant tamoxifen in fit elderly patients with operable breast cancer of high ER content. Breast 1999; 8:216.
32. Forouhi P, Walsh JS, Anderson TJ, Chetty U. Ultrasound as a method of measuring breast tumor size and monitoring response to primary systemic therapy. Br J Surg 1994; 81:223–225.
33. Preece PE, Wood RAB, Mackie CR, Cuschieri A. Tamoxifen as initial sole treatment of localised breast cancer in elderly women: a pilot study. BMJ 1982; 284:869–870.
34. Bradbeer JW, Kyndgon J. Primary treatment of breast cancer in elderly women with tamoxifen. Clin Oncol 1983; 9:31–34.
35. Allan SG, Rodger A, Smyth JF, Leonard RCF, Chetty U, Forrest APM. Tamoxifen as primary treatment of breast cancer in elderly or frail patients: a practical management. Br Med J 1985; 29:358.
36. Low SC, Dixon AR, Bell J, Ellis IO, Elston CW, Robertson JFR, Blamey RW. Tumor oestrogen receptor content allows selection of elderly patients with breast cancer for conservative tamoxifen treatment. Br J Surg 1992; 79:1314–1316.
37. McCarty KS Jr, Miller LS, Cox EB, Konrath J, McCarty KS Sr. Estrogen receptor analyses: correlation of biochemical and immunohistochemical methods using monoclonal antireceptor antibodies. Arch Pathol Lab Med 1985; 109:716–721.
38. McGuire WL, Carbonne PP, Sears ME et al. Estrogen receptor in human breast cancer: an overview. New York: Raven Press, 1975.
39. Furr BJ, Jordan VC. The pharmacology and clinical uses of tamoxifen. Pharmacol Ther 1984; 25:127–205.
40. Horobin JM, Preece PE, Dewar JA, wood RA, Cuschieri A. Long-term follow-up of elderly patients with locoregional breast cancer treated with tamoxifen only. Br J Surg 1991; 78:213–217.
41. Veronesi U, Bonadonna G, Zurrida S, Galimberti V, Greco M, Brambilla C, Luini A, Andreola S, Rilke F, Raselli R, Merson M, Sacchini V, Agresti R. Conservation surgery after primary chemotherapy in large carcinomas of the breast. Ann Surg 1995; 222:612–618.
42. Singletary SE, McNeese MD, Hortobagyi GN. Feasibility of breast-conservation surgery after induction chemotherapy for locally advanced breast carcinoma. Cancer 1992; 69:2849–2852.
43. Hunt KR, Fry DR, Bland KI. Breast carcinoma in the elderly patient: an assessment of operative risk, morbidity and mortality. Am J Surg 1980; 140:339–342.
44. Dixon JM, Renshaw L, Bellamy C, Cameron DA, Miller WR, Leonard RCF. Efficacy of anastrozole as neoadjuvant therapy in postmenopausal women with large operable breast cancers: reductions in tumor volume. Breast 1999; 8:215.

7

Adjuvant Endocrine Therapy

H. T. Mouridsen
Rigshospitalet
Copenhagen, Denmark

I. ABSTRACT

Many of the endocrine treatment modalities with demonstrated activity in advanced breast cancer have been applied in the adjuvant situation. These include ovarian ablation, additive therapy with diethylstilbestrol (DES) and medroxyprogesterone acetate (MPA), aromatase inhibitor therapy with aminoglutethimide, and competitive therapy with tamoxifen and toremifene.

Based on available data, tamoxifen given for 5 years is the established endocrine adjuvant therapy in patients with receptor-positive tumors either as a single agent in postmenopausal patients or added to chemotherapy in premenopausal patients. Although the data with tamoxifen are encouraging, continued research in the field is needed to improve further the therapeutic outcome.

New endocrine modalities are now being tested in the adjuvant setting, but a better understanding of the biology of breast cancer and a means to overcome endocrine resistance are major prerequisites to achieve a dramatic improvement of the prognosis of breast cancer.

II. INTRODUCTION

Endocrine therapy has been used in advanced breast cancer since it was realized that it was frequently a systemic disease. At the time of diagnosis, different endocrine modalities, which are active in advanced breast cancer, have been used as an adjuvant in primary breast cancer.

This chapter summarizes the available data from randomized trials looking at different endocrine modalities in the adjuvant setting and the criteria for patient selection in current standard endocrine therapy and describe the design of some major ongoing trials with different endocrine modalities in postmenopausal patients.

III. HISTORY OF ENDOCRINE THERAPY IN ADVANCED BREAST CANCER

Endocrine therapy of advanced breast cancer was introduced in 1896 when Beatson observed the regression of skin nodules following oophorectomy (1). Approximately 50 years later, the major ablative procedures, adrenalectomy (2) and hypophysectomy (3), were introduced.

The first experience of additive therapies using androgens (4), estrogens (5), and synthetic progestational compounds (6) were published in the 1950s, and these were followed in the early 1970s by tamoxifen (7), competing with the estrogen receptor (ER), and aminoglutethimide, the first aromatase inhibitor (8).

The past 10 years have seen the development of the new antiestrogens toremifene (9), raloxifene (10), and droloxifene (11), which have lower estrogenic properties compared to tamoxifen and the pure antiestrogen ICI 182.780 (12). In addition, second-generation aromatase inhibitors [fadrozole (13) and formestane (14)] and now third-generation aromatase inhibitors [anastrozole (15), letrozole (16), and exemestane (17)] provide even higher specificity with a close to complete suppression of circulating estrogens. Finally, we have seen the introduction of luteinizing hormone–releasing hormone (LHRH) analogues inhibiting ovarian synthesis of estrogens (18).

Although the modes of action differ, all the endocrine therapies aim to reduce estrogen-stimulated tumor growth and have largely similar efficacies, leading to a response in approximately one-third of unselected patients receiving endocrine therapy as first-line therapy for advanced breast cancer. It is generally accepted that the presence of estrogen and/or progesterone receptor in the tumor predicts the probability of response.

IV. ADJUVANT RANDOMIZED TRIALS WITH ENDOCRINE THERAPY

The concept of breast cancer frequently being that of a disseminated disease at the time of primary diagnosis has led to the introduction of systemic therapy as an adjuvant to the primary local approach. The benefits of the majority of advanced breast cancer endocrine therapies have been tested in this situation.

V. ABLATIVE THERAPY

The results of the randomized trials of ovarian ablation have been analyzed by the Early Breast Cancer Triallists' Collaborative Group (19). In the absence of chemotherapy, the relative reduction of recurrence rate at 15 years is 25 ± 7% and the relative reduction of mortality is 24 ± 7%. Thus, the effects are of the same order as that achieved with polychemotherapy (20). Similar benefit with the two modalities was confirmed in a randomized trial (21). Premenopausal patients with breast cancer stage T_0–T_3, N_0–N_2, M_0 and in whom axillary nodes were histologically proven to be affected were randomly allocated following primary local therapy to a 2 × 2 factorial designed trial with four options: ovarian ablation or chemotherapy alone, or ovarian ablation combined with long-term prednisolone, or chemotherapy combined with long-term predinisolone. Ovarian ablation was performed either through a surgical or irradiation procedure. The chemotherapy was CMF (cyclophosphamide 750 mg/m^2, methotrexate 50 mg/m^2, and 5-fluorouracil 600 mg/m^2), with all drugs being given intravenously eight times every 3 weeks. From December 1984, the regimen was modified to six three-weekly cycles for the remainder of the trial. Prednisolone was given orally, 7.5 mg daily, from the date of ovarian ablation or initiation of chemotherapy for 5 years' duration or until relapse if this occurred earlier. From March 1980 to May 1990, 332 patients were included. With a median follow-up of 5.9 years, there were no differences between the ovarian ablation and the CMF groups concerning recurrence-free survival (RFS) (relative risk [RR] = 1.08; 95% confidence interval [CI] = 0.77–1.51) and overall survival (OS) (RR = 1.12; CI = 0.76–1.63). Retrospectively, the data were analyzed according to the ER content with positivity defined ≥20 fmol/mg cytosol protein. In the receptor-positive cases, ovarian ablation was superior to CMF in RFS (RR = 0.65; CI = 0.37–1.15), but in the receptor negative cases, ablation was inferior to CMF (RR = 1.74; CI = 0.95–3.19).

This finding in the retrospective subgroup analysis, however, could not be confirmed in a prospective randomized trial conducted by the Danish Breast Cancer Cooperative Group (22). This trial allocated premenopausal patients with node-positive receptor-positive tumors to either ovarian ablation by irradiation or to CMF (cyclophosphamide 600, methotrexate 40, 5-fluorouracil 600 mg/m^2) administered intravenously every 3 weeks for a total of nine cycles. From January 1990 to June 1998, 732 patients were randomized and with a follow-up of 68 months RFS and OS in the ovarian ablation and CMF groups were 67 versus 66% and 78 versus 82%, respectively.

Table 1 Adjuvant Trials in Premenopausal Patients with Receptor-Positive, Node-Positive, or Node-Negative Tumors. Chemotherapy (CT) versus endocrine therapy (ET)

Ref.	CT	ET	N	FU	RFS	OS
23	CMF	G3 + T5	1.045	42 M	0.02	NS
24	CMF	G2 + T5	244	53 M	NS	NS

NS, not significant; G3, goserelin 3 years; G2, goserelin 2 years; T5, tamoxifen 5 years.

Preliminary data from two randomized trials comparing CMF with LHRH analogue (plus tamoxifen) have recently been reported (Table 1). In the Austrian trial, CMF (cyclophosphamide 600, methotrexate 40, 5-fluorouracil 600 mg/m^2) was administered intravenously on days 1 and 8 for six 28-day cycles to patients with receptor-positive, node-positive, or node-negative breast cancer (23). Dosage of goserelin was 3.6 mg every 4 weeks for 3 years and for tamoxifen 20 mg daily for 5 years. Concerning RFS, the results favored the combination endocrine therapy, with no difference being observed in survival. In the small Italian trial, patients received six classic cycles of CMF (cyclophosphamide 100 mg/m^2 po for 14 days, methotrexate 40 mg/m^2, 5-fluorouracil 600 mg/m^2 iv days 1 and 8 every 4 weeks (24). Goserelin was given for 2 years (3.6 mg every 4 weeks) and tamoxifen for 5 years (30 mg daily). With 53 months' follow-up, RFS were 69% in the CMF group compared to 72%. An identical OS was seen in the two groups (87%).

Another two trials compared CMF with 2 years of goserelin therapy (25,26), but so far no data are available. The International Breast Cancer Study Group (IBCSG) trial also analyzes the benefit with sequential chemotherapy and LHRH analogue, as does an American intergroup trial (27). The latter demonstrated no benefit when CAF (cyclophosphamide 100 mg/m^2 po for 14 days plus adriamycin 30 and 5-fluorouracil 500 mg/m^2 iv days 1 and 8) for six 28-day cycles was followed by goserelin (3.6 mg every 4 weeks for 5 years). However, combined goserelin and tamoxifen for 5 years was associated with a significant improvement in RFS (78 vs 67%).

Finally, an international trial (28) initially demonstrated improved RFS with goserelin compared to a control group; however, so far, these data are difficult to interpret, as the majority of patients had received prior tamoxifen or chemotherapy.

In conclusion, according to available data, ovarian ablation offers benefit in patients with receptor-positive tumors similar to that which can be achieved with chemotherapy with CMF, but whether additional benefit can be achieved with combination endocrine therapy remains to be investigated.

VI. ADDITIVE THERAPY

Data from published trials that use additive therapy as an adjuvant are summarized in Table 2. The first trial (29) recruited postmenopausal women with stage T_0–T_4, N_0–N_1, M_0 disease. Simple mastectomy was the standard surgical procedure at the time, and axillary surgery was not routinely performed; that is, the pathological nodal status is unknown. All patients received postoperative radiotherapy and were subsequently randomized to placebo versus tamoxifen 30 mg daily or diethylstilbestrol (DES) 3 mg daily for 2 years' duration. From March 1975 to March 1978, 154 patients were randomized. With a median follow-up of 9 years, RFS times were significantly superior in the DES group compared with the controls (RR = 0.57, CI = 0.33–0.99), but OS were not significantly different (RR = 0.69, CI = 0.34–1.18). Similar results were achieved with tamoxifen (RR = 0.60, CI = 0.35–1.04 and RR = 0.69, CI = 0.38–1.25, respectively). Side effects were frequently reported in the DES group, with the most important being metrorhagia, which occurred in 74% of the cases compared to 14% with tamoxifen and 0% with placebo.

In the second trial (30), postmenopausal patients with T_1-T_{3A}, N_0–N_2, Pn-positive tumors were randomized following radical and modified radical mastectomy to a control group or to oral treatment with medroxyprogesterone acetate (MPA). Dose was 1000 mg twice daily during 30 days followed by 500 mg twice daily for another 5 months. From April 1979 to March 1986, 138 patients entered the trial. With 36 months median follow-up, the estimated 5 years' RFS were not significantly different (24 vs. 50%, $P = .13$), nor were the survivals (48 vs 62%, $P = .34$). A significant proportion of the patients experienced side effects with the most frequent being vaginal spotting, tremors, sweating, and weight gain.

In the third and largest trial (31), premenopausal and postmenopausal patients with stage pN_0 (at least five examined nodes) and M_0 were randomized following primary surgery and optional radiotherapy (mandatory following

TABLE 2 Adjuvant Additive Endocrine Therapy (ET)

Ref.	ET	N	FU	RFS (%)	OS (%)
29	Placebo, 2 yrs	52	9 yrs	39	52
	DES 3 mg, 2 yrs	50		58	66
30	Control	66	3 yrs	24	48
	MPA HD 6 mons	69		50	62
31	Control	106	3 yrs	73	88
	MPA HD 6 mons	103		94	99

breast-conserving therapy) to a control group or to intramuscular treatment with MPA. The dose was 500 mg daily during 4 weeks or 500 mg daily 5 days a week for 5 weeks followed by 500 mg twice weekly for another 5 months. From 1982 to 1987, 240 patients were randomized. With a median follow-up of 3 years, additional RFS was significantly different in favor of MPA (94 vs 73%, $P =$ .0006). Overall survival in the two groups were 99 versus 88%, respectively ($P < .04$). Vaginal spotting, tremor, weight gain, sweating, and cramps occurred significantly more frequently in the MPA group.

Thus, according to the data from these small trials, adjuvant additive therapy with high doses of DES or MPA is associated with a prolongation of the time to recurrence or death. However, the trials are quite small, and in the two MPA trials, the follow-up time is very short, and unfortunately updated results have not been published. In addition, the therapies are associated with frequent unpleasant side effects.

VII. INHIBITIVE THERAPY

Two trials tested the value of aminoglutethimide in the adjuvant setting (Table 3). In the first (32), postmenopausal patients less than 75 years old with node-positive tumors were randomized to receive aminoglutethimide plus hydrocortisone for 2 years versus placebo. Initial dose of aminoglutethimide was 250 mg twice daily, which, if tolerated, was increased following 4 weeks to 250 mg three times daily and following 8 weeks to 250 mg four times daily. Dose of hydrocortisone was 20 mg twice daily throughout the treatment period. From 1979 to 1986, 354 patients entered the trial. With a median follow-up of 8 years, no significant differences were apparent in RFS and OS (Table 3). However, as reported earlier (33), there was a benefit for the treatment versus placebo group in RFS ($P =$ .005) and OS ($P = .05$) of up to approximately 4 years, which subsequently

Table 3 Adjuvant Endocrine Therapy (ET) with Aminoglutethimide (AG)

Ref.	ET	N	FU	RFS (%)	OS (%)
32	Placebo 2 yrs	179	8 yrs	36	45
	AG + HC 2 yrs	175		37	45
34	Tamoxifen 5 yrs		4 yrs	86	94
		2021			
	Tamoxifen 5 yrs +− AG 2 yrs			86	95

disappeared. The authors speculate whether the discrepancy between these data and tamoxifen data may be ascribable to the different biological properties of the two endocrine agents at a cellular level. The most frequent side effects included lethargy, skin rash, nausea, and ataxia.

Preliminary data from another trial were recently reported (34). Postmenopausal patients with receptor-positive, node-negative, or node-positive tumors were randomized to receive either tamoxifen 20 mg twice daily for 2 years and 20 mg daily for the subsequent 3 years or tamoxifen as above plus aminoglutethimide 250 mg twice daily for the first 2 years. From 1990 to 1996, 2021 patients were randomized, and with 4 years median follow-up, RFS and OS are similar in the two groups (see Table 3). Side effects prompted more withdrawals in the combination group.

In conclusion, the short-term benefit or lack of benefit combined with significant toxicity preclude further evaluation of aminoglutethimide in adjuvant therapy in breast cancer.

VIII. COMPETITIVE THERAPY

The available data from numerous randomized trials with tamoxifen including more than 30,000 women were reviewed in the early 1990s by the Early Breast Cancer Trialists' Collaborative Group (35). A highly significant reduction was achieved with tamoxifen at 10 years in both RFS (25 ± 2%) and OS (17 ± 2%). These results were confirmed in a recent update (36), which also analyzed in further detail the importance of treatment duration and the benefit according to different subgroups.

Indirect comparisons of the importance of treatment duration indicated that prolonged therapy is superior to shorter duration, and based on data from randomized trials, a 5-year treatment duration is now considered standard (37–39). Other major trials (aTTom and Atlas) are ongoing analyzing 2 versus 5 years tamoxifen (40) and tamoxifen beyond 5 years (41).

Subgroup analyses according to receptor status demonstrated that, for each tamoxifen duration, the proportional event reduction is greater for patients with ER-positive tumors than for ER-negative tumors. Thus, in receptor-positive tumors, the RFS reduction in trials of 1, 2, and about 5 years were all highly significant, 21, 28, and 50%, respectively, as was the mortality reduction (14, 18, and 28%, respectively). In contrast, in the ER-negative tumors' RFS reduction was significant only in trials of 2 years (13%) but nonsignificant in trials of 1 and approximately 5 years (4 and 6%, respectively), and with all durations, no significant mortality reductions were observed (6, 7, and −3%, respectively).

Other subgroups' analyses indicated similar event reduction according to nodal status. There are significant trends toward greater risk reduction in older women compared to younger women in trials of tamoxifen 1 or 2 years, but this trend is weaker in trials of tamoxifen of approximately 5 years.

A number of trials analyzing the beneficial effect of tamoxifen when added to chemotherapy showed an apparent risk reduction similar to that seen without the presence of chemotherapy.

Recently published preliminary data comparing tamoxifen with one of the newer SERMS, toremifen (42), reported no difference in RFS after a median follow-up of 3 years in 899 randomized patients.

Thus, standard endocrine adjuvant treatment today is tamoxifen for 5 years. This treatment significantly reduces the risk of relapse and death and the drug is well tolerated. Although long-term effects include significantly reduced contralateral breast cancer incidence, there is also an approximate doubling of endometrial cancer incidence (36).

IX. SELECTION OF PATIENTS FOR ADJUVANT ENDOCRINE THERAPY

The criteria for patient selection according to nodal status as defined at the International Consensus Conference on Primary Treatment of the Breast: Update 1998 (43) are presented in Tables 4 and 5. Tamoxifen is considered the endocrine treatment of choice in both premenopausal and postmenopausal patients with receptor-positive, node-negative, and node-positive tumors either alone (postmenopausal with node-negative intermediate and high-risk and node-positive tumors) or in combination with chemotherapy (premenopausal with node-positive tumors). In other subgroups, the role of endocrine therapy is still being tested in randomized trials.

Table 4 Adjuvant Treatment for Patients with Node-Negative, Receptor-Positive Tumors

Size, grade, receptor	Low risk $\leq$1, I, Pos	Intermediate risk $>$1–2, I–II, Pos	High risk $>$2, II–III, Neg
Premenopausal	None or Tamoxifen	Tamoxifen ± CT[a]	CT[a] ± Tamoxifen
Postmenopausal	None or Tamoxifen	Tamoxifen ± CT[a]	Tamoxifen ± CT[a]

[a] Considerations about low relative risk, costs, toxicity, age may justify the use of tamoxifen alone.

TABLE 5 Adjuvant Treatment for Patients with Node-Positive, Receptor-Positive Tumors

Premenopausal	CT + Tamoxifen
	Ovarian ablation/LHRHa ± Tamoxifen[a]
	CT ± Ovarian ablation/LHRHa ± Tamoxifen
Postmenopausal	Tamoxifen ± CT[b]

[a] Still being tested in randomized trials.
[b] Considerations about low relative benefit, costs, and toxicity with CT and age may justify the use of tamoxifen alone.

X. ONGOING RANDOMIZED TRIALS IN POSTMENOPAUSAL PATIENTS

Although the data with adjuvant tamoxifen are encouraging, we are still far from having achieved a dramatic improvement of prognosis in primary breast cancer, and there is a need to develop new and more active endocrine modalities. With this aim, recent years have seen the initiation of a number of major adjuvant trials analyzing the value of third-generation aromatase inhibitors and ICI 182.780 in the adjuvant setting. Many of these trials are company-sponsored, monitored, multi-institutional, and international studies.

In the adjuvant setting, two different designs have been adopted; head-to-head comparisons of tamoxifen versus the aromatase inhibitors or tamoxifen versus the sequential use of tamoxifen and aromatase inhibitors. In the neoadjuvant setting, the head-to-head designs have been used in trials comparing tamoxifen with aromatase inhibitor or ICI 182.780.

Study 1033 IL/0029 (ATAC) compares tamoxifen 5 years, or anastrozole 5 years or the combination of tamoxifen and anastrozole 5 years followed by 5 years combination therapy. The trial is double blind and recruits postmenopausal patients. The study started in 1996 and was closed in late 1999 with an entry of approximately 9000 patients.

The FEMTA 019 trial recruits postmenopausal patients with ER-positive, node-positive, or high-risk node-negative breast cancer, who are randomized in a double-blind trial to tamoxifen 5 years versus letrozole 5 years. Following the recruitment of approximately 2000 patients, the organization of the trial has been taken over by the IBCSG as part of a trial conducted by the Breast International Group (BIG). As a result, the trial design has been amended to a four-arm trial with tamoxifen 5 years versus letrozole 5 years versus tamoxifen 2 years followed by letrozole 3 years versus letrozole 2 years followed by tamoxifen 3 years. The study is planned to enter 7000 patients.

The study, NCIC CTG MA 17, randomizes postmenopausal patients with ER-positive tumors, who are recurrence-free following adjuvant treatment with tamoxifen for 5 years to either placebo or to another 5 years of treatment with letrozole. The planned number of patients to be randomized is 2380.

Another three trials also use a sequential design. Following 2 years of adjuvant tamoxifen, two identical trials, conducted by the German and Austrian Breast Groups, respectively, randomize postmenopausal patients to another 3 years of tamoxifen or anastrozole. The third trial (study OEXE 031) recruits patients who have received adjuvant tamoxifen for 2–3 years and then randomizes them to either tamoxifen or exemestane for another 2–3 years; that is, a total treatment duration of 5 years.

Two trials are conducted in the preoperative setting. One of these (study 024) randomized postmenopausal patients with ER-positive tumors to either tamoxifen or letrozole. The trial was closed August 1999 with an entry of 302 patients. The other trial to be activated shortly plans to recruit approximately 3500 premenopausal and postmenopausal patients and randomize them to placebo versus ICI 182.780.

Data from these important trials will be generated in the years to come, as will the data from adjoined studies with the objective to analyze long-term toxicities as concerns the potential consequences of therapy-induced alterations in bone and lipid metabolism.

REFERENCES

1. Beatson GT. On the treatment of inoperable cases of carcinoma of the mamma: suggestions for a new method of treatment with illustrative cases. Lancet 1896; 2: 162–165.
2. Huggins C, Bergenstal DM. Inhibition of human mammary and prostatic cancers by adrenalectomy. Cancer Res 1952; 12:134–141.
3. Luft R, Olivencrona H. Experiences with hypophysectomy in man. J Neurosurg 1953; 10:301–316.
4. Garton DAG. Androgen therapy in 70 cases of advanced mammary carcinoma. Br J Cancer 1950; 4:20–53.
5. Council on Drugs, Subcommittee on Breast and Genital Cancer, Committee on Research, American Medical Association. Androgens and estrogens in the treatment of disseminated mammary carcinoma. Retrospective study of 944 patients. JAMA 1960; 172:1271–1283.
6. Stoll BA. Progestin therapy of breast cancer: comparison of agents. BMJ 1967; 2: 338–341.
7. Cole MP, Jones CTA, Todd IDH. A new anti-estrogenic agent in late breast cancer. An early clinical appraisal of ICI 46474. Br J Cancer 1971; 25: 270–274.
8. Griffiths CT, Hall TC, Saba Z, Barlow JJ, Nevinny HB. Preliminary trial of aminoglutethimide in breast cancer. Cancer 1973; 32:31–37.
9. Valavaara R. Phase II trials with toremifene in advanced breast cancer: a review. Breast Cancer Res Treat 1990; 16:31–35.
10. Buzdar AU, Marcus C, Holmes F. Phase II evaluation of LY 156758 in metastatic breast cancer. Oncology 1988; 45:344–345.

11. Rauschning W, Pritchard KI. Droloxifene, a new anti-estrogen: its role in metastatic breast cancer. Breast Cancer Res Treat 1994; 31:83–94.
12. Howell A, DeFriend D, Robertson J, Blamey R, Walton P. Response to a specific anti-estrogen (ICI 182.870) in tamoxifen resistant breast cancer. Lancet 1995; 345: 29–30.
13. Falkson G, Raats JI, Falkson HC. Fadrozole hydrochloride, a new nontoxic aromatase inhibitor for the treatment of patients with metastatic breast cancer. J Steroid Biochem Mol Biol 1992; 43:161–165.
14. Goss PE, Powles TJ, Dowsett M. Treatment of advanced postmenopausal breast cancer with an aromatase inhibitor, 4-hydroxyandrostenedione: phase II report. Cancer Res 1986; 46:4823–4826.
15. Plourde PV, Dyroff M, Dukes M. Arimidex: a potent and selective fourth generation aromatase inhibitor. Breast Cancer Res Treat 1994; 30:103–111.
16. Demers LM. Effects of fadrozole (CGS 16949 A) and letrozole (CGS 20267) on the inhibition of aromatase activity in breast cancer patients. Breast Cancer Res Treat 1994; 30:95–102.
17. Thurlimann B, Paridaens R, Roche M, Bonneterre J, Zurlo MG, Lanzalone S, Arkhipov AI. Exemestane in postmenopausal pretreated breast cancer: a multicenter phase-II study in patients with aminoglutethimide failure. Ann Oncol 1994; 5(suppl 8):29.
18. Kaufmann M, Jonat W, Kleeberg U. Goserelin, a depot gonadotrophin-releasing hormone agonist in the treatment of premenopausal patients with metastatic breast cancer. J Clin Oncol 1989; 7:1113–1119.
19. Early Breast Cancer Triallists' Collaborative Group. Ovarian ablation in early breast cancer: overview of the randomized trials. Lancet 1996; 348:1189–1196.
20. Early Breast Cancer Triallists' Collaborative Group. Polychemotherapy for early breast cancer: an overview of the randomized trials. Lancet 1998; 358:930–942.
21. Scottish Cancer Trials Breast Group and ICRF Breast Unit, Guy's Hospital, London. Adjuvant ovarian ablation versus CMF polychemotherapy in premenopausal women with pathological stage II breast carcinoma: the Scottish trial. Lancet 1993; 341: 1293–1298.
22. Ejlertsen B, Dombernowsky P, Mouridsen HT, Kamby C, Kjaer M, Rose C, Andersen KW, Jensen MB, Bengtsson O, Bergh J. Comparable effect of ovarian ablation and CMF chemotherapy in premenopausal hormone receptor positive breast cancer patients. Proc ASCO 1999; 18:66a.
23. Jakesz R, Hausmaninger H, Samonigg H, Kubista E, Depisch D, Fridrik M, Stierer M, Gnant M, Steger G, Kalo R, Jatzko G, Hofbauer F, Reiner G, Luschin-Ebengreuth G. Comparison of adjuvant therapy with tamoxifen and goserelin vs CMF in premenopausal stage I and II hormone-responsive breast cancer patients: four-year results of Austrian Breast Cancer Study Group (ABCSG) Trial 5. Proc ASCO 1999; 18:67a.
24. Boccardo F, Rubagotti A, Amoroso D, Mesiti M, Minutoli N, Aldrigetti D, Bolognesi A, Genta F, Giai M, Irtelli L, Donati D, Pacini P, Farris A, Schieppati G. CMF versus tamoxifen plus goserelin as adjuvant treatment of ER-positive pre-perimenopausal breast cancer patients. Preliminary results of the GROCTA 02 Study. Proc ASCO 1998; 17:99a.
25. Jonat W, Study Group. The ZEBRA study: Zoladex (goserelin) vs CMF as adjuvant

therapy in the management of node positive stage II breast cancer in pre-perimenopausal women aged 50 years or less. Breast Cancer Res Treat 1998; 50:283.
26. Kaufmann M. Luteinizing hormone-releasing hormone analogues in early breast cancer: updated status of ongoing clinical trials. Br J Cancer 1998; 78:9–11.
27. Davidson N, O'Neill A, Vukov A, Osborne CK, Martino S, White D, Abeloff MD. Effect of chemohormonal therapy in premenopausal, node (+), receptor (+) breast cancer. An Eastern Cooperative Oncology Group Phase III intergroup Trial Proc ASCO 1999; 18:67a.
28. Rutquist LE. Zoladex and tamoxifen as adjuvant therapy in premenopausal breast cancer. A randomized trial by the Cancer Research Campaign (CRC). Breast Cancer Trials Group, the Stockholm Breast Cancer Study Group, the South East Sweden Breast Cancer Group and the Gruppo Interdisciplinare Valutazione Interventi in Oncologia (GIVIO). Proc ASCO 1999; 18:67a.
29. Palshof T. Adjuvant endocrine therapy in premenopausal and postmenopausal women with breast cancer. A report of the Copenhagen Breast Cancer Trials 1975–1987. Thesis, Medi-book, 1987.
30. Panutti F, Martoni A, Cilenti G, Camaggi CM, Fruet F. Adjuvant therapy for operable breast cancer with medroxyprogesterone acetate alone in postmenopausal patients or in combination with CMF in premenopausal patients. Eur J Cancer Clin Oncol 1988; 24:423–431.
31. Focan C, Baudoux A, Beauduin M, Bunesco U, Dehasque N, Dewasch L, Lobelle JP, Longeval E, Majois F, Mazy V, Nickers P, Salomon E, Tagnon A, Tytgat J, van Belle S, Vanderlinden B, Vandervellen R, Vindevoghel A. Adjuvant treatment with high dose medroxyprogesterone acetate in node-negative early breast cancer. Acta Oncol 1989; 28:237–240.
32. Jones AL, Powles TJ, Law M, Tidy A, Easton D, Coombes RC, Smith IE, Mckinna JA, Nash A, Ford HT, Gazet JC. Adjuvant aminoglutethimide for postmenopausal patients with primary breast cancer: analysis at 8 years. J Clin Oncol 1992; 10:1547–1552.
33. Coombes RC, Powles TJ, Easton D, Chilvers C, Ford HT, Smith IE, McKinna A, White H, Bradbeer J, Yarnold J, Nash A, Bettelheim R, Dowsett M, Gazet J-C. Adjuvant aminoglutethimide therapy for postmenopausal patients with primary breast cancer. Cancer Res 1987; 47:2494–2499.
34. Samonigg H, Jakesz R, Hausmaninger H, Depisch D, Fridrik M, Stierer M, Steger G, Kolb R, Hofbauer F, Reiner G, Schmid M. Tamoxifen vs aminoglutethimide for stage I and II receptor-positive postmenopausal node-negative or node-positive breast cancer patients: Four-year results of a randomized trial of the Austrian Breast Cancer Study Group (ABCSG). Proc ASCO 1999; 18:68a.
35. Early Breast Cancer Triallists' Collaborative Group. Systemic treatment of early breast cancer by hormonal, cytotoxic or immune therapy. Lancet 1992; 339:1–15.
36. Early Breast Cancer Triallists' Collaborative Group. Tamoxifen for early breast cancer: an overview of the randomized trials. Lancet 1998; 351:1451–1467.
37. Swedish Breast Cancer Cooperative Group. Randomized trial of two versus five years of adjuvant tamoxifen for postmenopausal early stage breast cancer. J Natl Cancer Inst 1996; 88:1543–1549.
38. Fisher B, Dignam J, Bryant J, DeCillis A, Wickerham DL, Wolmark N, Costantino

J, Redmund C, Fisher ER, Bowman DM, Deschenes L, Dimitrov NV, Margolese RG, Robidoux A, Shibata H, Terz J, Paterson AHG, Feldman MI, Farrar W, Evans J, Lickley HL. Five versus more than five years of tamoxifen therapy for breast cancer patients with negative lymph nodes and estrogen receptor-positive tumors. J Natl Cancer Inst 1996; 88:1529–1542.

39. Stewart HJ, Forrest AP, Everington D, McDonald CC, Dewar JA, Hawkins RA. Randomized comparison of 5 years of tamoxifen with continuous therapy for operable breast cancer. The Scottish Cancer Trials Breast Group. Br J Cancer 1996; 74: 297–299.
40. Gray R, Milligan K, Padmore L, Study Group. Tamoxifen: assessment of the balance of benefits and risks of long-term treatment (abst 031). Br J Cancer 1997; 76(suppl 1):24.
41. Davies C, Monaghan H, Peto R. Early breast cancer: How long should tamoxifen continue? (abstr 177). Eur J Cancer 1998; 34(suppl 5):S43. Abstr 177.
42. Holli K. Adjuvant trials of toremifene and tamoxifen: the European Experience. Oncology 1998; 12:23–27.
43. Goldhirsch A, Glick JH, Gelber RD, Senn H-J. International Consensus panel on the treatment of primary breast cancer V: Update 1998. In: Senn H-J, Gelber RH, Goldhirsch A, Thurlimann B, eds. Adjuvant Therapy of Primary Breast Cancer VI. Resent Results in Cancer Research. Berlin: Springer 1998:481–497.

Panel Discussion 2

Early Breast Cancer/Chemoprevention November 12, 1999

List of Participants

Michael Dixon Edinburgh, Scotland
Mitch Dowsett London, England
Paul Goss Toronto, Ontario, Canada
Harold Harvey Hershey, Pennsylvania
Stephen Johnston London, England
William Miller Edinburgh, Scotland
Henning Mouridsen Copenhagen, Denmark
Nam-Sum Paik Seoul, Korea
Carsten Rose Odense, Denmark
Ian Smith London, England
G. Stathopoulos Athens, Greece

Paul Goss: Thank you very much for an outstanding overview, Henning. I want to ask you a question that comes up in the clinic frequently, and also arose at the ASCO meeting. It's become our practice to use chemotherapy followed by tamoxifen in estrogen receptor (ER) patients who have high-risk node-negative and node-positive disease. At ASCO 1999, there were a number of papers that suggested that the outcome of patients after adjuvant chemotherapy is related to whether they enter menopause after chemotherapy. You mentioned an intergroup trial run by ECOG and Dr. Nancy Davidson. The randomization was cyclophosphamide, adriamycin, 5-fluorouracil (CAF) chemotherapy versus CAF + tamoxifen versus CAF + tamoxifen + Zoladex®. The question the trial addresses is does ovarian ablation medically add additional benefit to chemotherapy followed by tamoxifen? That is a question that our patients ask in the clinic when their chemotherapy is complete and they are about to start tamoxifen—is there now in 1999 a reason to add Zoladex® and ablate ovarian function if patients continue to menstruate after chemotherapy? What do you do in the clinic?

Henning Mouridsen: A difficult question. I think we need the data. Studies are ongoing to try and answer this question. I agree that an important question is whether part of the effect of chemotherapy in premenopausal women is mediated through medical castration. It is one of the areas where we made mistakes. The question has been elucidated in retrospective subgroup analyses that, however, are completely invalid. There is a major bias in all these retrospective analyses of the problem. Patients who have medical castration by the therapy have been claimed to do better—however, patients who might potentially benefit from medical castration are probably patients with receptor-positive tumors, who have a better prognosis than women with receptor-negative tumors. It needs to be demonstrated prospectively before we can definitively answer the question.

Paul Goss: I agree a prospective trial is very important. I think to date the analysis of the trial has shown an advantage of the two arms combined containing tamoxifen versus chemo alone—so that pooling the CAF + tamoxifen arms shows the disease-free survival advantage that you mentioned but not an overall survival (OS) advantage compared to chemo alone regarding the CAF + tamoxifen versus CAF + tamoxifen + Zoladex® arm, which has not yet shown a difference. So I think that the answer to the question is that we have to wait for further follow-up.

Nam-Sum Paik: What is the advantage of endocrine neoadjuvant treatment compared with conventional chemotherapeutic agents?

Michael Dixon: Although elderly patients' tumors are larger, if you compare size to size with younger women, they are less likely to be Neu positive, less likely to be grade 3, less likely to have a high proliferation, so it depends to some extent how you select your patients. You are right, some of these patients are suitable for neoadjuvant chemotherapy, and yet biologically their tumors are not that aggressive, and if you gave them preoperative chemotherapy, you probably wouldn't achieve much more than you would with neoadjuvant endocrine therapy apart from making them more ill.

So I think that the advantage of neoadjuvant endocrine therapy in these patients is that you achieve one of the aims—to do less aggressive surgery. When you come to operation, you find that these patients would have been selected to have postoperative adjuvant endocrine therapy rather than chemotherapy anyway. Obviously it is important to select patients carefully. Some of these larger tumors are large because the patients have had them for some time and they are not that aggressive tumors—often grade 1 or 2 on core biopsy, high ER positive and without clinically palpable nodes. In these patients, it is safe to treat with neoadjuvant endocrine therapy and you will get a substantial response, and if the patients want it, they can get away without mastectomy. There are obviously differences in different areas of the world, but in our center, of those patients we offer breast-conserving surgery, about 98% take it. In our society, a treatment which down stages the tumor and allows the patient to have less extensive surgery is an advantage.

Nam-Sum Paik: I agree that using neoadjuvant therapy for down staging is important. However, neoadjuvant therapy takes a long time and responses may not be as good as conventional therapy.

Michael Dixon: We've looked at time to half tumor volume and found it to be 40 days with neoadjuvant endocrine therapy. So that's not very long, although it is in a selected group of patients. One of the points is that when you reduce a tumor from 4 cm to 3 cm you are actually decreasing the amount of tumor you need to remove to get a wide excision by about 50%—so you are not taking 100 g, you are taking 50 g—which is a fairly dramatic reduction and makes breast conservation possible when it would otherwise be impossible. We have also looked at the histology of the responses in patients treated with either tamoxifen or letrozole, and this may provide some insight into the study reported by the group in the southeast of the UK who did the aminoglutethimide study. What we found was letrozole produces dramatic reductions in each individual tumor in the rate of proliferation. If you look at Ki67, it comes down in all patients treated with letrozole, which does not happen in all patients treated with tamoxifen. In about 50% of patients on both letrozole and tamoxifen, there is alteration in grade. In the letrozole patients, this is due to a decrease in proliferation, whereas in the tamoxifen group, there is a different histological pattern with more glandular differentiation, appearing almost as if it is dedifferentiating, possibly altering its subsequent behavior. This is why, 5 years tamoxifen may have a long-term effect, whereas aromatase inhibitors switch off proliferation, so that when they are stopped, proliferation may switch back on. The changes are really quite surprising and highly statistically significant on a relatively small number of patients, so I suspect that they are real.

William Miller: Can you tell us about the data in the literature comparing neoadjuvant endocrine therapy followed by surgery versus surgery followed by conventional adjuvant therapy?

Michael Dixon: There is only one randomized study that I know about—a small study done in Edinburgh showing no difference in survival. It did show that the median time to half tumor volume was 40 days and those patients which had a median reduction in their tumor volume of less than 40 days had a much better survival than those patients whose tumors did not respond as quickly—with a highly significant difference. This is also seen in chemotherapy studies; that patients with an early response seem to do better.

Mitch Dowsett: It takes perhaps 5 or 6 weeks for tamoxifen to get to steady state, whilst with aromatase inhibitors, steady state is achieved after only about 10 days. Do you think that clinical differences in your comparison of aromatase inhibitors and tamoxifen over relatively short periods; i.e., 3 months, could be explained by this difference?

Michael Dixon: Yes, I think it is a potential problem. When looking at data from Edinburgh, we noticed that a lot of the reduction in volume with tamoxifen occurs between 2 and 3 months. The problem with the model we use is that, because we do a wedge biopsy, there is some swelling initially, and although you can get a good idea of what happens between 1 and 2 months and between 2 and 3 months, it is always difficult in the first month. Now we are just doing core biopsies, it should be better. However, you are right—

if tamoxifen is not up to adequate levels in 5 weeks, then it may not be therapeutic, so in a 3-month study, one may be comparing 12 weeks of anastrozole or letrozole with 7–10 weeks tamoxifen.

Mitch Dowsett: In your analysis, Henning, you focused on ER positivity/negativity. But if you have an ER-negative, progesterone receptor (PgR)–positive tumor, would you give that patient tamoxifen?

Henning Mouridsen: Yes, of course. According to the overview analysis, the PgR status, irrespective of ER status, predicted the benefit with tamoxifen, as did the ER status.

Harold Harvey: Mr. Dixon, I think that your model is an absolutely wonderful in vivo model allowing us to answer a variety of biological questions. However, I think that the skeptic would have to argue that to have a facility to do this study in elderly patients with large receptor-positive tumors is, in fact, an indictment of a failure of detection of breast cancer in that population? Perhaps, as a practical matter, efforts should be concentrated on screening and earlier detection of these tumors and perhaps use your approach in much smaller tumors. How would you reply to the skeptic?

Michael Dixon: I think that the situation is changing. As the generations that have been involved in screening are now coming through—and Edinburgh was one of the places that started to screen first in the UK—the good news is that we are having great difficulty doing these studies anymore, because we don't see such large tumors. When we recruited for the first letrozole study, we got 12 large tumors in 14 weeks, now for the last exemestane study, to get 12 large tumors took us a year. So the good news is that over time the efforts to educate women, to get them along to screening, and get them more interested in their breasts has been successful. However, in all societies, you will always get some people who delay presenting for whatever reason.

William Miller: Mike, can I ask you what you would say to the other skeptic, who comes along and says, well you have a lot of information on the primary tumor, but what actually determines outcome is micrometastatic disease, and the primary tumor has a limited role to play in determining that.

Michael Dixon: We, and others, have looked at disease nodes as well as the primary cancer and there seems to be a reasonable correlation between ER positivity in nodes and primaries. Fortunately, of the 24 patients treated by letrozole, only 10 out of 24 were node positive despite the fact that they had large, locally advanced breast cancers. This comes down to the point I made earlier—this is a selected group of patients and who probably have less biologically aggressive disease, who have actually delayed treatment. But they have delayed treatment for a reason that they did not wish to have a mastectomy and they did not like the idea of having treatment, which is why they did not come to hospital. So actually by being able to get round that potential hurdle for them, they seem to have been very happy with the treatment we have given them.

Paul Goss: Mike, we have good randomized clinical trials of pre- and postoperative chemotherapy with long-term disease-free survival and OS data showing that there is no detriment to survival by delaying surgery with preoperative chemotherapy. I do not think there has been a good, well-powered trial with endocrine therapy that has been able to ask that same question. The patients that you have so carefully enrolled into your preoperative studies have an excellent response, because they are very highly selected. However, since the type of data you have presented has crept into the literature and into the public domain, we are seeing a lot of surgeons putting women onto endocrine therapy and delaying their cancer surgery. Their belief is that this clinical practice is comparable to delaying surgery for administrating neoadjuvant chemotherapy. I am very pleased to see how carefully you present the data and I am very careful to disseminate the news to my colleagues that it is clinical research data and not clinical practice data. It is obviously a very laudable goal to reduce mastectomy, but it wouldn't be if it came at the expense of increased mortality.

Michael Dixon: I think that is absolutely right. It is also important to make the point that it is a highly selected group of patients with high levels of ER.

Carsten Rose: Henning, I am always a bit puzzled when we talk about adding chemotherapy to tamoxifen in the postmenopausal patients being node positive or node negative. What is your opinion based on current data? Should we add it, in whom should we add it to, and what effect would you actually go for considering the added toxicity in these patients?

Henning Mouridsen: I think it is rather difficult to answer that question. If you go to the individual trials and look at the additional benefit with chemotherapy in these receptor positive patients, because these are the patients who had the tamoxifen, the additional benefit with chemotherapy is very very modest—a matter of 5% or something like that. So my personal opinion is that with the therapies that have been tested in the postmenopausal patients, and for the vast majority of patients it has been CMF (cyclophosphamide, methotrexate, fluorouracil), the additional benefit is too small to justify the general use of chemotherapy following tamoxifen.

Paul Goss: The NSABP 20 trial which randomized node-negative, receptor-positive pre- and postmenopausal patients to tamoxifen or tamoxifen plus CMF showed modest, but definite, improvements in disease-free survival and OS. In addition, Tannoch et al. polled North American women as to what survival benefit they would accept for a 6-month course of chemotherapy. Women answered that for a 1% improvement in survival, they would undergo a 6-month course of CMF chemotherapy. One of the speakers suggested that there is gap between what physicians think is worthwhile for patients and what patients think is worthwhile.

Carsten Rose: If I could just add to that, first of all, the NSABP paper actually demonstrated that the effect was very low in ER-rich tumors, and most of the effect was in those with intermediate ER values. That is one thing. The other is that I don't think these women that you are going to ask actually can perceive what they are saying yes to; 1% is actually not very much. We have to bear in mind that the absolute difference obtained between successful and unsuccessful chemotherapy is only around 4–5%. We are exploring very modest effects.

Henning Mouridsen: It is not just a matter of saying that, if there is a 1% chance of benefit, then we should offer the treatment to the patients. It is also a matter of keeping priority to different areas, because we can't afford to treat all patients in whom there is a 1% chance of improvement in their survival.

G. Stathopoulos: In the trial presented by Dr. Mouridsen on adjuvant treatment for 2 years with tamoxifen and then randomized to tamoxifen and aromatase inhibitors, if there is a difference in the outcome, then you can tell something, but if there is no difference how can you tell that the effect was not the 2 years of tamoxifen? Would it not be better to start the two arms from the beginning?

William Miller: Maybe I can rephrase what I think is the same question, because I was going to raise a general issue. What strikes me about the problems with adjuvant therapy vis-à-vis neoadjuvant therapy is that you need to study large numbers of patients for a long time to pick up sufficient events to observe differences between groups. We are talking about 9000 women and 7000 women. It may be controversial, but some of the study designs have gone ahead with aromatase inhibitors without the knowledge of whether the benefits of prior tamoxifen have been exhausted. It therefore is difficult to attribute benefits to tamoxifen or the aromatase inhibitor. In advanced disease, presumably you treat patients to exhaustion with one beneficial therapy before you start another. But in the adjuvant situation, you are blindly changing to another therapy without actually knowing whether you have exhausted the benefits of the first. So how do you know that the benefit is accruing specifically to an aromatase inhibitor rather than just continuing tamoxifen?

Mitch Dowsett: I think the FEMTA trial actually deals with that, because it has four arms, two of which persist with single-drug treatment.

William Miller: But the exemestane trial does not. Can I ask you, Henning, again, it is the same sort of drum that I am beating here, have we been carried away with enthusiasm, perhaps quite rightly, by the efficacy of the new aromatase inhibitors? For example, in the ATAC study, there is a combination arm which is being used adjuvantly but has never really been proven in the advanced situation. However, we are going ahead and using the combination of an aromatase inhibitor with tamoxifen without it necessarily being proven to be of clear benefit in advanced disease over the individual drugs. Is that true, and what are your thoughts about this if it is true?

Henning Mouridsen: To my knowledge there are no endocrine combination therapies which have proven superiority to single-modality therapies.

William Miller: So what's your thought about using a combination in an adjuvant situation?

Henning Mouridsen: It was very difficult for me to understand the rationale for the combined arm in the study. That is one of the reasons why we did not join the study.

Ian Smith: I just wanted to emphasize, as I see it, the key point of doing preoperative treatments, whether chemo or endocrine therapy, is not actually to reduce the need for mastectomy—although I believe that is a very useful spin-off. It relates to this terrible problem that we have of doing adjuvant trials that take many years to run to get answers. How do you introduce new therapies? How do you introduce Arimidex®, for example, as you have just asked? It also relates to the fact that we know from metastatic disease that individual patients are likely to respond differently. So it is the potential predictive value of preoperative neoadjuvant treatments that is really central. The problem is proving it, and that is why we need trials along the lines of which we have been discussing, where you can try to establish an early surrogate marker—whether clinical or pathological, or as many of us believe, biological—that will tell you in the long term whether there is going to be a survival benefit. I would like to ask Mike, he has shown very nicely that ultrasound is the best correlation, but is he aware of data that show that the extent of regression actually matters long-term? It would be very important if it is true.

Michael Dixon: I think that the size of the tumor has been used as a surrogate end marker. I agree with you that what matters is what you see under the microscope. Occasionally you can see a nidus on the ultrasound, and that has happened to some of our patients—a nidus which has very few tumor cells within it. So I accept that what matters most is what is happening pathologically. But when you are making clinical decisions as to whether a patient is suitable for a particular type of operation, you need to have some idea of what volumes you are trying to remove, so then it does become of practical importance how big the tumor is. From a surgical point of view, volume is important even if it may not have as much biological relevance. I agree with you that it is an excellent biological model, but it is very difficult to sell to patients that a biological model will just predict what therapy they are going to get later on, when we have such a highly selected group of patients. What do you do in patients who have had tamoxifen for 3 months and then you operate on them and they had a 10–20% reduction in tumor volume? Well you are going to do a wide excision, if possible, and if they are node negative, you are probably going to give them radiotherapy and continue them on the tamoxifen. So I think that part of the trouble is that, in the selected group of patients we have looked at, the response rates are so high that you are unlikely to change many of these people onto any other treatment. The other problem with your argument is, yes it is great that you should be able to give one type of chemotherapy, see if it responds, and if it doesn't try another type of chemotherapy, but in fact in practice, this rarely happens. There are very few groups that have actually looked to see if the first chemotherapy doesn't work and switching it for something else, and showing that if you switch to something else, you can get a reasonable response rate. So I think that is a great theoretical advantage, but what you really want to be able to do is give a shot of chemotherapy and know within a couple of weeks whether it has been effective. If it has not, then switch at that point rather than doing what seems to be happening now, which is waiting for a few months to see if the tumor responds clinically.

Ian Smith: I agree with your last point, the reason we don't switch is we don't know what the surrogate marker is. If there is one, we've got to find that first. If we could find it, then that would be the next generation study.

Michael Dixon: For the purposes of the study we've done, we have had to have a time for it, but in fact with tamoxifen, we actually treat patients for a very variable length of time, because we know from data that the median length of response is 9 months. So for the purposes of the studies I presented, it is 3 months, but actually in the clinical setting, we actually keep them going for much longer.

Paul Goss: It was a leap of faith to combine tamoxifen with an aromatase inhibitor in an adjuvant trial without any data of the combination in the metastatic setting. But so was to use a single arm of an aromatase inhibitor for 5 years, abandoning 5 years of tamoxifen, which had been so carefully demonstrated to be of benefit. In these trials, we did the equivalent of taking a chemotherapy drug which was new and had shown efficacy in second-line therapy in metastatic breast cancer and gone over to the adjuvant setting. We abandoned entirely our current chemotherapy (CMF, CAF, or whatever) and just replaced it with this second-line single agent chemotherapy drug. I think that would have been unthinkable in the chemotherapy setting. Perhaps it reflects peoples' beliefs in the indolent nature of hormone receptor–positive breast cancer that people did this, but it was a somewhat surprising thing to do prior to first-line therapy trials.

Stephen Johnston: Touching on the high response rates with letrozole shown by Mike Dixon. You may be stung by your success in not being able to be predictive, but you had two local recurrences there. Do you have any early information on whether there is anything happening that they did not respond very quickly, or that Ki-67 changes that you are seeing, like we are after 2 weeks, whether those two that relapsed locally had anything different?

Michael Dixon: I think it is simple—that they just did not get adequate local treatment. We just did a wide local excision; they did not have radiotherapy. About the time they were treated, a few years ago—it probably has happened quicker I think in many parts of the world than the UK—we tried to be less aggressive with elderly patients. We have even got a study in Scotland where patients are being randomized to have radiotherapy or not after wide excision, looking at quality of life issues. Some of my patients live some distance away from hospital, and for whatever reason, the radiotherapist decided not to give radiotherapy. Nowadays our practice has changed. They would all get radiotherapy unless they were very infirm. So I think the reason was that we excised the disease that was there, but wide excision alone is associated with a significant local recurrence rate. There were also some Italian data published about that time which said that if you are over 55 and you do a very wide excision without radiotherapy, you get adequate control. Other groups haven't shown this, and I think it is merely a factor of inadequate local treatment—so I don't think we are going to learn anything from these few patients.

8

Health Economics Aspects of Endocrine Therapy

Suzanne Wait
Université Louis Pasteur
Strasbourg, France

I. INTRODUCTION

Over the past few years, the high costs and increasing availability of new treatment options for breast cancer have channeled policymakers' and regulatory bodies' attention to determining the ''value for money'' of competing therapies. Many factors render breast cancer particularly amenable to economic evaluation. Given its high prevalence, breast cancer treatment is taking up an increasing share of health care resources, and this trend is likely to continue in view of the aging global population.

Breast cancer also raises important quality of life concerns. Every stage of treatment presents an important trade-off between clinical benefit and discomfort or toxicity. Breast cancer patient groups have become increasingly organized and involved in the drafting of treatment guidelines, in the dissemination of information, and the adoption of new treatment modalities. Patient groups have also driven the need for drug sponsors to describe the benefits of treatments in terms of outcomes that are relevant to the patient and address her concerns and needs.

Breast cancer management may be seen in a cascade of therapeutic decisions, with each decision having to balance evidence and expectations of clinical benefit, toxicity and side effects, impact on the patient's quality of life, and cost. It is within this framework that economic evaluation may play a role by providing an objective comparison in terms of costs and effects of treatment, thereby determining the value for money of different therapies (1,2).

The purpose of this chapter is to review economic aspects relevant to the endocrine treatment of breast cancer. Focus is limited to early breast cancer, with a particular emphasis on adjuvant endocrine therapy.

II. HEALTH ECONOMICS FRAMEWORK

A. Defining Economic Evaluation

Economic evaluation may be defined as "the comparative analysis of alternatives in terms of both costs and consequences" (1). The health economics framework for evaluation is depicted in Figure 1. Costs incurred are weighed against costs avoided and clinical benefits against side effects or associated morbidity in order to produce a final costs to effects ratio. The different methods for economic evaluation differ in how they value treatment effects. In cost-effectiveness analyses, the effects are expressed in physical units (number of deaths avoided, of life-years gained, of cancers detected), whereas in cost-utility analysis, these physical units are weighted by the quality of life of individuals, resulting in a cost per quality-adjusted life year (QALY) ratio. As is illustrated in Figure 1, the economic value of treatments is contingent on their clinical benefit. Clinical expecta-

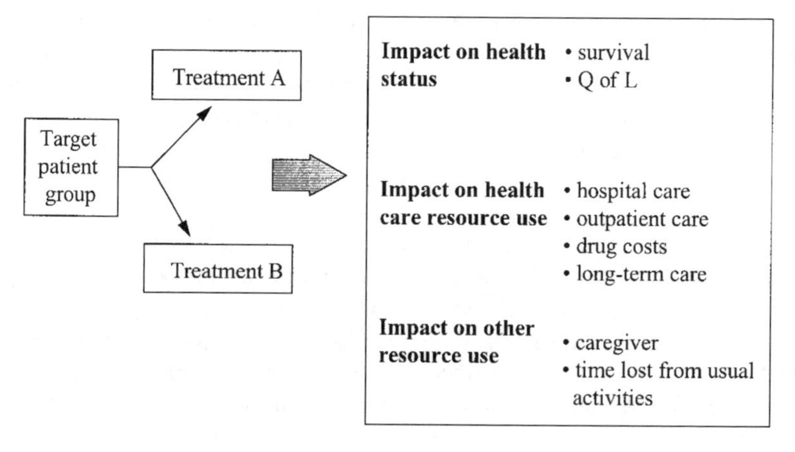

FIGURE 1 Health economics parameters.

tions of therapies drive economic expectations, and costs and quality of life effects of treatments are assessed in a continuum with clinical benefit. (For further reading on different methods of economic evaluation, please refer to Refs. 1 and 2.)

B. Quality-of-Life Evaluation

Quality of life is a term that has been appropriated by many academic disciplines, and the erroneous tendency in today's literature is to define anything other than response rate as "quality of life." As is illustrated in Figure 2, there are many components and influences on a patient's quality of life, of which treatment toxicity is just one component among many. When comparing two treatments, one may choose to focus evaluation on the symptoms of treatment or disease or choose a more comprehensive measurement of the impact of these factors on the patient's general well-being (health-related quality of life). For example, an instrument that specifically asks patients about menopausal symptoms may be the most sensitive to the differences between clinically equivalent endocrine treatments. In the adjuvant setting, a comprehensive evaluation of the patient's quality of life may be warranted over the entire duration of treatment and follow-up. In all cases, it is important to determine a priori the expected differences between treatments and to tailor the selection of the instrument to capture these differences. Careful attention must be paid to selecting an instrument which is patient-based, has validated psychometric properties, is acceptable to patients, and is most likely to capture differences between treatments (3).

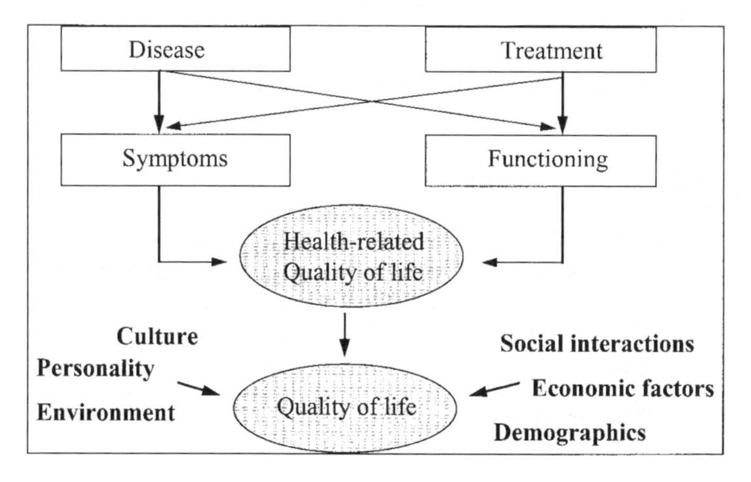

FIGURE 2 Influences on quality of life.

III. ADJUVANT ENDOCRINE THERAPY

A. Economic and Quality-of-Life Issues

The adjuvant setting for endocrine therapy is an obvious framework for decision making: First because there remains much controversy over the most appropriate treatment options for different patient groups, and second because one is incurring treatment costs and toxicity immediately in the hope of expected benefits years later. The balance between costs, side effects, and benefits of treatment will depend highly on clinical expectations—namely, the opportunity to reduce the risk ratio for recurrence and advanced disease. The age of the patient, prognostic factors, and treatment alternatives all enter into the equation of balancing costs to benefits.

The economic issues surrounding endocrine therapy depend on the reference treatment in each situation (4). If the treatment of reference is tamoxifen, the higher acquisition cost of the new endocrine therapy will need to be offset by higher efficacy, lower incidence of endocrine symptoms, and more favorable systemic effects. If the comparator is chemotherapy, endocrine therapy may present a quality of life advantage owing to lower toxicity; however, it must also offer an incremental clinical benefit as compared to chemotherapy. In the case of combination therapy, the addition of hormonal therapy to chemotherapy must provide sufficient additional clinical benefit to justify the increased therapy cost as compared to chemotherapy or hormonal therapy alone. Moreover, one must ascertain that the combination effects of treatments do not have detrimental effects on patients' quality of life. Postmenopausal women who would normally be receiving adjuvant tamoxifen may not be willing to accept the added toxicity incurred by chemotherapy if the marginal survival benefit is small. Similarly, some premenopausal women may not accept the induction of menopausal symptoms from treatment if their cancer is already of good prognosis. Each of these situations warrants careful decision making, which cannot be answered by existing clinical trial data alone.

B. Existing Literature

The economic literature on adjuvant endocrine therapy is mostly limited to comparisons of combination (chemo-endocrine) therapy, or either chemotherapy alone or tamoxifen alone. The two main questions addressed are (1) whether one should add tamoxifen to chemotherapy in premenopausal women, and (2) whether one should add chemotherapy to tamoxifen in postmenopausal women.

1. Chemoendocrine Therapy Versus Chemotherapy in Premenopausal Women

Smith and Hillner (5) conducted a cost-utility analysis based on the results from the 1992 EBCTCG meta-analysis in premenopausal women (6). They obtained utility values from focus groups of patients with breast cancer and derived costs from Medicare charges and from a single center's accounting data. The principal results are presented in Table 1. Based on their analysis, tamoxifen and chemotherapy are equally cost effective in premenopausal women who are estrogen receptor (ER) positive and node negative. Combination therapy may offer an attractive option for low-risk (node-negative) women, but it is of less value than either tamoxifen or chemotherapy alone. In node-negative, ER-negative and node-positive, ER-negative women, only chemotherapy provides a cost-effective treatment. Although in node-positive, ER-positive patients, tamoxifen is the most cost-effective option.

This model was the first to describe the relationship between costs and effects of adjuvant therapies in premenopausal women. The cost-utility results were highly sensitive to estimates of the relative risk of recurrence, the time frame considered, and the relative reduction in risk expected from treatment. Interestingly, changes in drug costs did not affect results significantly. The model is based on old data and would need to be updated in view of more recent trial results. Specifically, the model used efficacy estimates for tamoxifen that are significantly lower than the most recent efficacy data from the EBCTCG meta-analysis (7). Therefore, it is likely that the model underestimates the value of tamoxifen and combination therapy treatment scenarios as compared to chemotherapy in this population (8).

2. Chemoendocrine Therapy Versus Endocrine Therapy in Postmenopausal Women

Goldhirsch et al. (9) conducted an analysis of quality-time without symptoms and toxicity (Q-TWiST) to assess the balance between quality of life and survival in postmenopausal women receiving either adjuvant tamoxifen or combination therapy. In the Q-TWiST method, treatment duration is partitioned into phases according to the level of toxicity and clinical events: TOX (toxicity during adjuvant therapy), REL (relapse), and TWiST (time without symptoms and toxicity).

Table 1 Cost per QALY in Premenopausal Women

	Tamoxifen ($)	Cyclophosphamide ($)	Tamoxifen + cyclophosphamide ($)
N^-ER^+	11,440	11,370	33,100
N^-ER^-	214,000	4,970	186,200
N^+ER^+	4,330	9,230	14,750
N^+ER^-	57,800	4,890	80,700

Source: Adapted from Ref. 5.

A range of theoretical utility scores, from 0 to 1, is then assigned to each phase of treatment. The final analysis is expressed as an equation, in which:

$$\text{Q-TWiST} = \text{TOX}(u_t) + \text{REL}(u_r) + \text{TWiST}(u_w)$$

The authors based their clinical data on results from the Ludwig Trial III, which looked at postmenopausal node-positive women receiving either tamoxifen or tamoxifen combined with chemotherapy for 1 year after mastectomy. Results favored the chemoendocrine arm, and the authors concluded that the increased toxicity associated with combination therapy was outweighed by the survival benefit it allowed when compared to tamoxifen alone in this patient population.

Gelber and his colleagues updated the model in 1996 with results from nine trials involving 39,000 patients (10). After 7 years of follow-up, the toxic effects of chemotherapy were barely balanced by the modest increase in relapse-free and overall survival. None of the utility values tested yielded more Q-TWiST for patients undergoing chemotherapy plus hormonal therapy as compared to patients receiving tamoxifen alone. These results are much more ''humbling'' for chemotherapy advocates than the previous analyses had suggested (8).

The Q-TWiST analyses do have their inherent weaknesses. The model provides an oversimplification of the phases of treatment with regard to toxicity, and it equates ''utility'' or quality of life with toxicity, precluding any consideration for other impacts of treatment and disease on patients' quality of life. Also, no costs are included in the model. Despite these limitations, the Q-TWiST method is highly amenable to the study of adjuvant therapies, as it allows for a toxicity-adjusted survival analysis and provides easily interpretable data for decision making.

Another quality of life analysis in postmenopausal women receiving endocrine therapy was undertaken by Hürny et al. (11). The authors conducted a prospective quality-of-life assessment alongside the EBCTG Trial VII. They used a combination of linear analogue scales for five distinct quality-of-life items. The study found that patients receiving tamoxifen had higher scores compared to patients receiving chemoendocrine therapy. However, these differences disappeared over time as patients adapted to the toxicity of chemotherapy over the duration of the trial. Patient scores also reflected their anticipation of treatment. An important limitation of this study is the choice of quality-of-life instrument. Little is known of its psychometric properties, and extrapolation of results to different study settings is thus difficult.

The only economic evaluation found in the literature was a cost-utility analysis carried out by Hillner and Smith (12). The authors followed a similar modeling approach as their previous model (5) to assess the cost utility of adjuvant tamoxifen versus tamoxifen combined with chemotherapy in postmenopausal women. Using clinical data on the 1992 EBCTG trial results, the authors found a cost utility of $58,000/QALY for chemoendocrine therapy; however, this ratio was highly sensitive to clinical assumptions. Despite the model's strong conceptual basis, it is based on old data of short duration, precludes the comparison of anthracycline versus nonanthracycline regimens, and uses very low treatment costs, which probably do not reflect current agents being used. It needs to be updated with more current trial data.

IV. NEOADJUVANT APPLICATIONS FOR ENDOCRINE THERAPY

Neoadjuvant endocrine therapy may be offered as an alternative to neoadjuvant tamoxifen, neoadjuvant chemotherapy, or radical mastectomy and adjuvant hormonal therapy or chemotherapy. The intention of therapy is to decrease the need for radical surgery and aggressive adjuvant treatment through tumor shrinkage; therefore producing lower toxicity and better quality of life for the patient. Moreover, treatment costs may be decreased owing to fewer recurrences.

An economic evaluation of neoadjuvant administrated endocrine therapy is difficult within the current randomized clinical trial design for most neoadjuvant therapies. A cost-effectiveness comparison of two competing endocrine therapies would need to encompass all costs occurring over the full treatment period and compare them with survival in both groups. However, clinical trials may not be of a sufficient sample size to address all the hypotheses put forth over the full treatment period, and they may be concentrated on the neoadjuvant phase of treatment. Whereas a typical neoadjuvant trial may be powered to compare the endocrine therapy to tamoxifen (or placebo) in terms of response rates, tumor shrinkage, and frequencies of breast-conserving therapy, it may not be possible to address the most important questions that a breast cancer patient may have. These are:

- What are my chances if I take this treatment now as opposed to after surgery?
- How does the benefit of this treatment compare to chemotherapy, which is also being given before surgery?
- What will be the long-term impact of this treatment on my chances of survival and the probability of recurrence?

An experimental design to address each of these questions is perhaps unfeasible in terms of time and expense. Nonetheless, these questions can be addressed in a decision-analytic model that allows for comparison of treatment outcomes over the full course of breast cancer treatment. Such a broad conceptual model for breast cancer is currently being developed by MedTap UK with the support of Novartis (personal communication), and it may prove to be an excellent tool for patients and physicians facing these fundamental treatment decisions.

There exists an extensive literature on quality-of-life evaluations for women receiving mastectomy versus breast-conserving surgery (13–15). This pioneer research dismisses the myth that psychological adjustment is ''better'' in women receiving breast-conserving surgery as compared to mastectomy and highlights the individuality of patient preferences in this situation (16). Existing studies also suggest that women want more information on the outcomes of their surgical choices, and that the amount and thoroughness of information received affect their choices. However, not all women may want to exert the right to decide when offered that choice. These factors will be important in the comparison of benefits of neoadjuvant endocrine therapies that aim to reduce the occurrence of radical surgery.

V. CHEMOPREVENTION

Chemoprevention remains somewhat of an unexplored terrain for economic evaluation. The evaluation of chemoprevention warrants a public health paradigm in which a preventive agent is given to a large population of asymptomatic women in the hope of preventing eventual illness in a small number of individuals. The marginal effectiveness of chemoprevention depends not only on the number of cancers that are avoided owing to chemoprevention but also on the prognosis of women in the nonprevented group. In other words, if the control group is receiving aggressive mammographic screening and screening manages to prevent more poor-prognosis cancers than chemoprevention, the incremental number of deaths avoided and of life-years saved by chemoprevention will be very small. Much of the success of chemoprevention hinges on the identification of high-risk populations in which prevention may be most effective.

Currently there exist no economic evaluations of chemoprevention; however, an article by Chlebowski and colleagues (17) presents some tentative figures for cost effectiveness. Assuming that preventive endocrine therapy may reduce breast cancer incidence by 40% in a high-risk population of 8000 women, the authors suggest that 62 cancers would be prevented, of which a third (n = 21) would have resulted in death. Assuming a cost of nongeneric tamoxifen 20 mg per day of $1000 per year for 5 years, the cost per death prevented amounts to $645,000. Assuming an average survival of 20 years per cancer detected, this amounts to 420 life-years saved, and the cost per life-year saved becomes $95,000. By means of comparison, the cost effectiveness of breast cancer screening in women aged 50–69 years has been found to be in the order of $30,000–$50,000 per life-year saved in most studies (18). Although the data put forth by Chlebowski are purely hypothetical, they do suggest that the cost effectiveness of chemoprevention should be carefully scrutinized. Given that the cost of new endocrine therapies is always higher than tamoxifen, a comprehensive model that encompasses all costs (endocrine agents, all breast cancer treatment received and avoided) and effects (prevention of breast cancer incidence, morbidity and death, cardiovascular and other systemic effects of treatment, patient quality of life) will need to be developed to suitably address this question. Moreover, delicate questions remain in terms of the impact of prevention on women's expectation of disease, their psychological response to ineffective prevention, and their individual choices in terms of ultimate treatment decisions.

VI. COMPARISON BETWEEN DIFFERENT ENDOCRINE THERAPIES

Although the studies described above are helpful in providing comparative information on endocrine therapy versus chemotherapy or combination therapy, they do not shed light on the relative value of endocrine therapies among themselves. Moreover, all published studies pertain to tamoxifen, which merely reflects the current lack of published clinical data on the use of other endocrine therapies in the adjuvant, neoadjuvant, or preventive setting. Direct comparisons of endocrine therapies have been conducted in the metastatic setting. In a recent report (19), the cost effectiveness of letrozole versus megestrol acetate as a second-line hormonal therapy in postmenopausal women is modeled based on the results from the AR/BC2 trial and UK treatment patterns and cost data. Letrozole is associated with an incremental cost effectiveness of £3588 per life-year saved; thus, making it a very cost-effective treatment option in postmenopausal women with advanced breast cancer.

As new endocrine therapies, namely, aromatase inhibitors, move into the primary treatment of breast cancer, more information will be required on their respective impact on patients' quality of life, it is likely that the demand for direct comparative analyses of endocrine therapies will increase. A recent publication by Fallowfield et al. (20) explored the differences between endocrine therapies in the context of developing a new endocrine-specific (ES) quality-of-life instrument, the Functional Assessment of Cancer Therapy-ES (FACT-ES). This patient-based module should be administered in conjunction with the FACT for breast cancer patients (FACT-B), developed by David Cella and colleagues. It contains 30 items, is patient administered, and has demonstrated satisfactory psychometric properties in different populations of women (21). Its French version is currently being tested in a Phase IV open-label trial in postmenopausal metastatic breast cancer patients receiving either letrozole or anastrozole (Novartis, personal communication).

The publication of the FACT-ES is the first to report in detail the symptoms of patients receiving different endocrine therapies, although previous reports have suggested that patients' assessment of endocrine symptoms differ from that of their physicians (22). The authors compared FACT-ES scores among treatment groups as well as for different phases of treatment (advanced vs. adjuvant setting). FACT-ES scores showed high sensitivity to treatment effects over time. Differences between treatments were observed on two particular symptoms: vaginal dryness and hot flushes. Patients receiving megestrol acetate and adjuvant chemotherapy complained the most of vaginal dryness, whereas patients receiving tamoxifen reported the most hot flushes. No differences were observed in endocrine scores of advanced-disease patients with respect to patients with early breast cancer.

The interest of these results is that they allow a detailed comparison of the side effects of different endocrine treatments from the patient's perspective. Unfortunately, the study was limited to a restricted number of treatment comparisons, and no patients taking letrozole were included. Further studies using the FACT-ES may provide useful data to patients on the relative side effects of competing endocrine therapies, and may hopefully facilitate their selection among these treatments.

VII. CONCLUSIONS AND FUTURE CHALLENGES

As more treatment alternatives for early breast cancer become available and pressures on health care budgets force further choices between competing treatments, it is likely that more attention will be paid to the economic aspects of endocrine therapy. Endocrine therapy may offer a high-efficacy and low-toxicity treatment option to women with early breast cancer. However, the field of quality of life and health economics is still in its infancy in this area, and future analyses need to evaluate the value of endocrine therapies other than tamoxifen in the early breast cancer setting. If research questions are well articulated, appropriate methodologies are rigorously applied, and results from analyses well communicated, economic evaluation may play an important role toward improving the knowledge of the resource implications of clinical practice, as well as provide a comprehensive perspective on the value of clinical decisions, weighing factors of cost, survival, clinical benefit, and patient quality of life. Although decisions should by no means be based solely on the economic profile of competing therapies, factors of cost and quality of life should be weighed against survival and clinical benefit, individual patient preferences, and cultural and other considerations in order to arrive at the best care for breast cancer patients. The onus is on outcomes researchers to conduct studies following clear and rigorous methods, and always to position their findings within a relevant clinical context.

Hand-in-hand with the need for further research is the need for further sensitization of physicians to the potential role of health economics data in the early breast cancer setting. Recent surveys of oncologists would suggest that, although oncologists consider quality of life an obvious endpoint in the palliative setting, they do not consider it of importance in a curative setting. This perspective is regrettable, especially given the tendency of many oncologists and patients to overestimate the benefits derived from adjuvant chemotherapy (23,24). Moreover, as the quality of life literature in mastectomy has shown, the only valid assessment of quality of life is that directly reported by the patient, and breast cancer patients' perspectives on the impact of treatment on their quality of life may vary significantly from the perceptions of their physicians.

As patients become much more informed about the treatment options available to them, physicians will be under increasing pressure to better communicate evidence of the relative value of different treatment options. The ''value'' of treatments must balance the likelihood of local tumor control and survival, the burden of treatment on the patient, the impact of treatment on costs, and individual patient preferences. For health care researchers, the challenge is to communicate better the rationale, methods, and results from economic evaluations including quality-of-life studies, so that these findings may be better understood by physicians and incorporated into clinical decision making. Indeed, any advances in this growing field of research will require an implicit commitment from physicians and nurses, patient groups, policy makers, and researchers to strive constantly to find a better articulation between research, policy, and practice. Whichever the perspective, the objective must always be to improve the quality of care received by breast cancer patients at all states of their illness.

REFERENCES

1. Uyl-de Groot C, Touw CR. Economic evaluation of cancer treatments: methodological and practical issues. Anti-Cancer Drugs 1998; 9:835–841.
2. Williams C, Coyle D, Gray A, Hutton J, Jefferson T, Karlsson G, Kesteloot K, Uyl-de Groot C, Wait S. European School of Oncology Advisory Report to the Commission of the European Communities for the Europe Against Cancer Programme: Cost-effectiveness in Cancer Care. Eur J Cancer 1995; 31A(9):1410–1427.
3. Doward LC, McKenna SP. Evolution of quality-of-life assessment. In: Rajagopalan R, Sheretz EF, Anderson RT, eds. Care Management of Skin Diseases: Life Quality and Economic Impact. New York: Marcel Dekker, 1997:9–33.
4. Wait SH. Economic evaluation of endocrine therapy in the treatment of breast cancer. Anti-Cancer Drugs 1998; 9:849–857.
5. Smith TJ, Hillner BE, Desch CE. The efficacy and cost-effectiveness of adjuvant therapy of early breast cancer in premenopausal women. J Clin Oncol 1993; 11: 771–776.
6. Early Breast Cancer Trialists' Collaborative Group. Systemic treatment of early breast cancer by hormonal, cytotoxic or immune therapy. Lancet 1992; 339:1–15.
7. Early Breast Cancer Trialists' Collaborative Group: tamoxifen for early breast cancer: an overview of the randomized trials. Lancet 1998; 351:1451–1467.
8. Hillner BE. Review of cost-effectiveness assessments of chemotherapy in adjuvant and advanced breast cancer. Anti-Cancer Drugs 1998; 9:843–847.
9. Goldhirsch A, Gelber RD, Simes RJ, Glasziou P, Coates A. Costs and benefits of adjuvant therapy in breast cancer: a quality-adjusted survival analysis. J Clin Oncol 1989; 7:36–44.

10. Gelber RD, Cole BF, Goldhirsch A, et al. Adjuvant chemotherapy plus tamoxifen compared to adjuvant chemotherapy in post-menopausal node-positive breast cancer. Lancet 1996; 347:1066–1067.
11. Hürny C, Bernhard J, Coates AS, et al. Impact of adjuvant therapy on quality-of-life in women with node-positive operable breast cancer. Lancet 1996; 347:1279–1284.
12. Hillner BE, Smith TJ. Estimating the efficacy and cost-effectiveness of tamoxifen (TAM) versus TAM plus adjuvant chemotherapy in post-menopausal node-positive breast cancer. A decision-analysis model (abst 46). Proc Am Soc Clin Oncol 1992; 11:55.
13. Lasry JCM. Women's sexuality following breast cancer. In: Osoba D, ed. Effect of Cancer on Quality-of-Life. Boca Raton, FL: CRC Press, 1991:215–227.
14. Fallowfield LJ, Hall A. Psychosocial and sexual impact of diagnosis and treatment of breast cancer. Br Med Bull 1991; 47:388–399.
15. Kiebert GM, de Haes JCJM, van de Velve CJH. The impact of breast-conserving treatment and mastectomy on the quality-of-life of early-stage breast cancer patients: a review. J Clin Oncol 1991; 9:1059–1070.
16. Till JE, Sutherland HJ, Meslin EM. Is there a role for preference assessments in research on quality-of-life in oncology? Qual Life Res 1992; 1:31–40.
17. Chlebowski RT, Butler J, Nelson A, Lillington L. Breast cancer chemoprevention. Tamoxifen: current issues and future prospective. Cancer 1993; 72 (3 suppl):1032–1037.
18. Wait S. The cost-effectiveness of breast cancer screening in France: from research to policy. In: F. Calvo et al., eds. Breast Cancer Advances in Biology and Therapeutics. John Libbey Eurotext, 1996.
19. Nuijten M, Meester L, Waibel F, Wait S. Cost-effectiveness of letrozole in the treatment of advanced breast cancer in postmenopausal women in the UK. Pharmacoeconomics 1999; 16:379–397.
20. Fallowfield LJ, Leaity SK, Howell A, Benson S, Cella D. Assessment of quality-of-life in women undergoing hormonal therapy for breast cancer: validation of an endocrine symptom subscale for the FACT-B. Br Cancer Res Treat 1999; 55:189–199.
21. Brady MJ, Cella DF, Mo F, et al. Reliability and validity of the Functional Assessment of Cancer Therapy-Breast (FACT-B) quality-of-life instrument. J Clin Oncol 1997; 15:974–986.
22. Leonard RCF, Lee L, Harrison ME. Impact of side-effects associated with endocrine treatments for advanced breast cancer: clinicians' and patients' perceptions. T Breast 1996; 5:259–264.
23. Rajagopal S, Goodman PJ, Tannock IF. Adjuvant chemotherapy for breast cancer: discordance between physicians' perception of benefit and the results of clinical trials. J Clin Oncol 1994; 12:1296–1304.
24. Ravdin PM, Siminoff IA, Harvey JA. Survey of breast cancer patients concerning their knowledge and expectations of adjuvant therapy. J Clin Oncol 1998; 16:515–521.

Panel Discussion 3

Health Economics Aspects of Endocrine Therapy
November 12, 1999

List of Participants

Ajay Bhatnagar Basel, Switzerland
Michael Dixon Edinburgh, Scotland
Paul Goss Toronto, Ontario, Canada
Henning Mouridsen Copenhagen, Denmark
Ahn Myoung Ock Seoul, Korea
Nam-Sun Paik Seoul, Korea
Suzanne Wait Strasbourg, France

Michael Dixon: I would like to challenge a couple of things you said. First of all, there are some new data which are coming out now showing that patients treated for breast conservation do have a better quality of life. The concern about breast conservation has always been the increased risk of local recurrence, and that is what has held the quality of life back. What seems to be happening in the newer studies is that patients are becoming more convinced that breast conservation is equivalent to mastectomy. There are some studies that now are being presented showing improved quality of life by breast conservation is not necessarily better than that of patients undergoing mastectomy. Importantly for surgeons, there is now good evidence that cosmetic outcome after breast conservation also enhances quality of life with low levels of anxiety, depression with improved body image, and self-esteem in patients with a good cosmetic outcome. I think from the surgical perspective, it is not merely a matter anymore of chopping a lump out, and thinking that we have done a decent job at leaving half a breast there. I think one of the quality-of-life issues which has not been addressed greatly is the benefits patients get from breast conservation when it is well or poorly performed. We seem to be assessing endpoints and morbidity of chemotherapy in quite a lot of detail, but we have been fairly slow to look at surgical outcomes. Do you see any future in putting the surgeons under the microscope as well as the oncologists?

Suzanne Wait: I hope I did not give the message that I believe that mastectomy is better than breast conservation. That was certainly not my intention. What I tried to present was just existing data. I completely agree with your comments, and I would like to point out what was not illustrated in a slide—that these data are actually quite old. They date back to 1991, which I think is old in the history of mastectomy and of quality-of-life research. I think one of the unfortunate things with economic evaluations and quality-of-life analyses is that they always are a step behind in a sense what the clinical results may be, because we obviously rely on clinical trial results. It is once you have the clinical trial results in hand that we can start to look at economic and quality-of-life aspects. To my knowledge, there are not very many recently published quality-of-life studies that explore this question. It is definitely an area where I think that the analyses that are conducted need to be much more sophisticated, going far beyond looking at the short-term ramifications of the extent of surgery. A much more long-term perspective needs to be adopted.

Paul Goss: Actually we have just had a paper accepted for publication between the MD Anderson Hospital and Princess Margaret on quality of life following mastectomy versus breast-conserving surgery: To our surprise, the data which emerged from our study were that quality of life was markedly superior in women receiving breast-conserving surgery for the initial 24 months of follow-up. And thereafter the curves completely reversed, and the quality of life in mastectomy patients long-term was superior to conservative surgery patients. We do not know the explanation for this, but we hypothesize that it may be driven by concern regarding recurrence in women treated conservatively and, once adjustment to body image has taken place over time, in women treated by mastectomy.

Suzanne Wait: In one of the earlier papers published in the late 1980s or the early 1990s, the authors suggest that it was different aspects that were more favorable early than later depending on the type of surgery. Also, women adapted very differently to the disease, but the main issues in terms of the really psychological impact of having a disease remain there regardless of the type of surgery received. There may be other confounding factors as well, which may affect the results seen.

Henning Mouridsen: Just before we leave the discussion about this subject, may I ask you again which fraction of the patients with mastectomy had a reconstruction later? And were they still included in the analysis of quality of life following mastectomy?

Paul Goss: We actually did not include any patients with reconstruction in the study for that very purpose. We did not want to confound the quality-of-life analysis. We believe that is a separate study that needs to be done carefully.

Henning Mouridsen: But, if you did not include patients with secondary reconstruction, your mastectomy group is a highly selective group?

Paul Goss: We included long-term follow-up patients. Reconstructive surgery was not being practiced widely in Canada in the time period we studied.

Suzanne Wait: Another argument I did not bring up in my talk is that the results of quality-of-life analyses depend on what instruments you are using. I do not know what instrument you used in your analysis, but quality-of-life analyses that focus on specific symptoms, for example, that may be very relevant 24 months after having mastectomy but completely irrelevant 5 years later, will obviously give very different results, or may present different advantages to one surgical option versus the other as compared to a more comprehensive instrument.

Ahn Myoung Ock: I would like to ask both Dr. Wait and Dr. Goss. As you indicated the quality-of-life issues really differ in terms of any aspects of quality of life—the economic terminology will differ from psychological aspects of quality of life. So I would like to know whether there were only the psychological aspects of quality of life?

Suzanne Wait: Dr. Goss was saying that he was only looking at the quality of life aspect. There are different methodologies used to address very different questions. The cost–utility analysis is basically an economic evaluation where survival is weighted by a quality-of-life parameter, that of utility. It does not constitute a quality-of-life analysis per se.

Ahn Myoung Ock: Both of you have addressed different aspects of quality of life. So I would like to point out that it is important to make the distinction between analysis of the psychological impact of treatment on the patient (quality of life), and an economic analysis which includes a cost component.

Ajay Bhatnagar: You said a couple of things which give rise to the following issue. You said that publications from 1991 and 1992 are very old, and you also said that during these quality-of-life assessments you are trying to match relatively hard scientific data with perceptions which are fairly soft, because they are the subjective view of either patients or people. In that case, quality-of-life data are certainly not going to be very time resilient. Since society is changing very rapidly today, you are going to have very different results when society's perceptions change 2 years from now compared to today. So you will have to do these studies repetitively maybe on a year-by-year basis or maybe on an every two-year basis?

Suzanne Wait: I am going to challenge you there, on your use of the terms *hard* and *soft* data. The separation between hard and soft data is a very relative one, and what one often refers to as hard data depends on perceptions as well. With quality-of-life data, you are looking at the perceptions of patients. You are not trying to replicate an assay over time. It is a different perspective, it is definitely challengeable, sometimes from a methodological standpoint, but I would argue so are clinical data at times. In terms of the data being "old," when I said old, I meant especially in quality-of-life research. This is definitely an evolving science, and the instruments that are being used today are still being validated, and some are very well validated, but others are not. No, of course, you do not have to redo the study every year, just as you do not have to redo clinical studies every year. However, the research question that was addressed in these "older" studies, the ones that I was quoting anyway, was fairly simple, and they were really looking at mastectomy versus breast-conserving surgery and in a fairly short time frame. And, I think now the audience for this type of data is more sophisticated, and they don't just want to know the impact of the type of surgery, but the long-term consequences of surgical choices. So, it is perhaps a more sophisticated paradigm and a fairly new one, I would argue.

Ajay Bhatnagar: If I just could follow that up, Suzanne. I think you misunderstood the question I was asking. What I was saying is that, since perceptions change, and perceptions are based on the information that one has, then the amount of information one has changes with time. Today there is a tremendous explosion in information, with much more information being available more widely to the general population from which the patients come. Thus, the analysis you make which relates value to the patient's perception of value is also going to change. So you have continuously changing parameters, and these very variable parameters may not be the same 2 years from now and could not have been 4 years previously.

Suzanne Wait: Yes, of course, and the results of the studies, again like clinical studies, depend on the population, the time frame, and the quality of the outcomes you are measuring. And, by no means should the results of cost analyses and quality-of-life analyses ever replace decision making. However, I think it is important to provide this information in a scientific manner that is in an experimental context. You cannot say obviously that, because 90% of women in this study preferred mastectomy, that all women prefer mastectomy. These are individual, patient preferences, and also physician preferences will always enter into play. And I think that it will always be the challenge of how to use these data and bring them from scientific papers to the bedside with the patient.

Henning Mouridsen: Can I just ask you about the economics? You gave figures for specific subgroups of breast cancer patients having had different treatments, tamoxifen or chemotherapy or the combination, and the figures were approximately from $4000 to $200,000 per life-year gained. How is this economic level for treatment of breast cancer patient compared to other areas of medicine?

Suzanne Wait: When cost-effectiveness analyses started, the idea was that you had this wonderful laundry list of all the possible therapeutic interventions, and then you would be able to say; ''Oh, there is my threshold, I take all these, and I reject all these.'' There is a publication that is often disputed, but everyone quotes it. It is a Canadian publication, in which they try to set some kind of framework for what are acceptable cost-effectiveness ratios, and the cut-off level is $30,000 per quality-adjusted life-year (QALY). This publication appeared in 1995, I believe. Again, it is always difficult as a health economist to present the data. Whenever you put dollar signs down there, people expect them to be very precise and definite to provide an answer. The whole idea behind cost-effectiveness analyses, and economic evaluation in general is that you are comparing things in relative terms. So, the idea is that it is not really important whether it is $30,000 per QALY. What is important, if, for example adjuvant endocrine therapy is $30,000 per QALY and treatment of hypertension is $2000. So in this case you could say the value of treating hypertension is ''better'' than adjuvant endocrine therapy. But to answer your question, the threshold is usually $30,000. Anything under $30,000 is obviously great value, but anything above $100,000 per QALY is considered the area where you start asking yourself questions about the relative value of your investment.

Nam-Sun Paik: But your instances in the 1989 data are not proper for your lecture, because the oncologists consider the patient in terms of patient survival, and the costs, and also the quality of life. But, in 1989, too many doctors questioned the proper treatment. In 1997, for breast cancer treatment, patients can have chemotherapy and hormonal therapy depending on the estrogen receptor (ER) status. Nowadays we have to take some good instances to lecture the quality of life or the cost?

Suzanne Wait: I agree with you. The data I presented unfortunately reflect what is available in the economic literature on endocrine therapy. Obviously there are more updated clinical data. However, economic analyses and quality-of-life analyses based on these data today have not been published, otherwise I would have presented them.

9

Chemoprevention with Aromatase Inhibitors

Paul E. Goss
Princess Margaret Hospital
University Health Network
Toronto, Ontario, Canada

I. ABSTRACT

Aromatase inhibitors have established themselves in the treatment of advanced breast cancer. Recent data suggest that they may supplant tamoxifen as first-line endocrine therapy in breast cancer patients. Their use as alternative therapy, or in combination, or in sequence with tamoxifen is now the subject of adjuvant therapy clinical trials. In addition to their effect on recurrent local and distant disease in this setting, data will also be obtained on their ability to reduce contralateral primary tumors and also on their effects on other end organs and on the general and menopause-specific quality of life. These data may support a potential role for this class of drugs as chemopreventatives of breast cancer. In preclinical experiments, the third-generation inhibitors have shown a potent ability to prevent tumor induction and growth in carcinogen-induced rat mammary tumor models. Pilot studies of letrozole are underway in healthy postmenopausal women to ascertain its effect on breast density as a surrogate for breast cancer risk and on other target tissues. Surrogate markers of breast cancer risk such as dense mammogram, dense bones, and elevated plasma estradiol may be one way of selecting specific subsets of women for chemoprevention with aromatase inhibitors.

II. INTRODUCTION

Breast cancer remains a serious public health concern. In Canada alone, approximately 20,000 women and their families are affected by the diagnosis annually, and approximately 5000 will die from their disease (1). Reductions in mortality from breast cancer have been achieved by earlier diagnosis and treatment as a result of increased public awareness and mammographic surveillance, improved surgical and radiation techniques, and the implementation of increasingly effective chemotherapy and endocrine therapy. However, because of the systemic nature of the disease from the time of earliest detection, gains from these strategies are likely to remain modest. Chemoprevention of breast cancer is therefore an important goal, because, if achievable, it is most likely to lead to the greatest reduction in breast cancer mortality. In this chapter, the association of estrogen and breast cancer risk is outlined and the current status of chemoprevention of breast cancer with endocrine therapies is reviewed. The potential application of aromatase inhibitors specifically as chemopreventatives of breast cancer is then discussed.

III. ESTROGEN AND BREAST CANCER RISK

Preclinical, clinical, and epidemiological data strongly support the role of estrogens in the development and growth of breast cancer. However, estrogen-induced carcinogenesis has not yet been explained at a molecular level (2).

The potential mechanisms involved include the alkylation of cellular molecules and/or the generation of active radicals that may damage DNA (3). In addition, estrogen itself or some of its reactive metabolites (e.g., the catechol estrogens) may have genotoxic effects (4).

Based on epidemiological data, it has been proposed that breast cancer risk is determined by the cumulative exposure of breast tissue to bioavailable estrogens and the resulting mitotic activity (5,6).

Exposure to estrogens and other sex hormones in a woman occurs mainly in the reproductive period taking into account her pregnancies. This is supported by the finding that early age at menarche, late age at first birth, and exogenous estrogen replacement therapy (ERT) all are associated with increased risk of breast cancer (6,7), whereas early menopause lowers the risk (8) (Fig. 1).

Estrogen homeostasis is complex and dependent on its synthesis, tissue responsiveness, and catabolism. Any of these three mechanisms are potential targets for an antiestrogen mechanism. Recent clinical trials have provided us

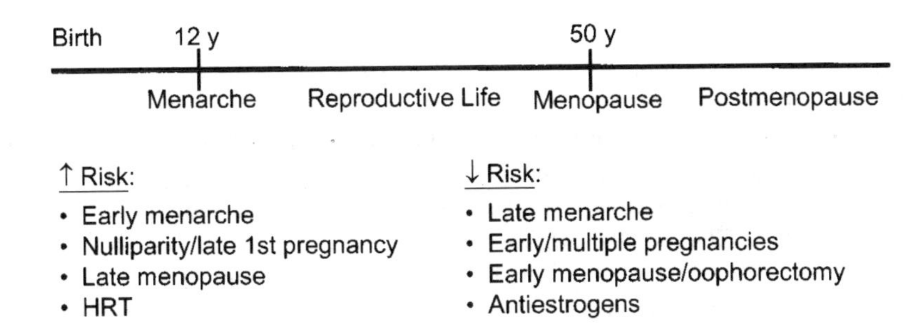

Figure 1 Estrogen and breast cancer risk.

with data regarding the chemopreventative effects of two selective estrogen receptor modulators (SERMS), tamoxifen and raloxifene. These compounds are partial agonists/antagonists on target tissues. It is important to review the reduction in breast cancer risk as well as other end organ effects of these SERMS in order to compare and contrast the potential role of aromatase inhibitors in breast cancer prevention to them.

IV. ANTIESTROGEN CHEMOPREVENTION OF BREAST CANCER WITH SERMS

The most widely studied SERM is tamoxifen, a nonsteroidal antiestrogen, which has been in clinical use for over 20 years. Tamoxifen has been shown to be effective in the treatment of metastatic breast cancer, as adjuvant therapy for early-stage breast cancer, and, most recently, as a preventative agent (9–16).

Importantly, from the adjuvant tamoxifen overview data and from the NSABP B-14 trial, reduction of contralateral breast cancer gave the first clue that tamoxifen might be a potential chemopreventative (10). Reduction in contralateral breast cancer rates was also seen in the Early Breast Cancer Trialists Collaborative Group (EBCTCG) meta-analysis which showed a 13, 26, and a 47% reduction after 1, 2, and 5 years of tamoxifen, respectively (17). In 1992, the NSABP initiated a chemoprevention trial with tamoxifen. Their P1 trial of tamoxifen versus placebo enrolled 13,388 well women at increased risk of breast cancer; defined as women over the age of 60 years or women aged 35–59 years whose 5-year risk of developing breast cancer (by the Gail model) (18) equaled or exceeded that of a 60-year-old woman. A highly significant ($P = .00001$) 45% decrease in the relative risk of developing cancer in the tamoxifen treatment group (14,15,19) was observed. Recently published data show that tamoxifen reduced the risk of invasive breast cancer by 49% ($P < .00001$) with a cumulative incidence (over 69 months) of 43.4 cases versus 22.0 cases per 1000 women in the placebo and tamoxifen groups, respectively. This risk reduction occurred in all age groups <49 (44%), 50–59 (51%), and 60+ (55%). Tamoxifen was also shown to reduce the incidence of noninvasive breast cancer by 50% ($P < .002$). There is no survival difference reported in this trial population to date. Tamoxifen reduced the occurrence of estrogen receptor (ER)–positive tumors by 69% (RR = 0.31; 95% CI = 0.22–0.45) but did not affect the incidence of ER-negative tumors. The natural history and outcome of the cancers that occurred in women in this trial is as yet unknown. Tamoxifen was shown to have a beneficial effect on hip, radius, and spinal fractures with a 19% reduction in fractures that almost reached statistical significance (RR = 0.81; 95% CI = 0.63–1.05). However, tamoxifen was found to increase the risk of endometrial cancer (RR = 2.53; 95% CI = 1.35–4.97). This increased risk was observed predominantly in women aged >50 years. The rates of stroke (risk ratio 1.59; 95% CI = 0.93–2.77), pulmonary embolism (risk ratio 3.01; 95% CI = 1.15–9.27), and deep vein thrombosis (risk ratio 1.60; 95% CI = 0.91–2.86) were also elevated in the tamoxifen treatment arm; again predominantly in women age >50 years (16).

Two European tamoxifen chemoprevention trials (the Italian and the British) (20,21) were also conducted, neither of which showed a reduction in risk with tamoxifen. Several things might explain the differences between the European trials and the P-1 study (22,23). Since the chemopreventative effects of tamoxifen were greater in older women and with longer treatment duration in P-1, the young, low-risk study population (only 12% were over 60 years of age, 48% were hysterectomized, no risk factors were required) and the problems with compliance in the Italian trial might have accounted for differences. Dissimilarities in study cohorts could also have contributed to the different outcomes of the British study from P-1. The relatively young study population with a strong family history in the British trial may not be as susceptible to a chemopreventive effect of tamoxifen as the women in P-1, who were selected according to more nongenetic risk factors. In addition, the shorter duration of follow-up in P-1 as compared to the British trial might have also played a role. Overall, the results of P-1 seem to be robust for the cohort observed, particularly because they are consistent with the preventive effect of tamoxifen on contralateral breast cancer in the adjuvant setting.

A detailed analysis of health-related quality of life in women on P-1 was published recently (24). Women on tamoxifen experienced significantly more vasomotor and gynecological symptoms than women on placebo, but this did not contribute to a reduction in quality of life or an increase in depression as measured by the MOS SF-36 and CES-D quality of life questionnaire instruments, respectively. This important clinical trial established the proof of principle that breast cancer risk reduction could be achieved with the use of a SERM. However, in reviewing the NSABP P-1 study results, it was decided that although efficacy had been demonstrated in all subsets of women in terms of breast cancer reduction, the therapeutic index of the drug was such that its use should be recommended outside of a clinical trial only in women at increased risk of breast cancer.

In the multiple outcomes of raloxifene evaluation (MORE) trial, another SERM, raloxifene, was evaluated. Raloxifene is a benzothiophene derivative that appears to act as an estrogen antagonist in breast tissues but as an estrogen agonist with respect to its effects on circulating lipids and bone. In the uterus, raloxifene causes minimal endometrial thickening after prolonged therapy and no increase in uterine cancer as is seen with tamoxifen (25–29).

In the MORE trial, the patient population (n = 7705) included postmenopausal women up to 80 years of age who had osteoporosis (defined as hip or spine bone mineral density [BMD] T score $\leq$ 2.5 or presence of vertebral fractures) and no history of breast or endometrial cancer. Women were randomized to receive raloxifene (60 or 120 mg) or placebo daily. This trial was designed to test the hypothesis that raloxifene would lower the risk of fractures in the study population. At a median follow-up of 40 months, 54 cases of breast cancer were included in the analysis: 12 classified as ductal carcinoma in situ and 40 as invasive (2 had insufficient data to classify). Of the 40 cases of invasive cancer, 13 occurred on raloxifene and 27 on placebo (RR 0.24; 95% CI = 0.13–0.44; $P < .001$). This difference is statistically significant in showing that raloxifene markedly reduced the incidence of breast cancer, although the study was not designed to test this hypothesis (30).

Raloxifene was also shown to decrease the risk of ER-positive breast cancer by 90% (RR 0.10; 95% CI = 0.04–0.24) but not ER-negative tumors (RR 0.88; 95% CI = 0.26–3.0). Although raloxifene did not increase the risk of endometrial cancer (RR 0.8; 95% CI = 0.2–2.7), in the women who underwent transvaginal ultrasound, endometrial thickness was increased by 0.01 mm in the raloxifene arm and decreased by 0.27 mm in the placebo group ($P < .01$). In addition, 14.2% in the raloxifene group and 10.1% in the placebo arm had an endometrial thickness >5 mm ($P = .02$) and 3.3% (raloxifene) and 1.5% (placebo) had an endometrial thickness that had increased by more than 5 mm compared with baseline ($P = .03$). Raloxifene also increased the risk of venous thromboembolic disease (RR 3.1; 95% CI = 1.5–6.2). The findings of the MORE trial are important, since raloxifene appears only to be slightly uterotrophic, but it has beneficial effects on the breasts, on bones, and on the lipid profile (28–31).

When evaluating the benefits versus risks, that is, the therapeutic index of these interventions, all effects on peripheral tissues have to be taken into account. This is also of particular importance if considering the use of aromatase inhibitors in chemoprevention, and this will be emphasized below.

V. POTENTIAL APPLICATION OF AROMATASE INHIBITORS IN BREAST CANCER PREVENTION

Aromatase inhibitors block the enzyme complex responsible for the final step in estrogen synthesis: the conversion of androgens to estrogens (Fig. 2). In considering the use of aromatase inhibitors in the chemoprevention of breast cancer, a number of important parameters need to be considered. These are shown in Table 1.

A. Therapeutic Index

Antiestrogen strategies for the chemoprevention of breast cancer are intimately related to the function of estrogen in other target tissues. This point has been highlighted in the review above of the SERMS as chemopreventatives, or "chemoprotectors." Because of the apparent "pure" antiestrogenic effects of aromatase inhibitors, these agents may at first seem unsuitable as breast cancer

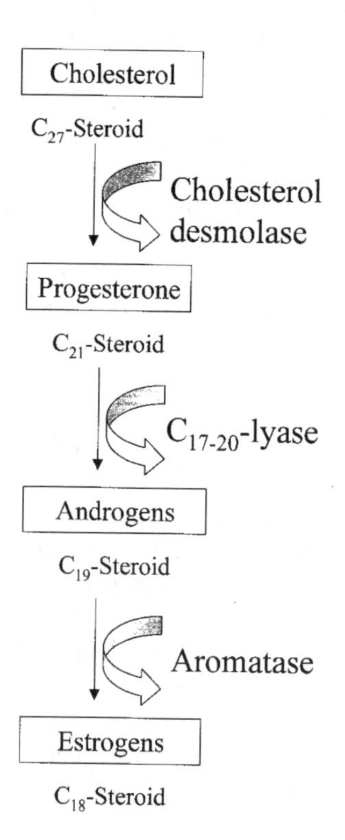

FIGURE 2 The role of aromatase in estrogen biosynthesis.

TABLE 1 Issues Related to Aromatase Inhibitors as Chemopreventatives

Therapeutic index of chemopreventatives
Design of adjuvant aromatase inhibitor trials
Dose/duration issues
Cohorts of high-risk women for pilot studies

preventatives. However, careful analysis of the multiorgan issues involved may not support this view. An event of breast cancer is an all or none phenomenon effecting only a small minority of women exposed, whereas effects on bone and lipid metabolism and on quality of life may potentially impact on all women exposed to the intervention (Table 2).

B. Chemoprevention in Preclinical Models with Aromatase Inhibitors

A number of experiments have been conducted to determine the efficacy of aromatase inhibitors in preclinical rat mammary tumor models. For example, Lubet et al. tested 4-hydroxyandrostenedione (4-OHA) and vorozole in the methylnitrosourea (MNU)–induced rat mammary tumor model. Animals were treated by gavage for 7 days before the administration of MNU, which typically produces highly hormone-responsive tumors. It is known that 4-OHA is conjugated extensively by first-pass metabolism and therefore, not unexpectedly, had minimal

TABLE 2 Therapeutic Index of Chemopreventatives

	SERMS				
	TAM	RAL	AI	AI + SERM	AI + supportive care
Breast cancer	✓✓	✓✓	✓✓	✓✓✓	✓✓
Bone metabolism	✓	✓	x	—	✓
Cardiovascular	✓	✓	x	—	✓
Endometrial cancer	x	?x	✓	—	✓
Thromboembolism	x	x	✓	—	✓
Quality of life	—	—	—	—	—
Menopausal Q.O.C.	x	x	x	x	x
Cognitive function	?	?	?	?	?
Skin	?	?	?	?	?
Other	?	?	?	?	?

antitumor effect. In contrast, vorozole decreased tumor incidence from 100 to 10% and tumor multiplicity from 5.0 tumors per animal to 0.1 tumors per animal (32) (Fig. 3). These data suggest a potent chemopreventative effect in this hormone-dependent model.

C. Review of Current Clinical Trials with Aromatase Inhibitors

Letrozole, anastrozole, and vorozole have all shown clinical efficacy when tested as second-line therapy in postmenopausal women with receptor-positive advanced breast cancer in progression on tamoxifen. For example, letrozole produced significant response rates (23%), duration of response, and time to treatment failure. Detailed data on the adverse events are available from these trials. In general, these agents are well tolerated and, at least in these trials, discontinuation of therapy due to side effects was low (33–35) (Table 3).

A far more relevant and detailed assessment of issues related to efficacy and toxicity of aromatase inhibitors is being addressed in the current generation of adjuvant breast cancer trials. These trials are exploring aromatase inhibitors as alternatives to tamoxifen in combination with tamoxifen and in sequence with

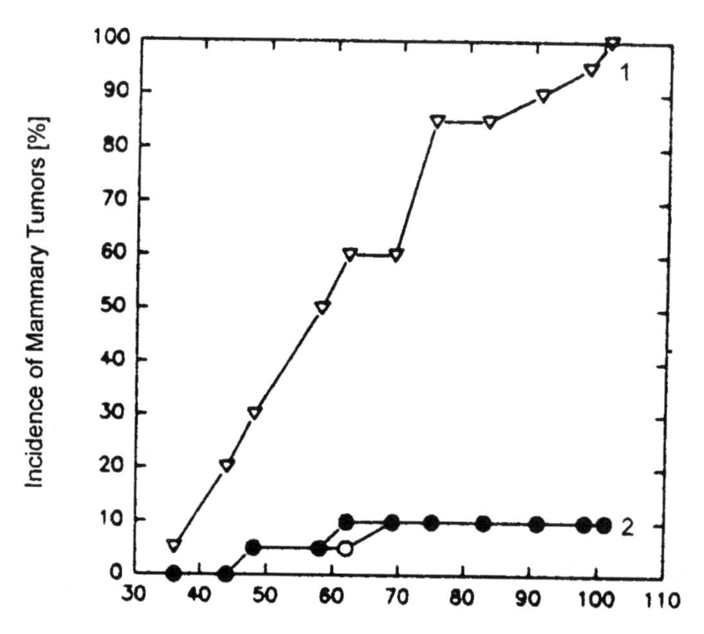

FIGURE 3 Effect of vorozole on mammary tumor incidence.

TABLE 3 Aromatase Inhibitors Versus Megestrol Acetate

	Anastrozole (1 mg) (combined analysis)		Letrozole 2.5 mg		Vorozole (2.5 mg)	
	ANA	MA	LET	MA	VOR	MA
RR	10.3%	7.9%	23.6%	16.4%	10.5%	7.6%
			$p = 0.04$			
Duration of response	—	—	Not reached	17.9	18.2	12.5
			$p = 0.02$			
TTP (mo)	4.8	4.6	5.6	5.5	2.7	3.6
TTF (mo)	6.0	5.0	5.1	3.9	—	—
			$p = 0.04$			
Survival (mo)	26.7	22.5	25.3	21.5	26.0	28.7
	$p < 0.025$					
QoL	—		—		VOR > MA (select subscales)	
Discontinuation due to side effects	2.7		3		3.1	

RR: response rate; TTP: time to progression; TTF: time to treatment failure; ANA: anastrozole; MA: megestrol acetate; LET: letrozole; VOR: vorozole; QoL: quality of life.

tamoxifen. The design of these trials is outlined in Table 4. Importantly, rigorous companion studies are being conducted evaluating organ effects other than on breast cancer. For example, bone density, bone biomarkers and lipid metabolism, quality of life (QoL) and menopausal QoL are being carefully evaluated in the Breast International Group (BIG) and National Cancer Institute of Canada–Clinical Trials Group (NCIC-CTG) adjuvant letrozole trials. Evaluation of all-cause morbidity and mortality is essential in determining the ultimate clinical utility of these applications.

D. Alternative Therapeutic Strategies for Aromatase Inhibitor Chemoprevention Trials

1. Monotherapy

In Sequence with Tamoxifen. Aromatase inhibitors have been shown to have efficacy after tamoxifen in women with advanced metastatic receptor-positive disease. Their efficacy may be enhanced by prior treatment with tamoxifen. Preclinical experimental data have indicated alterations of the estrogen receptor in prolonged tamoxifen-exposed breast cancer cells, resulting in an estrogen

Table 4 Planned and Ongoing Adjuvant Trials with Aromatase Inhibitors

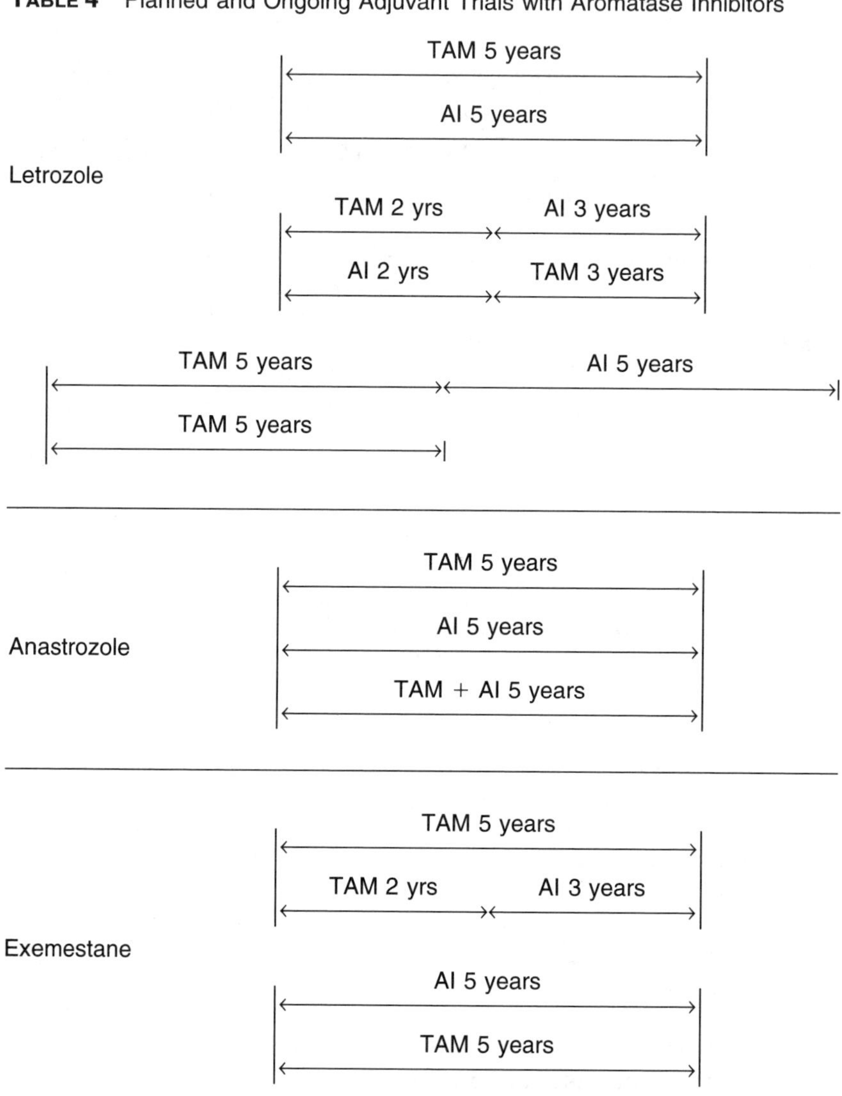

''hypersensitivity'' state (36). In addition, MCF-7 cells grown in an estrogen-deprived medium analogous to tamoxifen blockade have been shown to upregulate aromatase expression producing estrogen directly for autocrine and paracrine use and becoming ''supersensitive'' to estrogen stimulation (37). Concurrent ''tamoxifen dependence'' has been demonstrated. Taken together these data suggest that tamoxifen-exposed breast cancer cells may become ''exquisitely'' and uniquely vulnerable to aromatase inhibitor therapy and make the results of the ''sequenced'' adjuvant trials the most appropriate use of these agents. This principle is being tested, for example, in the international adjuvant trial of letrozole versus placebo in receptor-positive women completing 5 years of tamoxifen treatment.

The impact of estrogen inhibition on other target tissues may well be influenced by prior exposure to tamoxifen. Thus, in postmenopausal women who have undergone 5 years of tamoxifen treatment, the impact on bone, lipid metabolism, and cardiovascular risk may be less than in treatment-naïve women. The effect on vasomotor, gynecological, and other symptoms may also be less as the women coming off tamoxifen may be more remote from menopausal symptoms.

Finally, some of the residual adverse effects of tamoxifen, for example, endometrial and thromboembolic effects, may be partially reversed by aromatase inhibition.

As First-Line Therapy. Preliminary data from a recent study suggest the superiority of anastrozole over tamoxifen as first-line therapy for metastatic disease (38). Results from a comparable study of letrozole are expected. Similarly, a study comparing letrozole to anastrozole as first-line therapy will help to select the optimal inhibitor. If the aromatase inhibitors appear to be superior to tamoxifen, they might be considered as initial chemopreventative therapy. On the other hand, the sequence strategy may overall still be superior and results from Breast International Group Femara and Tamoxifen (BIGFEMTA), Anastrozole, Tamoxifen and Combined (ATAC), and the MA17 trials should help to address this.

In addition, the effects on other end organs such as bone and the cardiovascular system may be more favorable if the compounds are given in sequence. Administration of an aromatase inhibitor in the ''early'' postmenopausal period may aggravate menopausal symptoms significantly. Results from the head-to-head aromatase inhibitor versus tamoxifen trials will provide data in this regard.

2. Combination Therapy

Aromatase Inhibitor Plus a SERM. Preclinical data from Brodie et al. have suggested that combining an aromatase inhibitor with a SERM may be less efficacious than using either compound alone (39). Ingle et al. (40) assessed the pharmacokinetics in patients receiving tamoxifen prior to the addition of letrozole and showed no decrease in tamoxifen or its metabolites. However, evidence does exist for increased clearance of letrozole when combined with tamoxifen (41). No such effect has been observed for anastrozole (42). Nevertheless, one of the three arms of the ATAC trial is a combination of anastrozole and tamoxifen. Not only will efficacy be assessed compared to tamoxifen and anastrozole alone but also other endpoint evaluation; for example, QoL, bone and lipid metabolism, and others. It is possible that efficacy for breast cancer prevention may be demonstrated but that a negative effect on bone and lipid metabolism may occur. These negative effects may be overcome by concurrent administration of a SERM. The potential effects of endocrine chemopreventatives on different end organs are shown in Table 2.

Aromatase Inhibitor Plus Other Supportive Therapy. In the event that aromatase therapy reduces breast cancer risk but exerts a negative effect on bone and lipid metabolism, it is possible that other supportive therapy might override these effects. For example, the use of vitamin D and calcium as well as bisphosphonate therapy may overcome any bone problems, and lipid-lowering agents may have a similar role in the cardiovascular system. Data related to these concomitant therapies will also be forthcoming from the adjuvant trials.

E. Optimal Dose

An important consideration when choosing a chemopreventative in healthy women is the dose selected, because it too will determine the therapeutic index of the intervention. No dose effect has been demonstrated in the anastrozole studies between 1 and 10 mg (33). Preliminary results from the letrozole studies suggested 2.5 mg to be superior to 0.5 mg (34). However, for chemopreventative purposes, lower doses might be desirable. Each end organ may have its own level of estrogen homeostasis and sensitivity to aromatase inhibition. Thus, if estrogen synthesis by epithelial or interstitial cells in the breast is important in the pathogenesis of breast cancer, it may be favorable to inhibit this selectively. A very small dose of aromatase inhibitor may suffice. In this regard, letrozole seems to be superior compared to the other aromatase inhibitors. In preclinical studies of human breast adipocyte aromatization, letrozole was superior to anastrozole in penetrating the cell and inhibiting aromatase (43). This may allow very small doses of letrozole to be given as a chemopreventative even to premenopausal women without perturbing their other end organs. It is challenging to select an appropriate dose for chemoprevention. To ensure efficacy it is safest to select a dose which has shown activity in breast cancer treatment. This is particularly true if one is treating incident cases, but if efficacy is determined at the level of a premalignant lesion, a smaller dose may be adequate. Testing of various doses in breast cancer patients or in lesions with a useful surrogate for efficacy may be the way to determine the minimal effective dose for chemoprevention.

F. Optimal Duration

Duration of chemoprevention is another important but difficult decision. Five years of tamoxifen and raloxifene treatment have been chosen because of the data pertaining to the optimal duration of therapy for adjuvant breast cancer being 5 years. This 5-year "rule" may apply only to the "treatment" of occult incident established tumors and not to premalignant lesions. In addition, this rule applies to tamoxifen, and the phenomenon of aromatase inhibitor resistance may not exist as it does for SERMS. However, the data generated by Santen (37) indicate that, under depleting concentrations of estrogen, MCF-7 cells in culture become increasingly sensitive to diminishing doses of estradiol. In other words, breast cancer cells will, in the same way as they do when exposed to SERMS, adjust their growth ability and overcome the therapeutic effect. However, if hydroxylated estrogens are indeed carcinogenic, then sustained depletion of these may cause true prevention of breast cancer and perhaps result in indefinite therapy being the optimal chemopreventative strategy. As discussed previously, the efficacy of breast cancer reduction will need to be balanced against the potential negative effects in other important target tissues.

VI. PILOT STUDIES OF CHEMOPREVENTION

It is useful when assessing the potential of a novel chemopreventative agent to consider selecting cohorts of women at high risk for breast cancer with "surrogate biomarkers" of risk for pilot chemoprevention studies. The risk factors mentioned above in Section III are epidemiological rather than individual and hence not appropriate for selection of women at high risk. The Gail risk assessment has limitations, particularly in postmenopausal women where its discriminatory power between "all risk" and "high risk" is limited. A number of clinical markers have recently been identified which may allow identification of more specific high-risk postmenopausal women for chemopreventative studies. Three of these will be discussed below.

A. Postmenopausal Plasma Estrogen Levels

In several large prospective, nested, case-control studies, higher free estradiol levels in postmenopausal women have been associated with an increased risk of developing breast cancer compared to women with lower estrogen levels (44,45).

The osteoporotic fracture study (46) confirmed this relationship by showing a relative risk (RR) of subsequent breast cancer of 3.2 (95% CI = 1.4–7.0) between women in extreme quartiles of estrone levels. Thomas et al. (47) also compared serum estradiol concentrations in postmenopausal women and found an almost fivefold increase in breast cancer risk (95% CI = 2–12) in the upper third of the distribution compared to the lower third (Fig. 4).

B. Breast Density on Screening Mammogram

The radiological appearance of the female breast depends on the relative amount of fat and connective and epithelial tissues present and varies during the lifetime and from woman to woman. As dense breasts are associated with higher breast cancer incidence (44), mammography can be used as a means to assess breast cancer risk. It has been shown that the relative risks of breast cancer range from 4 to 5 between extreme categories of mammographic density (48) (Fig. 5).

A reduction of serum estrogens in women with dense breasts reduces mammographic density, as has been shown by using tamoxifen (49), dietary modifications (50), and gonadotropin-releasing hormones (51). Further evaluation of antiestrogenic strategies in reducing breast density and their association with subsequent breast cancer risk should be pursued.

C. Bone Mineral Density and Breast Cancer Risk

The hypothesis that women with a history of osteoporotic bone fractures have a relatively low risk of breast cancer has been supported by two studies (52,53). On the other hand, Cauley et al. (54) showed that increased bone mineral density (BMD) is significantly associated with higher risk of breast cancer. Between

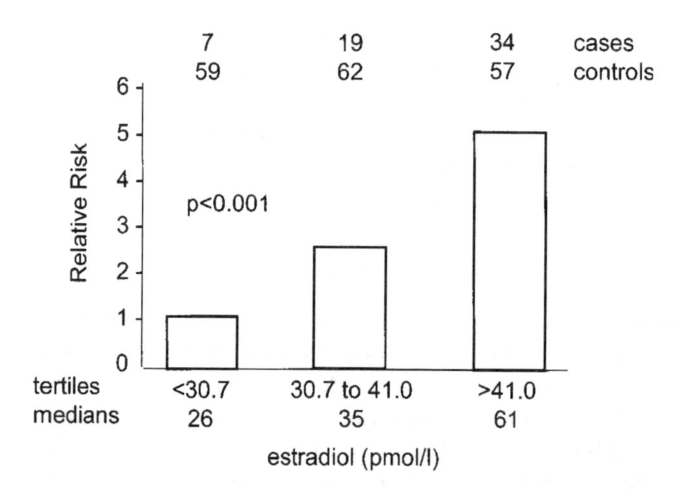

Figure 4 Odds ratios for breast cancer in relation to postmenopausal plasma estradiol concentration.

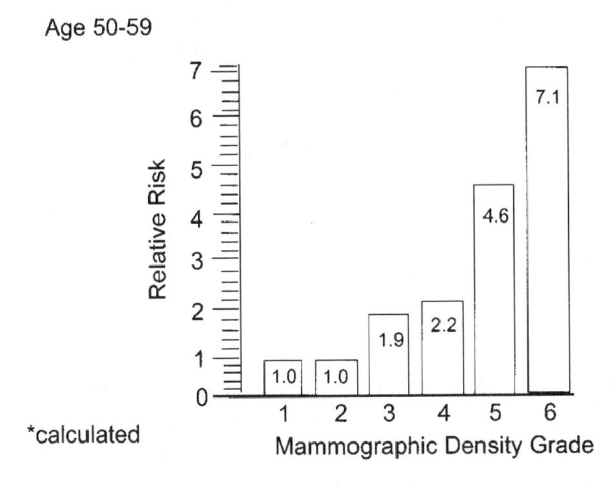

FIGURE 5 Breast cancer risk and mammographic density.

women in extreme quartiles of BMD, a greater than twofold increase of risk was noted. In a study of Zhang et al., (55), BMD in the highest quartile was associated with an increased incidence rate of breast cancer of sevenfold and a relative risk of 3.5 compared to women in the lowest quartile (Fig. 6). These studies emphasize the relationship between high bone mass as a marker for postmenopausal estrogen exposure (and perhaps longer exposure) and breast cancer risk (Fig. 7).

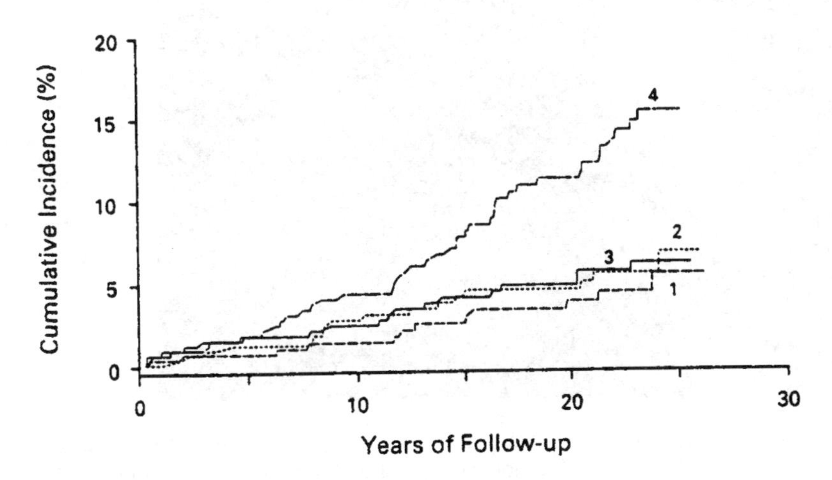

FIGURE 6 Cumulative incidence of breast cancer.

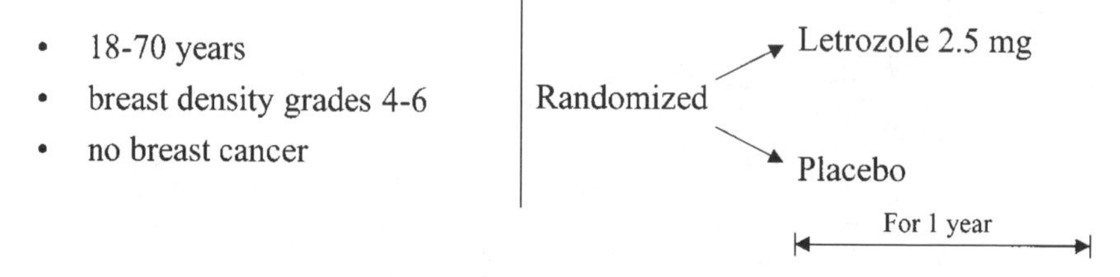

Figure 7 Design of dense mammogram study.

Selection of women from any of these ''risk factor'' groups may be appropriate for chemopreventative pilot studies. We are conducting a randomized trial of letrozole versus placebo in women with increased density on their screening mammogram (Fig. 8). The participants in this study are a cohort of postmenopausal women with an elevated risk of breast cancer (48). Careful evaluation of bone and lipid metabolism, markers of coagulation, quality of life, and menopause-specific quality of life are being undertaken. Treatment duration is for one year, and the surrogate findings of breast density reduction together with other

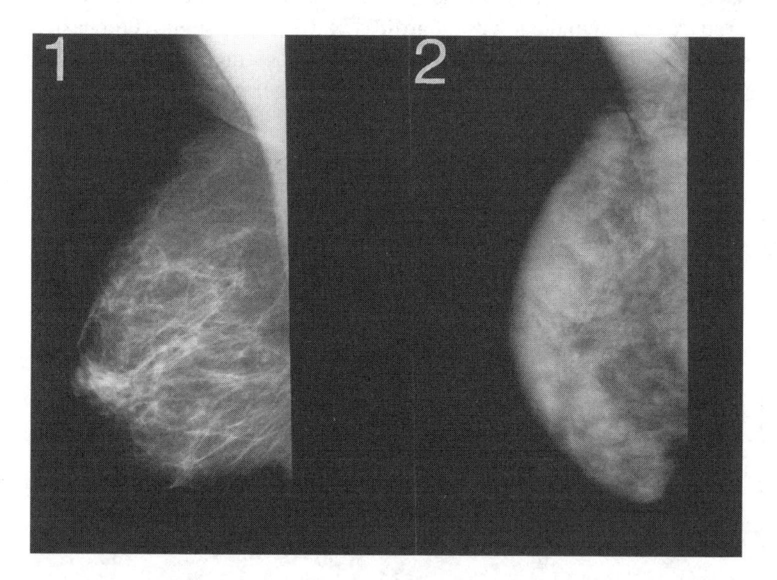

Figure 8 Mammograms from two patients showing: (1) normal breast and (2) breast with extensive areas of radiologically dense breast tissue. (Courtesy of K. Bukhanov).

end organ effects will help to decide whether this class of agents may play a future role in breast cancer chemoprevention.

VII. CONCLUSIONS

The efficacy of aromatase inhibitors in the treatment of breast cancer has been established. Preclinical and clinical data suggest that inhibition of estrogen synthesis should result in a reduction in breast cancer risk. Detailed evaluation in terms of breast cancer recurrence and reduction in contralateral breast cancer will be forthcoming from the current adjuvant trials. These trials will also provide comparative data against tamoxifen and in combination and sequence with tamoxifen. Detailed evaluation of other organ endpoints will also give important data regarding the toxicity and therefore potential therapeutic index of aromatase inhibitors in healthy postmenopausal women. Results from ongoing pilot studies examining the effects on surrogate markers of breast cancer risk and other endpoints will also soon be available from our dense mammogram study. Measures to protect potential negative effects on organs such as bone may be overcome by combination therapy with SERMS or concurrent use of agents such as bisphosphonates. Initial studies of aromatase inhibitors as chemopreventatives will probably employ standard doses and limit the duration of therapy to 5 years, although pilot studies exploring lower doses and different durations of therapy would be useful. Finally, although not addressed in this chapter, application of aromatase inhibitors in selected high-risk premenopausal women could also be explored in conjunction with either surgical or medical ovarian ablation and along similar lines to those discussed for postmenopausal women.

REFERENCES

1. National Cancer Institute of Canada. Canadian Cancer Statistics 1999, Toronto, Canada, 1999.
2. Lupulescu A. Estrogen use and cancer incidence: a review. Cancer Invest 1995; 13: 287–295.
3. Nandi S, Guzman RC, Yang J. Hormones and mammary carcinogenesis in mice, rats, and humans: a unifying hypothesis. Proc Natl Acad Sci USA 1995; 92(9):3650–3657.
4. Yager JD, Liehr JG. Molecular mechanisms of estrogen carcinogenesis. Annu Rev Pharmacol Toxicol 1996; 36:203–232.
5. Pike MC, Spicer DV, Dahmoush L, Press MF. Estrogens, progestogens, normal breast cell proliferation, and breast cancer risk. Epidemiol Rev 1993; 15(1):17–35.
6. Rosner B, Colditz GA. Nurses' health study: log-incidence mathematical model of breast cancer incidence. J Natl Cancer Inst 1996; 88(6):359–364.
7. Colditz GA. Relationship between estrogen levels, use of hormone replacement therapy, and breast cancer. J Natl Cancer Inst 1998; 90(11):814–823.

8. Paffenbarger RS Jr, Kampert JB, Chang HG. Characteristics that predict risk of breast cancer before and after the menopause. Am J Epidemiol 1980; 112(2):258–268.
9. Litherland S, Jackson IM. Antioestrogens in the management of hormone dependent cancer. Cancer Treat Rep 1988; 15:183–194.
10. Fisher B, Costantino J, Redmond C, Poisson R, Bowman D, Couture J, Dimitrov NV, Wolmark N, Wickerham DL, Fisher ER, Margolese R, Robidoux A, Shibata H, Terz J, Paterson AHG, Feldman MI, Farrar W, Evans J, Lickley HL, Ketner M, et al. A randomized clinical trial evaluating tamoxifen in the treatment of patients with node negative breast cancer who have estrogen-receptor-positive tumors. N Engl J Med 1989; 320:479–484.
11. Early Breast Cancer Trialists Collaborative Group. Systemic treatment of early breast cancer by hormonal, cytotoxic or immune therapy. 133 randomized trials involving 31,000 recurrences and 24,000 deaths among 75,000 women. Lancet 1992; 339:1–15.
12. Clarke RB, Laidlaw IJ, Jones LJ, Howell A, Anderson E. Effect of tamoxifen on Ki67 labeling index in human breast tumours and its relationship to oestrogen and progesterone receptor status. Br J Cancer 1993; 67:606–611.
13. Rutqvist LE, Wilking N. Analysis of hormone receptors and proliferation fraction in fine-needle aspirates from primary breast carcinomas during chemotherapy or tamoxifen treatment. Acta Oncol 1992; 31:139–141.
14. Smigel K. Breast cancer prevention trial shows major benefit, some risk. J Natl Cancer Inst 1998; 90:647–648.
15. Josefson D. Breast cancer trial stopped early. BMJ 1998; 316:1187.
16. Fisher B, Costantino JP, Wickerham DL, Costantino JP, Wickerham DL, Redmond CK, Kavanah M, Cronin WM, Vogel V, Robidoux A, Dimitrov N, Atkins J, Daly M, Wieand S, Tan-Chiu E, Ford L, Wolmark N, and other National Surgical Adjuvant Breast and Bowel Project Investigators. Tamoxifen for prevention of breast cancer: report of the National Surgical Adjuvant Breast and Bowel Project P-1 study. J Natl Cancer Inst 1998; 90:1371–1388.
17. Early Breast Cancer Trialists Collaborative Group. Tamoxifen for early breast cancer: an overview of the randomized trials. Lancet 1998; 351:1451–1466.
18. Gail MH, Brinton LA, Byar DP, Corle DK, Green SB, Schairer C, Mulvihill JJ. Projecting individualized probabilities of developing breast cancer for white females who are being examined annually. J Natl Cancer Inst 1989; 81:1879–1886.
19. Bush TL, Heizisouer KJ. Tamoxifen for the primary prevention of breast cancer: a review and critique of the concept and trial. Epidemiol Rev 1993; 15:233–243.
20. Veronesi U, Maisonneuve P, Costa A, Sacchini V, Maltoni C, Robertson C, et al. Prevention of breast cancer with tamoxifen: preliminary findings from the Italian randomised trial among hysterectomised women. Italian Tamoxifen Prevention Study. Lancet 1998; 352:93–97.
21. Powles T, Eeles R, Ashley S, Easton D, Chang J, Dowsett M, et al. Interim analysis of the incidence of breast cancer in the Royal Marsden Hospital tamoxifen randomised chemoprevention trial. Lancet 1998; 352:98–101.
22. Chlebowski RT, Collyar DE, Somerfield MR, et al. American Society of Clinical

Oncology Technology Assessment on Breast Cancer Risk Reduction Strategies: Tamoxifen and Raloxifene. J Clin Oncol 1999; 17(6):1939–1955.
23. Pritchard KI. Is tamoxifen effective in prevention of breast cancer? Lancet 1998; 352:80–81.
24. Day R, Ganz P, Costantino J, Cronin W, Wickerham D, Fisher B. Health-related quality of life and tamoxifen in breast cancer prevention: a report from the National Surgical Adjuvant Breast and Bowel Project P-1 Study. J Clin Oncol 1999; 17:2659–2669.
25. Clemens JA, Bennett DR, Black LJ, Jones CD. Effects of a new antiestrogen, keoxifene (LY 156758) on growth of carcinogen-induced mammary tumors and on LH and prolactin levels. Life Sci 1983; 32:2869–2875.
26. Black LJ, Masahiko S, Rowley ER, Magee DE, Bekele A, Williams DC, Cullinan GJ, Bendele R, Kauffman RF, Bensch WR, Frolik CA, Termine JD, Bryant HU. Raloxifene (LY139481 HCl) prevents bone loss and reduces serum cholesterol without causing uterine hypertrophy in ovariectomized rats. J Clin Invest 1994; 93:63–69.
27. Ashby J, Odum J, Foster JR. Activity of raloxifene in immature and ovariectomized rat uterotrophic assays. Reg Toxicol Pharmacol 1997; 25:226–231.
28. Draper MW, Flowers DE, Huster WJ, Neild JA, Harper KD, Arnaud C. A controlled trial of raloxifene (LY139481) HCl: impact on bone turnover and serum lipid profile in healthy postmenopausal women. J Bone Miner Res 1996; 11(6):835–842.
29. Delmas PD, Bjarnason NH, Mitlak BH, Ravoux A-C, Shah AS, Huster WJ, Draper M, Christiansen C. Effects of raloxifene on bone mineral density, serum cholesterol concentrations, and uterine endometrium in postmenopausal women. N Engl J Med 1997; 337:1641–1647.
30. Cummings SR, Eckert S, Krueger KA, Grady D, Powles TJ, Cauley JA, Norton L, Nickelsen T, Bjarnason NH, Morrow M, Lippman ME, Black D, Glusman JE, Costa A, Jordan VC. The effect of raloxifene on risk of breast cancer in postmenopausal women: results from the MORE Randomized Trial. JAMA 1999; 281(23):2189–2197.
31. Cummings S. European Congress on Osteoporosis, Berlin, September 1998.
32. Lubet RA, Steele VE, Casebolt TL, Eto I, Kelloff GJ, Grubbs CJ. Chemopreventive effects of the aromatase inhibitors vorozole (R-83842) and 4-hydroxyandrostenedione in the methylnitrosourea (MNU)–induced mammary tumor model in Sprague-Dawley rats. Carcinogenesis 1994; 15(12):2775–2780.
33. Buzdar A, Jonat W, Howell A, Jones SE, Blomqvist C, Vogel CL, Eiermann W, Wolter JM, Azab M, Webster A, Plourde PV for the Arimidex International Study Group. Anastrozole, a potent and selective aromatase inhibitor, versus megestrol acetate in postmenopausal women with advanced breast cancer: results of overview analysis of two phase III trials. J Clin Oncol 1996; 14(7):2000–2011.
34. Dombernowsky P, Smith I, Falkson G, Leonard R, Panasci L, Bellmunt J, Bezwoda W, Gardin G, Gudgeon A, Morgan M, Fornasiero A, Hoffmann W, Michel J, Hatschek T, Tjabbes T, Chaudri HA, Hornberger U, Trunet PF. Letrozole, a new oral aromatase inhibitor for advanced breast cancer: double-blind randomized trial showing a dose effect and improved efficacy and tolerability compared with megestrol acetate. J Clin Oncol 1998; 16(2):453–461.

35. Goss PE, Winer EP, Tannock IF, Schwartz LH, on behalf of the North American Vorozole Study Group. Randomized phase III trial comparing the new potent and selective third-generation aromatase inhibitor vorozole with megestrol acetate in postmenopausal advanced breast cancer patients. J Clin Oncol 1999; 17(1):52–63.
36. Masamura S, Santner SJ, Heitjan DF, Santen RJ. Estrogen deprivation causes estradiol hypersensitivity in human breast cancer cells. J Clin Endocrinol Metab 1995; 80(10):2918–2925.
37. Yue W, Santen R. Aromatase inhibitors: rationale for use following anti-estrogen therapy. Semin Oncol 1996; 23(4 Suppl 9):21–27.
38. Goss PE. Aromatase inhibitors in the treatment and prevention of breast cancer. Curr Oncol 1999; 6(3):138–143.
39. Brodie A, Lu Q, Liu Y, Long B, Wang JP, Yue W. Preclinical studies using the intratumoral aromatase model for postmenopausal breast cancer. Oncology 1998; 12(3 Suppl 5):36–40.
40. Ingle JN, Suman VJ, Johnson PA, Krook JE, Mailliard JA, Wheeler RH, Loprinzi CL, Perez EA, Jordan VC, Dowsett M. Evaluation of tamoxifen plus letrozole with assessment of pharmacokinetic interaction in postmenopausal women with metastatic breast cancer. Clin Cancer Res 1999; 5(7):1642–1649.
41. Dowsett M, Pfister C, Johnston SRD, Miles DW, Houston SJ, Verbeek JA, Gundacker H, Sioufi A, Smith IE. Impact of Tamoxifen on the pharmacokinetics and endocrine effects of the aromatase inhibitor letrozole in postmenopausal women with breast cancer. Clin Cancer Res 1999; 5(9):2338–2343.
42. Dowsett M, Tobias JS, Howell A, Blackman GM, Welch H, King N, Ponzone R, von Euler M, Baum M. The effect of anastrozole on the pharmacokinetics of tamoxifen in post-menopausal women with early breast cancer. Br J Cancer 1999; 79(2):311–315.
43. Miller W. Qualitative effects of letrozole as primary medical therapy on in situ oestrogen synthesis and endogenous oestrogen levels within the breast. Breast 1997; 6(4):a0–12.
44. Toniolo PG, Levitz M, Zeleniuch-Jacquotte A, Banerjee S, Koenig KL, Shore RE, Strax P. A prospective study of endogenous estrogens and breast cancer in postmenopausal women. J Natl Cancer Inst 1995; 87(3):190–197.
45. Cauley JA, Lucas FL, Kuller LH, Stone K, Browner W, Cummings SR. Elevated serum estradiol and testosterone concentrations are associated with a high risk for breast cancer. Ann Intern Med 1999; 130(4 part 1):270–277.
46. Cummings SR, Browner WS, Bauer D, Stone K, Ensrud K, Jamal S, Ettinger B. Endogenous hormones and the risk of hip and vertebral fractures among older women. Study of Osteoporotic Fractures Research Group. N Engl J Med 1998; 339(11):733–738.
47. Thomas HV, Key TJ, Allen DS, Moore JW, Dowsett M, Fentiman IS, Wang DY. A prospective study of endogenous serum hormone concentrations and breast cancer risk in premenopausal women on the island of Guernsey. Br J Cancer 1997; 75(7): 1075–1079.
48. Boyd NF, Byng JW, Jong RA, Fishell EK, Little LE, Miller AB, Lockwood GA, Tritchler DL, Yaffe MJ. Quantitative classification of mammographic densities and

breast cancer risk: results from the Canadian National Breast Screening Study. J Natl Cancer Inst 1995; 87(9):670–675.

49. Ursin G, Pike MC, Spicer DV, Porrath SA, Reitherman RW. Can mammographic densities predict effects of tamoxifen on the breast? (letter). J Natl Cancer Inst 1996; 88(2):128–129.
50. Knight JA, Martin LJ, Greenberg CV, Lockwood GA, Byng JW, Yaffe MJ, Tritchler DL, Boyd NF. Macronutrient intake and change in mammographic density at menopause: results from a randomized trial. Cancer Epidemiol Biomarkers Prev 1999; 8(2):123–128.
51. Ursin G, Astrahan MA, Salane M, Parisky YR, Pearce JG, Daniels JR, Pike MC, Spicer DV. The detection of changes in mammographic densities. Cancer Epidemiol Biomarkers Prev 1998; 7(1):43–47.
52. Persson I, Adami HO, McLaughlin JK, Naessen T, Fraumeni JF Jr. Reduced risk of breast and endometrial cancer among women with hip fractures (Sweden). Cancer Causes Control 1994; 5(6):523–528.
53. Olsson H, Hagglund G. Reduced cancer morbidity and mortality in a prospective cohort of women with distal forearm fractures. Am J Epidemiol 1992; 136(4):422–427.
54. Cauley JA, Lucas FL, Kuller LH, Vogt MT, Browner WS, Cummings SR. Bone mineral density and risk of breast cancer in older women: the study of osteoporotic fractures. Study of Osteoporotic Fractures Research Group. JAMA 1996; 276(17): 1404–1408.
55. Zhang Y, Kiel DP, Kreger BE, Cupples LA, Ellison RC, Dorgan JF, Schatzkin A, Levy D, Felson DT. Bone mass and the risk of breast cancer among postmenopausal women. New Engl J Med 1997; 336(9):611–617.

Panel Discussion 4

Chemoprevention
November 12, 1999

List of Participants

Mitch Dowsett London, England
Paul Goss Toronto, Ontario, Canada
Henning Mouridsen Copenhagen, Denmark
Ian Smith London, England
Arnold Verbeek Basel, Switzerland
Suzanne Wait Strasbourg, France

Arnold Verbeek: This was a very exciting talk on an interesting topic. As you know, in a commercial organization, there are always decisions to make regarding the priorities for investments we should make and in which indication areas we ought to be active. I think the prevention trials are extremely exciting, also for the field of aromatase inhibitors. But then I look at a full package of data we have currently available. I think there are lots of exciting data in second-line therapy, but very minimal data in first-line. There are no data in a different setting. There are exciting data in preoperative, but very small trials. My question to you is: Is it now time to start prevention trials with aromatase inhibitors? If yes, why? And if no, when is the right time?

Paul Goss: My personal view is that the efficacy of letrozole has been established. I do not think I need to be convinced—as I said on one of the slides—that aromatase inhibitors would reduce the risk of breast cancer in an analogous way to tamoxifen. I find this hard to imagine. So, if I were making a decision, I would not be hesitating on the efficacy end of the scale. What would make me hesitate right now is the therapeutic index of the drug and the lack of data on toxicity both short- and long-term. Dr. Larry Riggs has presented a poster at an endocrine society meeting on letrozole's effect on bone biomarkers. But I do not think we know nearly enough about its effects in that regard. I do not know whether the adjuvant trial data on biomarkers will become available before the primary endpoint data. I do not know whether these data are blinded. We will get some of these data from the small chemoprevention trial I am doing. So, that is the second point. And the third point that is difficult is in whom should we try this. Should we actually go and select women with dense mammograms, or should we give women with genetic risk a gonadotrophin inhibitor plus aromatase inhibitor. You have to look at the event rate in the population and whether the trial is feasible and justifiable on the potential therapeutic index of the drug.

Mitch Dowsett: Paul, you were looking at me in the terms of biomarker data. Certainly within the ATAC (Anastrozole, Tamoxifen, and Combined) trial there is only a subprotocol in which bone density itself is being measured, and therefore bone density data will only be available on between 300 and 500 patients. I cannot remember the exact numbers. At the moment, there is no plan actually to analyze those before the main endpoint of the trial comes about. But I believe that the issues you alluded to are too important that we should, both in the ATAC trial and the FEMTA trials, be pressing for some way in which statistical firewalls can be built, so that we can get these early data on bone lipids.

One of the findings that came out of the raloxifene studies was the early bone biomarker data; the bone fracture data rather than the bone density data predicted fractures better than bone density. And, therefore, we probably can get data on the impact of the aromatase inhibitors much earlier than we had initially anticipated. So, I think we really have to press for these data. And it is entirely possible to build these statistical firewalls which will not unblind the eventual impact.

Paul Goss: I don't think we need large numbers of patients to see the effect on bone biomarkers. If it is going to have a negative effect, it is going to be demonstrable on a small number of patients. The data are very reliable at 3 and 6 months for c-telopeptide or cross-lab measurements. Also, we are giving calcium and vitamins as well as placebo in the treatment arms in our prevention studies, but they do not really influence the bone biomarker significantly enough to obscure the negative effect you might see from the aromatase inhibitors.

Ian Smith: These were very persuasive accounts of how we should be thinking about aromatase inhibition. There is one possible problem—one potential advantage that seems to me selective estrogen receptor modulators (SERMS) may have and you may have alluded to it. And that is the problem of menopausal flushes, which is a real big problem as everyone here knows. It seems to me theoretically quite possible that you could give hormone replacement therapy (HRT) with SERMS. But I cannot see how there could be any logic in fitting that together with aromatase inhibition, and just wonder whether you have thoughts on that.

Paul Goss: I think that is an extremely important point. You know that the majority of women still take HRT for the alleviation of vasomotor symptoms, although the pendulum is gradually swinging in the United States toward taking it for preventive measures. The interesting thing about vasomotor symptoms is that in the three trials (anastrozole, letrozole, and vorozole studies) the incidence of hot flushes across the trials was markedly different. I think it reflected the way these data were collected. So, the first thing is that I am not sure exactly what the level of hot flash provoked by aromatase inhibitors actually is. I think it is surprisingly low, although in theory it should not be. The interesting thing is that the paper just published on the quality of life follow-up on the NSABP-P1 study showed that vasomotor symptoms and sexual dysfunction were statistically significantly worsened by tamoxifen. But they did not impact on the quality of life as reported by women. Therefore, I believe you have to distinguish between side effects and quality-of-life impact. They are different.

Suzanne Wait: To comment on that, I think again it depends on what you are measuring. If you are looking at symptoms, then I agree with you Paul, these do not equate with quality of life. A lot of these quality-of-life instruments were designed for sick women. And most of them were designed to look at the effects of chemotherapy, usually in a metastatic setting. Using these instruments in populations of women with early breast cancer or with a risk of breast cancer may therefore not be appropriate. But I have another question for you. In a long-term randomized controlled aromatase inhibitor chemoprevention trial where you would be looking at the impact on mortality, what would your comparator be—it would be placebo I would assume—but what confounding effects would you have to think of? I am thinking specifically of how you could account for different kinds of mammography use.

Paul Goss: I think the first question is in whom you decide to study prevention, and where you conduct the study. For example, now in high-risk women you could not study a chemopreventive intervention in the United States against a placebo. You would have to do it against tamoxifen, whereas in Europe, you definitely could. The second point is, we actually approached various countries to do a chemoprevention study with another agent in the last 2 years. When we asked the Scandinavian countries, they made the point that if you want to study an intervention, such as a drug to reduce breast cancer mortality, you would effectively do that if you did it within the context of an intervention already proven to reduce mortality from breast cancer (for example, a national breast cancer screening program). In other words, do not alter the mammography screening program, just add the current intervention.

Henning Mouridsen: Paul, now you and Suzanne started to discuss together, could you just comment about the estimate of the economic implications using preventive therapy? The figures were 100-fold higher than for treatment of breast cancer. What would your comments be to that?

Paul Goss: First of all, I do not understand cost analysis well enough to give a good answer. The problem I see is that you cannot count hard events only. How do you count in cost analysis terms the hot flashes and urogenital or sexual dysfunctions in dollar amounts?

Suzanne Wait: It is also beyond the purview of the economists. The idea is not to translate absolutely every single outcome into a cost term. Another thing is that in the data I presented on chemoprevention, the only costs the authors actually took into account were the actual costs of tamoxifen. There was not even an outlook to what the added costs would be. They looked at one effect, and that was mortality, and they looked at one cost, and that was the cost of tamoxifen. So, it was a really, really simplistic model. And as you suggested, a comprehensive economic evaluation that would really look at chemoprevention in its full impact would have to be very, very complex.

Paul Goss: One should try to build a model and study it. You should look at a best- and worst-case scenario for each endpoint. For example, will aromatase inhibitors cause accelerated fractures? In the worst-case scenario, you get a group of bone experts to say what is the worst impact that that could have in this age group. Every endpoint should be answered like this.

Part IV

NEW DIRECTIONS FOR AROMATASE RESEARCH

N. Harada, Chair
D. Evans, Chair

10

Immunohistochemistry of Aromatase: A Recent New Development

Hironobu Sasano, Takashi Suzuki, and Takuya Moriya
Tohoku University School of Medicine
Sendai, Japan

I. ABSTRACT

It has become very important to identify factors that might predict the likely response of breast cancer patients to aromatase inhibitor treatment before treatment starts. In postmenopausal women, aromatase inhibitors are thought to exert their effects on breast cancer by inhibiting intratumoral aromatase activity or expression. Therefore, an assessment of intratumoral aromatase activity and/or expression in surgically resected human breast carcinoma specimens is considered as a promising predictive marker, which could possibly be used in combination with the analysis of the presence of estrogen receptor (ER) α.

Biochemical measurement of aromatase activity in resected specimens is generally considered to be the gold standard method of assessing aromatase activity, but it is associated with technical problems, including the time-consuming and laborious nature of the assay. The reverse transcriptase–polymerase chain reaction (RT-PCR) method can demonstrate the presence of aromatase mRNA in clinical breast cancer specimens, and it requires much smaller amounts of breast cancer tissue. However, the results of these assays are greatly influenced by the ratio of carcinoma cells and cannot be correlated with the nature of the resected specimens. On the other hand, immunohistochemical analysis of aromatase in surgical pathology specimens is relatively easy and straightforward, and can localize the sites of aromatization in the tissue. Therefore, immunolocalization of aromatase in surgical pathology breast cancer specimens is considered the most promising method of assessing intratumoral aromatase. However, this method is also associated with several problems. First, results may be influenced by the nature of the antibodies used and specimen preparation. Second, a scoring system needs to be introduced to provide objective and qualitative analysis of the immunoreactivity. Therefore, the development of antibodies against aromatase which can recognize the epitopes of archival materials and a straightforward and reproducible scoring system of immunoreactivity to provide biologically inert findings are required to establish aromatase immunohistochemistry as the routine method of evaluating intratumoral aromatase in clinical breast cancer specimens.

II. INTRODUCTION

Estrogens are considered to play important roles in the development and progression of a large proportion of human breast cancer. Increased aromatase expression and activity have been reported in human breast cancer compared with cells in normal breast tissue, with the overexpression of aromatase being considered to play an important role in the estrogen-related development and progression of at least some human breast cancers (1–5). Aromatase inhibitor therapy is one of the endocrine treatments available to breast cancer patients. It has, therefore, become very important to study which patients may respond when aromatase inhibitor therapy before it is initiated.

At present, aromatase inhibitors are considered primarily to exert their effects through the reduction of intratumoral aromatase expression or activity, especially in postmenopausal women. Some studies have demonstrated that the response of breast cancers to aromatase inhibitors is greater when the cancer tissues have detectable aromatase activity; that is, a positive correlation of intratumoral aromatase activity with response to treatment with various aromatase inhibitors including aminoglutethimide and hydrocortisone (6,7). If aromatase inhibitors work by reducing in situ estrogen biosynthesis and concentration, the cancer cases that respond to treatment should express estrogen receptors (ER). In human breast cancer, ERα was demonstrated to be the predominant form and not ERβ, a newly identified ER isoform (8–10). We have also demonstrated a statistically significant correlation between clinical response to aromatase inhibitors in the patients of human breast carcinoma and the presence of both ERα and aromatase (unpublished observations). Therefore, it is very important to assess intratumoral aromatase activity and expression in clinical breast cancer specimens with accu-

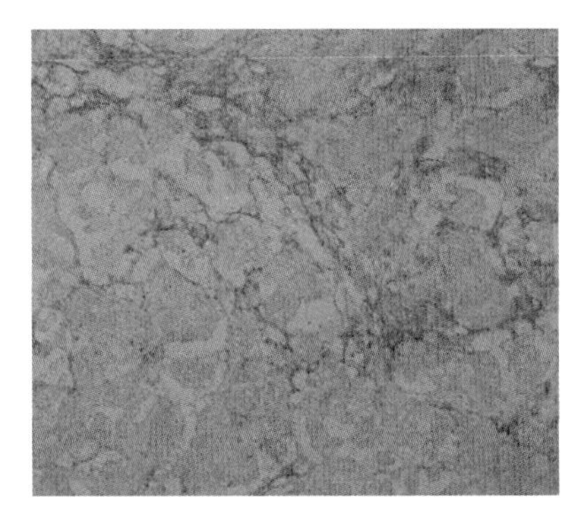

FIGURE 1 Immunohistochemistry of aromatase in breast cancer using polyclonal antibody against the enzyme generated in Dr. Nobuhiro Harada's laboratory. Aromatase immunoreactivity was detected in the stromal cells.

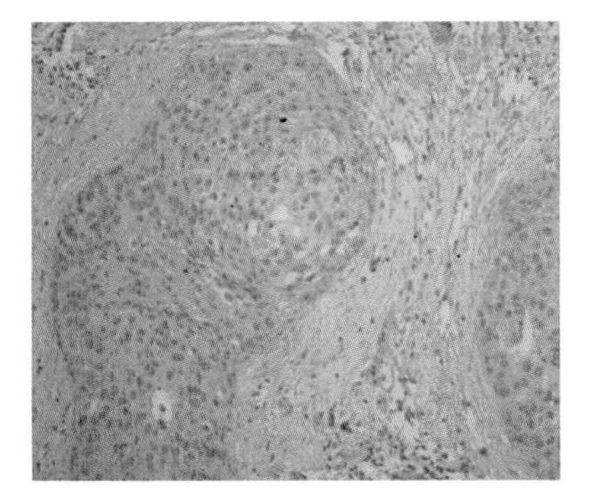

FIGURE 2 Immunohistochemistry of aromatase in breast cancer using monoclonal antibody against the enzyme generated in Dr. Evan Simpson's laboratory. Aromatase immunoreactivity was detected in the carcinoma cells.

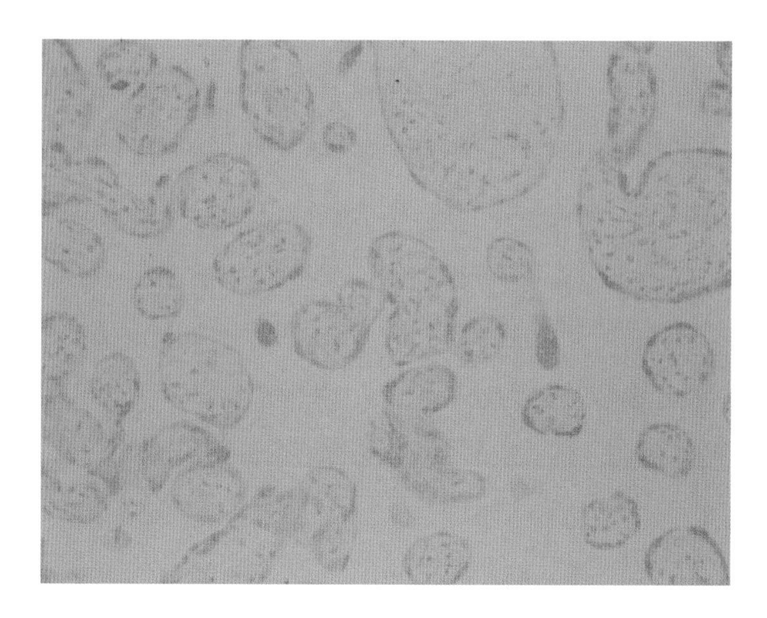

FIGURE 3 Immunohistochemistry of aromatase in normal full-term human placenta using newly developed monoclonal antibody. Aromatase immunoreactivity was detected in syncytiotrophoblasts.

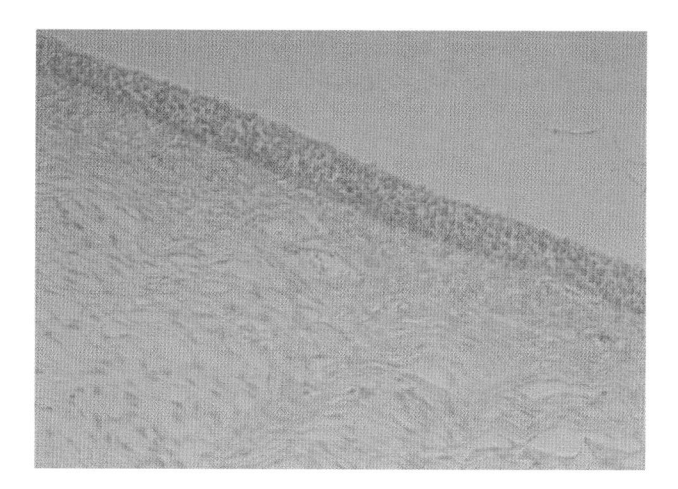

FIGURE 4 Immunohistochemistry of aromatase in normal cycling human ovary using newly developed monoclonal antibody. Aromatase immunoreactivity was detected in the membrana granulosa but not in theca interna cells.

racy and reproducibility to establish whether intratumoral aromatase with or without other factors can serve as a predictor of aromatase inhibitor therapy. In this chapter, the advantages and disadvantages of assessment methods for intratumoral aromatase in clinical breast cancer specimens will be reviewed with an emphasis on immunohistochemical methods.

III. BIOCHEMICAL ASSESSMENT OF AROMATASE

Biochemical measurement of aromatase activity is generally considered as the gold standard method for quantitative assessment of aromatase activity (11). However, this method has several technical disadvantages. First, a relatively large volume (more than 0.5 g) of fresh or freshly frozen breast cancer tissue is required. As the average size of the breast cancer clinically detected has become smaller owing to improvements in mammography or other radiological diagnostic tools, it is increasingly difficult to obtain the volume of breast cancer tissue needed. In addition, aromatase activity is very labile, so tissues need to be frozen in liquid nitrogen immediately after removal to avoid possible degradation. Second, the assays are cumbersome and time consuming. Therefore, it is relatively difficult to apply this method to breast cancer patients worldwide as a means of examining intratumoral aromatase—despite its theoretical importance.

Reverse transcriptase–polymerase chain reaction (RT-PCR) can demonstrate the presence of aromatase mRNA in clinical breast cancer specimens. This method is relatively fast and straightforward, requiring much smaller amounts of breast cancer tissue, although specimens still need to be frozen immediately after resection and stored at −80°C or in liquid nitrogen. The RT-PCR method is not a quantitative assay in itself, but it can demonstrate relative abundance of aromatase mRNA in clinical specimens using a competitive method for comparing results to those of internal controls such as β-actin. In addition, alternative splicing of aromatase mRNA, which is considered to play an important role in the regulation of aromatase expression in breast cancer tissues (12, 13), can also be examined from the same preparation of specimens.

However, biochemical assessment methods of intratumoral aromatase are associated with serious disadvantages, which can greatly influence the results of clinical breast cancer specimen examinations. Although human breast cancer tissue is composed of tumor and nontumor cells (which include interstitial or stromal cells, vasculature, and adipose cells), these methods treat the tissue as one mass. In addition, there are diverse histological types of breast cancer, such as ductal carcinoma, lobular carcinoma, medullary carcinoma, and others. The results of biochemical analyses are, therefore, easily influenced by the ratio of carcinoma to noncarcinoma cells in specimens. Thus, in order to obtain a better understanding of intratumoral aromatization in human breast cancers, it is extremely important to correlate the morphological features of lesions with the findings (14). Although biochemical studies of resected breast cancer specimens, including assays for tumor aromatase activity, have provided important information on intratumoral aromatase status in human breast disorders, it is nearly impossible to determine localization of aromatase (14).

IV. IMMUNOHISTOCHEMICAL ANALYSIS OF AROMATASE

The development of antibodies against aromatase and of an immunohistochemical staining system has made it possible to detect aromatase immunoreactivity in tissue sections in situ (3–5,14–17). The advantages and disadvantages of immunohistochemical aromatase analysis in clinical breast cancer specimens are summarized in Table 1.

A. Advantages of Aromatase Immunohistochemistry for Assessment of Intratumoral Aromatase

Owing to marked improvements in immunostaining methods and the antibodies used, it is now possible to immunolocalize the increasing number of antigens in routinely processed specimens (i.e., 10% formalin-fixed and paraffin embedded tissue). Furthermore, the technique is no longer a cumbersome, laborious, and time-consuming process, but rather, owing to the technical improvement of immunostaining methods, it now forms part of routine diagnostic methods which are incorporated in anatomical pathology laboratories. The process can now be performed rapidly and without many technical difficulties. Another advantage is that resected breast cancer specimens, which are fixed in 10% formalin and embedded in paraffin, are stored as archives in the greater majority of hospitals and institutions in many countries. This method has the enormous potential to provide information about breast cancer patients worldwide, allowing us to assess intratumoral aromatase both prospectively and retrospectively.

As expected, immunohistochemical analysis of aromatase can accurately demonstrate the presence of aromatase regardless of the nature of the cancer

TABLE 1 Advantages and Disadvantages of Immunohistochemical Analysis of Aromatase in Clinical Specimens of Human Breast Carcinoma

Advantages	Disadvantages
Relatively easy and can be done in 6–18 h	Results greatly influenced by the characteristics of antibodies
Can be widely applicable	Interobserver and intraobserver differences and reproducibility
Localize the site of aromatization and correlation with histological features	Qualitative but not quantitative

specimens obtained and the ratio of carcinoma to noncarcinoma or stromal cells present in the cancer tissues. However, although this method can be applied to needle biopsy specimens, immunohistochemical analysis does have some disadvantages.

B. Disadvantages of Aromatase Immunohistochemistry for Assessment of Intratumoral Aromatase

1. Antibodies

Immunohistochemistry results can be greatly influenced by the nature or characterization of antibodies employed as a primary antibody of the immunostain. In previous reports on aromatase immunohistochemistry of breast cancer specimens, there have been controversies regarding the location of aromatase immunoreactivity in the tissues. Some groups reported aromatase in stromal cells, including adipocytes (3–5,17) (Fig. 1) (see color insert for Figs. 1–4), whereas others reported immunoreactivity in carcinoma cells (15,16) (Fig. 2). These results may be due to differences in the epitopes recognized by polyclonal and monoclonal antibodies. In addition, aromatase antibodies need to be able to recognize processed tissue (i.e., 10% formalin fixation–resistant epitopes of aromatase molecules), because this method of sample preparation is routinely used.

2. Fixation and Tissue Processing

As in the immunohistochemistry of other antigens, results are also greatly influenced by the quality of specimen preparations. Delayed fixation usually results in the degradation of immunoreactivity, leading to misinterpretation of data as false-negative findings. The time between removal of specimens and fixation should be as brief as possible, although this is not as critical as with biochemical studies analyzing aromatase activity and mRNA. The specimens should be trimmed appropriately to ensure sufficient permeation of fixatives into the tissue specimens. Over fixation should be avoided, as this usually results in the masking of the epitopes, due to formalin-induced excessive cross-linking of the protein, which also results in a false-negative reaction. However, these types of false-negative results may be unmasked by recent developments in antigen-retrieval methods, such as the autoclave method, microwave irradiation, and others. In any event, prompt and brief fixation is ideal for the accurate assessment of intratumoral aromatase if using immunohistochemistry.

3. Interpretation of Aromatase Immunoreactivity in Breast Cancer Specimens—Scoring System or Semiquantitative Approaches

The most serious disadvantage of using aromatase immunohistochemistry as a method of assessing intratumoral aromatase may be the lack of quantitative data produced compared to biochemical studies. Whether in cancer or stromal cells, aromatase immunoreactivity is located in the cytoplasm of cells, making it extremely difficult to obtain the number, or ratio, of ER-positive cells or labeling index of aromatase, in immunostained slides. It is relatively easy to obtain the labeling index of nuclear antigens, such as ERα or Ki-67 (one of the proliferation markers associated with nuclear antigens) either by routine light microscopic examination or computer-assisted image analysis (18).

Semiquantitative analyses of aromatase immunoreactivity in tissue specimens have been performed using two different approaches. One is the use of the CAS200 computed image analysis system (19–21). Using this program, the total optical density (OD) of aromatase visualized by diaminobenzidine in the cytoplasm was calculated from the 500-nm channel, and the number of nuclei stained by ethyl green was calculated from the 620-nm sensor. The percentage of positively immunostained areas of the tissue sections per case were then subsequently calculated by computer. The correlation between aromatase labeling index and the percentage of aromatase-positive areas was significant in the specimens examined by CAS 200 (19). However, this approach requires relatively expensive image analyzers and supporting computer programs; has limited areas of examination in tissue sections; observers have to define the threshold of detection and the areas of examination, making the assay subjective; and the procedure makes it relatively difficult to obtain results compared to nuclear antigens.

The other approach is to determine the percentage of stromal cells with aromatase immunoreactivity using routine light microscopy. In this model, aromatase positive stromal cells were divided into three groups: 0–5%, 5–25%, and >25% cells positive for aromatase. There was a significant correlation ($P < .01$) between the aromatase labeling index and amount of aromatase mRNA determined by RT-PCR in common epithelial ovarian cancer (20). Aromatase immunoreactivity determined by polyclonal antibody was also demonstrated to be correlated with mRNA expression as determined by in situ mRNA hybridization study in endometrioid endometrial cancer (21). This approach is considered more promising for widespread application, as it requires no special instruments or equipment. However, as in any morphological or histological classification or criteria, establishing of intraobserver and interobserver standards is crucial to make results as subjective and reproducible as possible, making considerable experience and familiarity necessary. In addition, this scoring of aromatase immunoreactivity should be straightforward and easily applicable for any pathologists with reasonable experience of breast pathology and interpretation of immunohistochemistry results. When introducing this scoring system into the laboratory, immunohistochemical slides should first be screened together by investigators using double-headed or multiheaded light microscopy to determine the reproducible criteria, and then they should be independently reviewed by different investigators. Disconcordant cases should then be reevaluated together to ensure consistency in the scoring system.

V. SUMMARY AND FUTURE PROSPECTS

Aromatase immunohistochemistry could undoubtedly be the most suitable and widely applicable method of studying the status of intratumoral aromatase in clinical breast cancer specimens, especially with respect to predicting the response of aromatase inhibitors. However, two factors need to be clarified first. First, the introduction of widely available and reliable monoclonal antibodies, which recognize aromatase epitopes in 10% formalin-fixed and paraffin-embedded tissue sections, was demonstrated in the immunohistochemistry of human full-term placenta (Fig. 3) and human normal cycling ovary (Fig. 4). These monoclonal antibodies certainly have great potential to clarify the problems associated with antibodies when assessing intratumoral aromatase using immunohistochemistry. And second, a straightforward and reproducible scoring system of aromatase immunoreactivity using these new monoclonal antibodies needs to be established. With new antibodies and the associated scoring system, the correlation between intratumoral aromatase and the clinical response to aromatase inhibitors can possibly be established in combination with ERα expression.

REFERENCES

1. Miller WR, Forrest APM. Oestradiol synthesis from C19 steroids by human breast cancer. Br J Cancer 1974; 33:905–911.
2. Miller WR, Hawkins RA, Forrest APM. Significance of aromatase activity in human breast cancer. Cancer Res 1982; 42:3365–3368.
3. Sasano H, Nagura H, Harada N, Goukon Y, Kimura M. Immunolocalization of aromatase and other steroidogenic enzymes in human breast disorders. Hum Pathol 1994; 25:530–535.
4. Sasano H, Harada N. Intratumoral aromatase in human breast, endometrial, and ovarian malignancies. Endocrine Rev 1998; 19:593–607.
5. Sasano H, Frost AR, Saitoh R, Harada N, Poutanen M, Vihko R, Bulun SE, Silverberg SG. Nagura H. Aromatase and 17 beta-hydroxysteroid dehydrogenase type 1 in human breast carcinoma. J Clin Endocrinol Metabol 1996; 81:4042–4046.
6. Miller WR, O'Neil J. The importance of local synthesis of estrogen with the breast. Steroids 1987; 50:537–548.
7. Bezwoda WR, Mansoor N, Dansey R. Correlation of breast tumor aromatase activity and response to aromatase inhibition with aminoglutethimide. Oncology 1987; 44: 345–349.
8. Sasano H, Suzuki T, Matsuzaki Y, Fukaya T, Endoh M, Nagura H, Kimura M. Messenger ribonucleic acid in situ hybridization analysis of estrogen receptors alpha and beta in human breast carcinoma. J Clin Endocrinol Metab 1999; 84:781–785.

9. Dotzlaw H, Leygue E, Watson PH, Murphy LC. Expression of estrogen receptor-beta in human breast tumors. J Clin Endocrinol Metab 1997; 82:2371–2374.
10. Vladusic EA, Homby AE, Guerra-Vladusic FK, Lupu R. Expression of estrogen receptor α messenger RNA variant in breast cancer. Cancer Res 1998; 58:210–214.
11. Shenton KC, Dowsett M, Lu Q, Brodie A, Sasano H, Sacks NP, Rowlands MG. Comparison of biochemical aromatase activity with aromatase immunohistochemistry in human breast carcinomas. Breast Cancer Res Treat 1998; 49(suppl 1):S101–107; Discussion S109–119.
12. Bulun SE, Simpson ER. Regulation of aromatase expression in human tissues. Breast Cancer Treat 1994; 30:19–29.
13. Harada N. Novel properties of human placental aromatase as cytochrome P450: Purification and characterization of a unique form of aromatase. J Biochem 1988; 103: 106–112.
14. Sasano H, Murakami H. Immunolocalization of aromatase in human breast disorders using different antibodies. Breast Cancer Res Treat 1998; 49(suppl 1):S79–84; Discussion S109–119.
15. Esteban JM, Warsi Z, Haniu M, Chen S. Detection of intratumoral aromatase in breast carcinomas. An immunohistochemical study with clinicopathologic correlation. Am J Pathol 1992; 940:337–343.
16. Lu Q, Nakamura J, Savinov A, Yue W, Weisz J, Dabbs DJ, Wolz G, Brodie A. Expression of aromatase protein and messenger ribonucleic acid in tumor epithelial cells and evidence of functional significance of locally produced estrogen in human breast cancers. Endocrinology 1996; 137:3061–3068.
17. Santen RJ, Martel J, Hoagland M. Stromal spindle cells contain aromatase in human breast tumors. J Clin Endocrinol Metab 1994; 79:627–632.
18. Sasano H, Kimura M, Shizawa S, Kimura N, Nagura H. Aromatase and steroid receptors in gynecomastia and male breast carcinoma: an immunohistochemical study. J Clin Endocrinol Metab 1996; 81:3063–3067.
19. Suzuki T, Sasano H, Sasaki H, Fukaya T, Nagura H. Quantitation of P450 aromatase immunoreactivity in human ovary during the menstrual cycle: relationship between the enzymes activity and immunointensity. J Histochem Cytochem 1994; 42:1565–1573.
20. Kaga K, Sasano H, Harada N, Ozaki M, Sato S, Yajima A. Aromatase in human epithelial ovarian neoplasms. Am J Pathol 1996; 149:45–51.
21. Watanabe K, Sasano H, Harada N, Ozaki M, Niikura H, Sato S, Yajima A. Aromatase in human endometrial carcinoma an hyperplasia: immunohistochmical, in situ hybridization, and biochemical studies. Am J Pathol 1995; 146:491–500.

11

Molecular Epidemiology of Aromatase Expression in 1182 Primary Breast Cancers

Urs Eppenberger, S. Levano, F. Schoumacher, H. Müller, and S. Eppenberger-Castori
Universitäts-Frauenklinik
Basel, Switzerland

D. Evans
Novartis Pharma AG
Basel, Switzerland

I. ABSTRACT

Today, potent selective aromatase inhibitors like letrozole are in clinical trials. In order to optimize individual first- and second-line treatment modalities, it will be important to measure intratumoral aromatase expression levels or its enzyme activity as drug targets in primary breast cancer biopsies. We assessed the mRNA expression levels by real-time polymerase chain reaction (PCR) in 1182 primary breast cancer tissue samples and in 86 corresponding normal tissue samples, assuming a positive correlation between aromatase mRNA expression, protein, and enzyme activity levels.

The intratumoral aromatase mRNA expression levels measured varied from negative expression values to very high expression levels (over 1 million copies per total RNA). These expression levels were independent of patients' ages and menopausal, nodal, or steroid hormone receptor status. In 40% of the matched normal adjacent tissue (NAT) samples, aromatase expression levels were higher compared to the intratumoral aromatase expression levels. According to these results, aromatase expression levels should also be measured in matching NAT to evaluate the paracrine impact of estrogen on tumor growth. Survival analysis revealed a poor clinical outcome with respect to relapse and death in a subset of patients with high intratumoral expression levels.

II. INTRODUCTION

One of the most challenging areas in breast cancer is the management of patients with metastatic disease, as even very potent adjuvant chemotherapy produces only a modest reduction in tumor burden. As a result, treatment goals for advanced breast cancer are defined as palliation and/or prevention of disease progression by improving the quality of life and extending survival. In this context, the choice of hormonal manipulations or cytotoxic agents is often individualized with respect to a specific subset of patients with advanced disease. Premenopausal patients receive systemic chemotherapy, whereas postmenopausal women who are estrogen receptor (ER) positive get tamoxifen as a first-line therapy. Progress was made in managing hormonal manipulation by the assessment of the ER and progesterone receptors (PgR), since those patients with positive receptors have a hormonal therapy response rate of 60–70% (1,2). This was an important step forward, as it was the first time that the use of intratumoral proteins as targets for drug response was demonstrated.

Estrogen tumor availability remains one of the risk factors involved in the progression of breast cancer. However, today, the treatment of metastatic breast cancer by the ablation of endocrine organs (e.g., ovaries) has been largely abandoned because of the availability of effective pharmacological hormonal agents. These include the progestins, aminoglutethimide and luteinizing hormone–releasing hormone (LHRH) agonists, which block secondary estrogen synthesis. Estrogen production declines with age, as shown in postmenopausal women. After menopause estrogens are mainly produced in peripheral tissues like adipose and the adrenals (3). This peripheral production is carried out by the cytochrome P450 enzyme complex known as aromatase or estrogen synthetase, which mediates the conversion of androstenedione and testosterone to estrone and estradiol, respectively. In addition to aromatase activity in peripheral tissue, several reports show that aromatase is expressed at a higher level in human breast cancer as compared to normal breast tissue (4). However, it is not clear if all breast tumors express aromatase or if this enzyme activity is modified owing to specific post-translational events. Tumor aromatase has also been shown to stimulate breast cancer growth in both autocrine and paracrine manners. Various methods were applied to assess intratumoral aromatase levels either by biochemically measuring enzyme activity or identifying protein expression by immunohistochemistry in paraffin-embedded tumor tissue or by determining aromatase mRNA aromatase expression levels by real-time polymerase chain reaction (RT-PCR). Gene transcriptional studies revealed that aromatase expression in breast tissue is regulated by the use of alternative promoters. Several studies indicated that aromatase promoter switches are tissue specific (5–8). In breast cancer, a promoter switch from a glucocorticoid-stimulated promoter I.4 in normal tissue to cAMP-stimulated promoters I.3 and II in neoplastic tissue has been reported (9,10).

Suppression of in situ estrogen biosynthesis can be achieved by the prevention of aromatase expression or by the inhibition of aromatase activity in breast tumors. Although the control mechanism of aromatase expression in breast cancer tissue is not yet fully understood, aromatase-inhibitor therapy is already applied as second-line treatment in patients who fail antiestrogen therapy. According to several clinical studies, 20–30% of these patients respond to aromatase-inhibitor treatment. Today, several potent and selective aromatase inhibitors are being used in clinical studies to compare their activity with that of antiestrogens in first-line endocrine therapy for metastatic breast cancer (11). Other clinical trials are investigating the possibility of using aromatase inhibitors in chemoprevention.

In this context, we initiated a study to assess retrospectively aromatase mRNA expression levels in the tumors of primary breast cancer patients. The goal was to establish a precise and reproducible technique in a large number of breast tumors to assess the intratumoral distribution pattern of aromatase mRNA expression levels. This knowledge, combined with ER and PgR protein levels, will be useful in the future to predict patient response to selective aromatase inhibitors like letrozole who may not respond to antiestrogens at the time of primary surgery, identifying those who might profit from sequential endocrine therapy (12, 13).

III. MATERIAL AND METHODS

A. Patients

This retrospective study was performed on tumor samples from 1182 patients with untreated operable-primary breast cancer. The majority of these patients underwent curative surgery between 1997 and 1998. The patient collective was not randomized but selectively chosen according to lymph node involvement and menopausal and ER status. A total of 739 patients (63%) exhibited a lymph node–negative status and were postmenopausal. Of these, 583 (49% of total) patients had ER-positive tumors (ER $>$ 20 fmol/mg protein) (group 1) and 156 (13% of total) ER-negative tumors (group 2). A further 443 (37%) patients had lymph node–positive and ER-positive tumors. Of these patients, 111 (10% of total) were premenopausal (group 3) and 332 (28% of total) were postmenopausal (group 4). In 86 cases, the matching normal adjacent tissue (NAT) was available and analyzed for mRNA aromatase expression levels.

B. Follow-Up

In a subset of 119 patients, follow-up information was obtained. This small subset of patients, who underwent primary surgery in 1992–1996, had a median duration of follow-up of 38 months (range 10–83 months). Recurrence-free survival (RFS) and overall survival (OS) were calculated from surgery until the date of the first recurrence or death. Recurrence was defined as the first evidence of relapse at the locoregional or distant site. Recurrence and death were documented in 34 and 9 cases, respectively.

C. Tissue Preparation and Set-Up of the RT-PCR

Tissues were shock frozen in liquid nitrogen and stored at −70°C until further processing. Stored tissues were pulverized in liquid nitrogen and total RNA was isolated from 30 mg of tissue using the QIAshredder and RN-easy system (Quiagen). The RNA was stored at −20°C. RNA isolation steps were performed in a different laboratory unit to the tissue handling and PCR. Tubes for quantitative PCR were handled only and tightly closed in the designated RNA laboratory. The concentration of RNA was determined by a fluorescence method (SYBR-Green II) in 96-well plates with yeast RNA as the standard. Taq-Man (Perkin Elmer) was used to perform RT-PCR. The primers and probe were designed using the program ''Primer Express'' from Perkin Elmer. The forward primer was complementary to the exon 4, whereas the reverse primer was complementary to the exon 5, and the probe spans one exon/intron. The probe was labeled at the 5′ end with the reporter dye FAM and at the 3′ end with the quencher TAMRA. For the evaluation of the primers, a common PCR with cDNA from placental tissue was performed. Placental RNA was commercially obtained and purified according with the Quiagen protocol. The thermal cycling parameters included 2 min at 94°C, 35 cycles at 94°C for 20 s, and at 60°C for 1 min. The reaction was stopped at 4°C. The PCR-product was mixed with a loading buffer and loaded onto an agarose gel. After electrophoresis, a single band of approximately 150 base pairs (bp) was observed. The PCR product was purified and sequenced on a ABI 310 sequencer (Perkin Elmer) and corresponded to the aromatase mRNA sequences published (14).

D. Validation of the Assay

We used the tested primers for the total placental RNA as templates. The total placental RNA was diluted to yield samples ranging from 2.7 ng to 21.6 pg (2.7,

0.54, 0.18, and 0.0216 ng). These samples were measured with the Taq-Man RT-PCR assay. The thermal cycling parameters were 2 min at 50°C, 30 min at 60°C, 5 min at 95°C for 40 or 45 cycles and then at 95°C for 20 s and 60°C for 1 min. Each sample was measured eight times in three different plates. Standard deviations were 2–3%, demonstrating that the method was reproducible.

E. Standards

Placental RNA was chosen as the standard for determining aromatase RNA in tumor samples. This was in order to be able to treat the standard RNA exactly the same way as the tumor RNA (same purification protocol) and avoid any contamination of the valuable tumor samples with PCR product. The quantified PCR product was used to calibrate the placental RNA and calculate the amount of apparent aromatase RNA in the total RNA. The number of molecules in the PCR product solution was determined by the fluorescence method (SYBR Green I) and dilutions from 300,000 to 480 molecules were made. Using these dilutions as a standard curve, the number of aromatase RNA copies was determined in total placental RNA (84,433 molecules aromatase RNA/ng total RNA). Consequently, by using this number and the placental RNA as standard, we are able to calculate the number of aromatase copies in the respective tissue samples. The same stock of placental RNA was used for all standard curves and diluted from 2700.0 to 21.6 ng, as shown in Figure 1.

F. Tissue Samples Measurements

Where possible, 300 ng total RNA isolated from the tissue samples was used for the Taq-Man measurements (Fig. 2). The threshold cycle number (C_T) was set

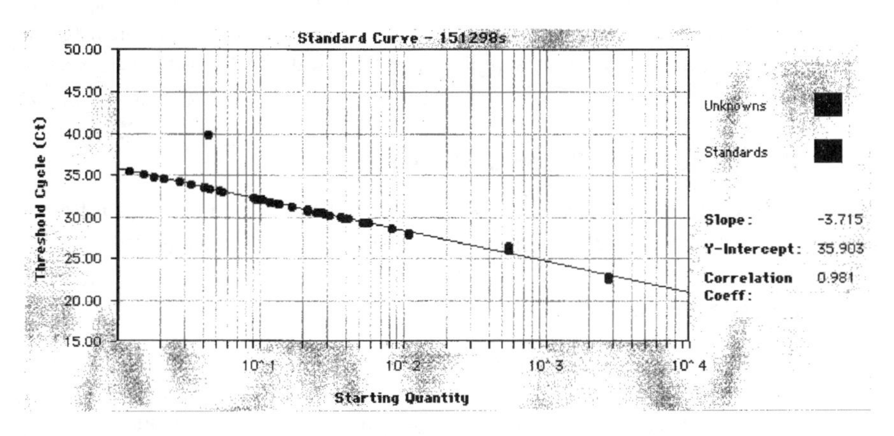

FIGURE 1 Standard curve for RT-PCR.

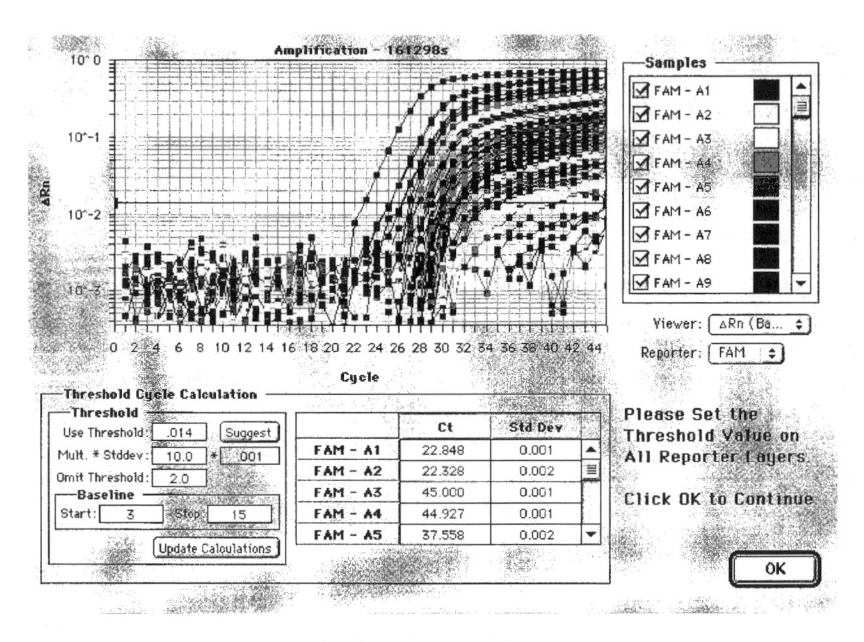

FIGURE 2 An example of RT-PCR measurements.

above the background noise (black horizontal line). If we could not detect any signal using 300 ng total RNA in the RT-PCR set-up, we repeated the measurements with an increased amount of template (500 ng). This protocol allowed us to identify samples containing no aromatase RNA and thus eliminate false-negative values. All determinations were performed in duplicates, and their mean value used as the final RNA copy number. Measurements were repeated if the duplicate determinations differed by more than 15% (for samples with high aromatase RNA expression values) or 50% (for samples with very low RNA aromatase expression).

G. Steroid Hormone Receptor Status

Estrogen receptor and PgR were routinely measured in our laboratory on fresh/frozen tissue samples by electroimmunoassay (EIA) as previously described (15).

H. Statistical Analysis

The relationships and interactions of aromatase mRNA levels and ER and PgR, as well other prognostic factors, were tested using the Mann-Whitney test, the Spearman rank correlation coefficient (r_s), and the linear regression method. The concentration levels of erbB-2 (continuous values) were first tested in univariate regression analysis versus the rate of relapse. In view of the significant prognostic outcome, a cut-off value was searched by means of Classification and Regression Trees (CART) analysis (16,17). Recurrence-free survival and OS probabilities were calculated by the Kaplan-Meier method. The univariate relationships between prognostic factors and survival were assessed by means of log-rank analysis (18,19). All *P* values were two-tailed. All statistical analyses were performed with S-plus 4.5 (MathSoft Inc., Seattle, WA).

IV. RESULTS

Highly selective aromatase inhibitors, such as letrozole, are becoming promising alternatives in second-line therapy for postmenopausal women with advanced breast cancer whose disease has progressed or recurred under tamoxifen treatment (20), since the inhibition of aromatase is more likely to accomplish an intratumoral estrogen blockade, as demonstrated in a nude mouse model (21). Therefore, the assessment of intratumoral aromatase activity as a predictor for drug response will become important in improving individual treatment modalities for postmenopausal women with advanced disease (22).

In 1182 primary breast tumors, the expression levels of intratumoral aromatase ranged from 0 to 1,219,000 mRNA copy number/ng total RNA. The mean and median values were 38,980 and 18,340 mRNA copy number/ng total RNA. In a small subset of 36 tumors (3%), no intratumoral aromatase expression could be found despite ER and PgR distribution levels similar to any other breast cancer tissue. This indicates that intratumoral estrogen production, catalyzed by aromatase, is not necessarily decisive for tumor proliferation and may be independent of ER expression. In general, tumors with no aromatase expression were accompanied with high aromatase expression levels in their respective matching NAT samples. The inverse situation was also observed: 12% of the analyzed NAT samples showed no aromatase expression, whereas very high aromatase levels were expressed in the corresponding tumor tissues. However, in two premenopausal and lymph node–negative patients, no aromatase expression was found in either the tumor or respective NAT.

The histogram of the calculated mean values of aromatase mRNA expression levels shows that the left-tailed distribution pattern could be normalized by logarithmic transformation (Fig. 3). Although the maximum aromatase mRNA levels detected in the 86 matching NAT were only 318,000 mRNA copy number/ng total, the mean and median values of 33,200 and 18,140 mRNA copy number/ng total RNA were comparable (no significant difference) with the values of the matching tumor samples or of the total number of tumor samples analyzed.

No correlation was found between the menopausal status or age (mean 67 years; range 26–90 years) of the patients and the intratumoral aromatase expression levels. However, the aromatase levels in NAT were significantly lower in

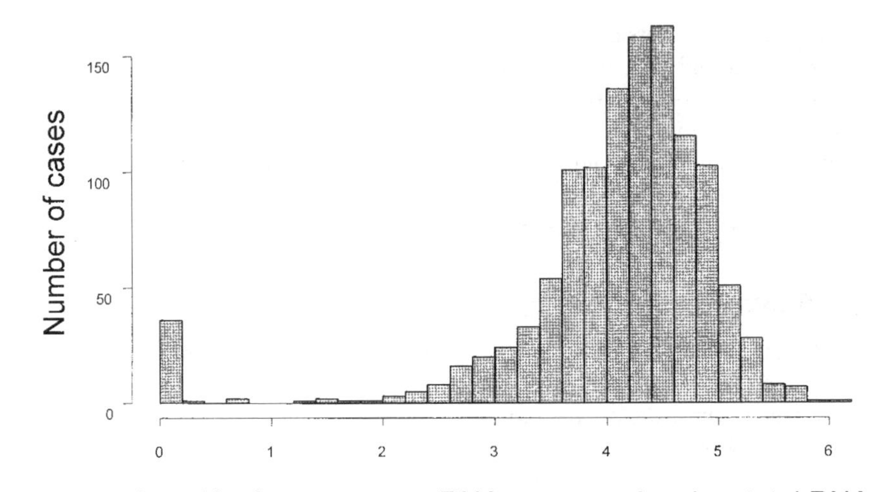

FIGURE 3 Distribution of aromatase mRNA expression levels in 1182 primary breast tumors.

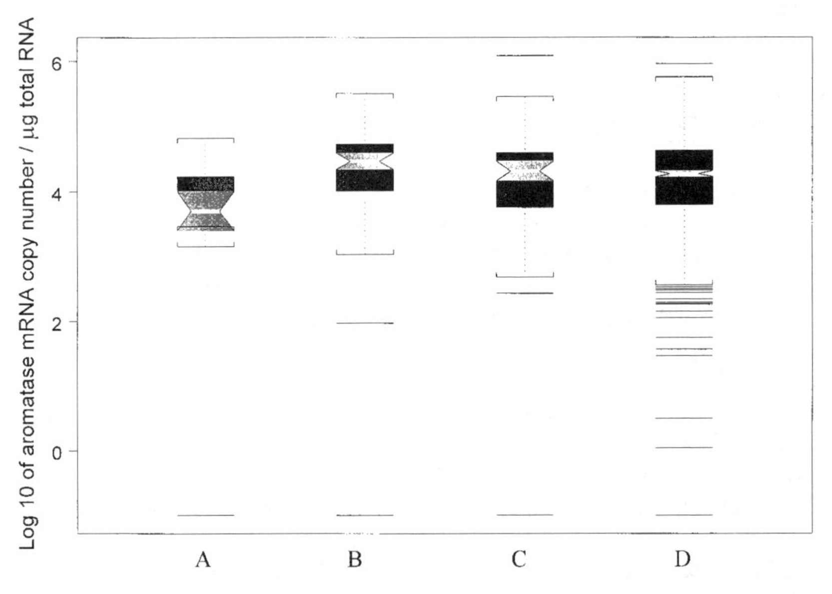

FIGURE 4 Notch-boxplots of aromatase expression levels in: (A) NAT of premenopause, (B) NAT of postmenopausal, (C) intratumoral levels of premenopausal, and (D) intratumoral levels of postmenopausal patients.

premenopausal women compared to postmenopausal patients ($P < .005$), as shown in Fig. 4. This indicates indirectly that the induction of aromatase may be influenced by follicle-stimulating hormone (FSH) serum levels or other factors having an impact on the menopausal status. The comparison of the intratumoral aromatase expression levels with the respective NAT levels revealed a heterogeneous pattern. In fact, in 39 (45%) of the matching 86 cases, intratumoral aromatase expression (median 79850 mRNA copy number/ng total RNA) was significantly higher as with the respective NAT (median 15780 mRNA copy number/ng total RNA) ($P < .0001$). The intratumoral aromatase values could be 3- to 100-fold higher as compared to the respective matching NAT. In 15% of the paired tissue samples, the analyzed aromatase expression levels were comparable ($\pm$ 10%). However, in contrast to previous reports, we found that, in 33 (40%) matched cases, the aromatase expression levels (median 51950 mRNA copy number/ng total RNA) of NAT were significantly higher compared to the respective intratumoral aromatase values (median 12480 mRNA copy number/ng total RNA) ($P < .0001$).

One of the goals of this study was to investigate the impact of local estradiol biosynthesis by aromatase on the ER protein expression levels, since it is known that high estrogen plasma levels downmodulate the ER levels in breast tumor tissue. No correlation was found between aromatase expression levels and ER or PgR levels. In postmenopausal patients, aromatase expression levels greater than 40,000 mRNA copy number/ng total RNA (approximate median value) are found in tumors with ER-positive status with the same distribution pattern also being seen in ER-negative tumors (Fig. 5). Moreover, the majority of tumors with high aromatase expression levels and ER-negative status also exhibited PgR protein levels <20 fmol/mg protein (Fig. 6, group 2). This small subset of patients with hormone-independent tumors but very high aromatase expression levels could also be considered for aromatase inhibitor treatment modalities. As shown in Figure 5, the ER levels of group 3 (premenopausal patients with positive node and ER) are lower compared to the ER levels of all postmenopausal patients. Also, in this subset of patients, no statistically significant correlation could be found between ER and intratumoral aromatase expression levels. In all NAT samples tested, ER and PgR levels were very low (median values 10 and 12 fmol/mg protein, respectively), as expected for normal breast epithelial tissue and completely independent from the aromatase expression levels assessed.

In a subset of 119 patients with known clinical outcome, the intratumoral aromatase expression levels were investigated for their prognostic value. As shown by Kaplan-Meier survival curves (Fig. 7), patients whose tumors expressed higher aromatase levels (>100,000 mRNA copy number/ng total RNA) significantly correlated with increased risk of relapse and death. These results reconfirm data reported by other investigators that the intratumoral amount of aromatase is decisive for the in situ estrogen promotion of breast cancer cell growth.

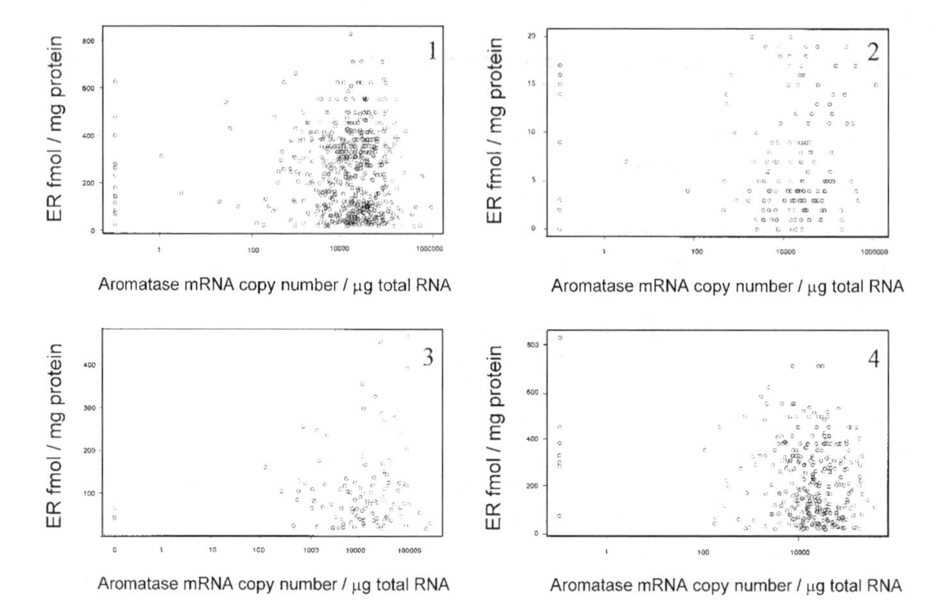

FIGURE 5 Intratumoral aromatase expression levels versus ER protein levels. Groups 1 and 2 are postmenopausal pN0 patients: 1 with ER positive and 2 with ER negative tumors; groups 3 and 4 are pN1 patients with ER-positive tumors: 3 premenopausal and 4 postmenopausal.

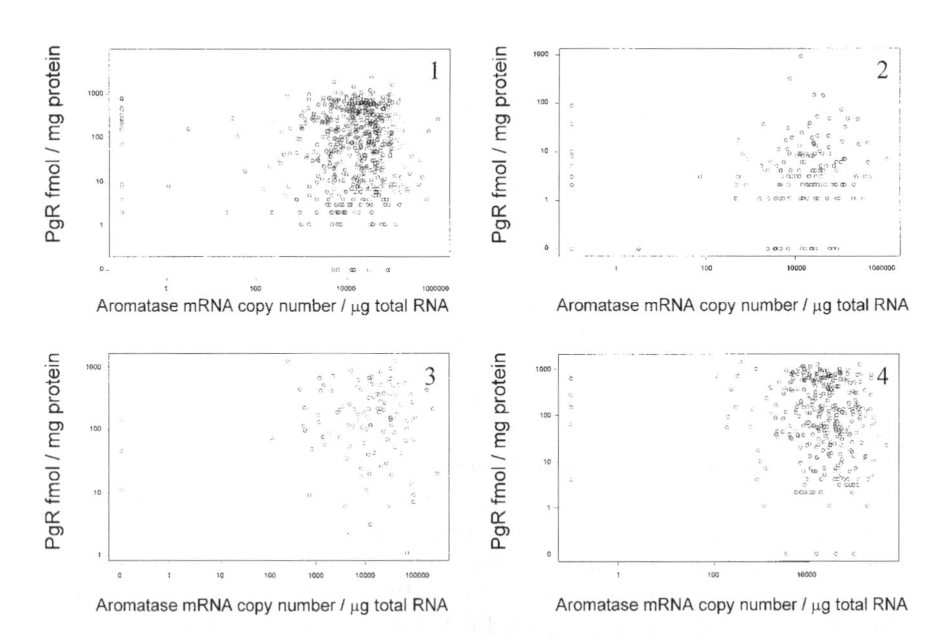

FIGURE 6 Intratumoral aromatase expression levels versus PgR protein levels. Groups 1 and 2 are postmenopausal pN0 patients: 1 with ER positive and 2 with ER negative tumors; groups 3 and 4 are pN1 patients with ER-positive tumors: 3 premenopausal and 4 postmenopausal.

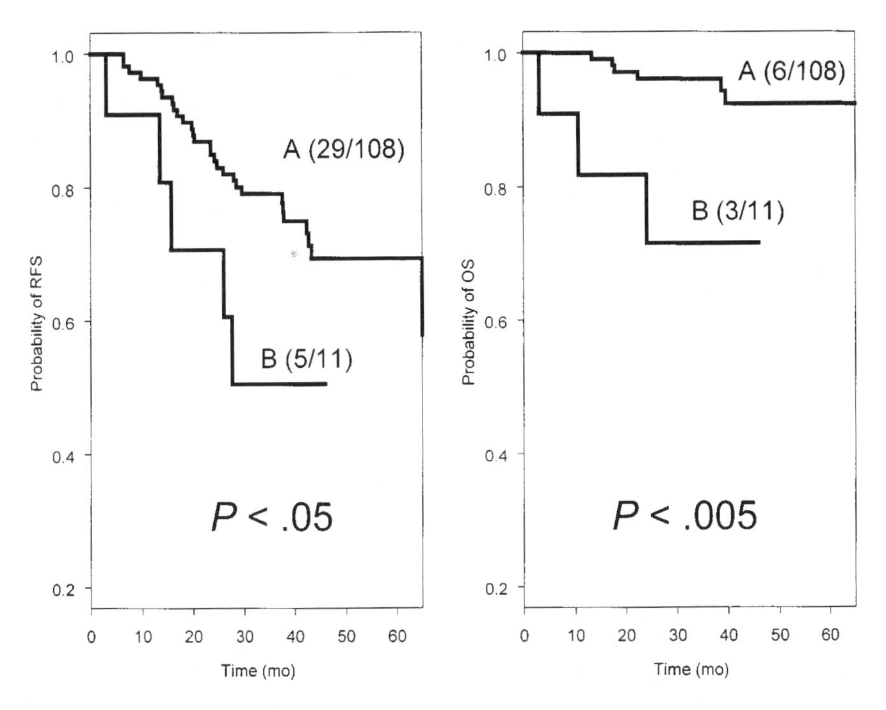

FIGURE 7 RFS and OS Kaplan-Meier curves as a function of intratumoral aromatase expression levels. A and B indicate aromatase values < and ≥ 100,000 copy number/μg total RNA, respectively. Numbers in parentheses indicate numbers of relapses/total in each group.

V. DISCUSSION

It has been known for many years that aromatase is one of the key enzymes involved in estrogen biosynthesis in human reproduction, but it only recently became evident that it is also distributed in tumors and extragonadal tissues. According to previous reports (5,6,8), aromatase is tissue specific and regulated by a set of molecular factors which either induce or suppress its expression. Such a complex molecular mechanism can be explained by an alternative use of tissue-specific exon I promotors within the aromatase gene and helps to define estrogen-dependent physiological functions mediated through a paracrine and/or autocrine estrogen production in the same tissue. It appears that estrogen function is a mitogenic factor in neoplastic tissue (23,24). Such mitogenic and proliferative effects of estrogens were recognized and are believed to correlate with ER and PgR status in breast tumors. Interestingly, the proportion of patients with hormone-sensitive tumors is greater among postmenopausal than premenopausal patients (25). In postmenopausal women, plasma levels of estrogen are greatly decreased and estrogens synthesized in the periphery virtually become the only source of plasma estrogen (26). However, postmenopausal women maintain a higher intratumoral estrogen content owing to the presence of aromatase even though the plasma concentration of estrogen decreases to lower values after menopause (27). This is in agreement with our observation that aromatase mRNA expression levels in NAT significantly increase in postmenopausal patients (see Fig. 4) as well as tumor ER levels (see Fig. 5, group 3 versus group 1 or 4). Unfortunately, overproduction of estrogen may promote the risk of estrogen-dependent cancers and deficient production may promote the risk of osteoporosis or arteriosclerosis.

According to our data, intratumoral aromatase expression levels are independent from the menopausal and the ER/PgR status. However, high levels of intratumoral aromatase expression correlate with poor clinical outcome, as shown in Figure 7. Therefore, one may speculate that there is another important biological role for tumor aromatase: Excessive intratumoral estrogen levels may propagate neovascularization and therefore the spread of metastatic cells. This information is important to design prospective clinical trials.

In a first step, the correlation between mRNA expression levels and its posttranslation protein levels and its enzyme activity have to be evaluated in a prospective clinical trial, because the functionality of the enzyme is decisive for a positive drug response. An ongoing collaborative study will clarify this important aspect. If a positive correlation is found between mRNA expression levels and enzyme activity, one may suggest that RT-PCR is the appropriate methodology to measure routinely the quantitative levels of this aromatase as a predictive factor in prospective clinical studies.

Based on our results, the aromatase expression levels also have to be assessed in NAT to investigate the possible paracrine impact of estrogen on tumor growth in those tumors expressing very low intratumoral aromatase levels but with high mRNA aromatase levels in their normal adjacent tissues. According to these findings, it will be an important step forward in the disease management of hormone-dependent cancers if the intratumoral aromatase expression or enzyme activity levels can be established as predictive factors for response to selective type II aromatase inhibitors.

ACKNOWLEDGMENTS

This work was supported by the Swiss National Science Foundation grant Nr. 3100-49505.96 (U. Eppenberger), the Tumorbank Foundation of Basel, and the Oncology Research of Novartis Pharma AG, Basel, Switzerland. We thank W.

Kueng, N. Tognoni, K. Paris, A. Takahashi, F. David, H. Bolliger, and D. Rehm for their technical assistance.

REFERENCES

1. McGuire WL, Chamness GC. Studies on the estrogen receptor in breast cancer. Adv Exp Med Biol 1973; 36:113–136.
2. Agarwal VR, Bulun SE, Leitch M, et al. Use of alternative promoters to express the aromatase cytochrome P450 (CYP19) gene in breast adipose tissues of cancer-free and breast cancer patients. J Clin Endocrinol Metab 1996; 81:3843–3849.
3. Bolufer P, Ricart E, Lluch A, et al. Aromatase activity and estradiol in human breast cancer: its relationship to estradiol and epidermal growth factor receptors and to tumor-node-metastasis staging. J Clin Oncol 1992; 10:438–446.
4. Sasano H, Harada N. Intratumoral aromatase in human breast, endometrial, and ovarian malignancies. Endocr Rev 1998; 19:593–607.
5. Means GD, Kilgore MW, Mahendroo MS, et al. Tissue-specific promoters regulate aromatase cytochrome P450 gene expression in human ovary and fetal tissues. Mol Endocrinol 1991; 5:2005–2013.
6. Mahendroo MS, Means GD, Mendelson CR, et al. Tissue-specific expression of human P-450AROM. The promoter responsible for expression in adipose tissue is different from that utilized in placenta. J Biol Chem 1991; 266:11276–11281.
7. Simpson ER, Mahendroo MS, Means GD, et al. Tissue-specific promoters regulate aromatase cytochrome P450 expression. J Steroid Biochem Mol Biol 1993; 44:321–330.
8. Harada N, Utsumi T, Takagi Y. Tissue-specific expression of the human aromatase cytochrome P-450 gene by alternative use of multiple exons 1 and promoters, and switching of tissue-specific exons 1 in carcinogenesis. Proc Natl Acad Sci USA 1993; 90:11312–11316.
9. Chen S, Zhou D, Okubo T, et al. Breast tumor aromatase: functional role and transcriptional role. Endocr Rel Cancer 1999; 6:149–156.
10. Harada N. Aromatase and intracrinology estrogen in hormone dependent tumors. Oncology 1999; 57:7–16.
11. Harvey HA. Emerging role of aromatase inhibitors in the treatment of breast cancer. Oncology (Huntingt) 1998; 12:32–35.
12. Brodie A, Lu Q, Long B. Aromatase and its inhibitors. J Steroid Biochem Mol Biol 1999; 69:205–210.
13. Dowsett M, Pfister C, Johnston SR, et al. Impact of tamoxifen on the pharmacokinetics and endocrine effects of the aromatase inhibitor letrozole in postmenopausal women with breast cancer. Clin Cancer Res 1999; 5:2338–2343.
14. Simpson ER, Evans CT, Corbin CJ, et al. Sequencing of cDNA inserts encoding aromatase cytochrome P-450 (P-450AROM). Mol Cell Endocrinol 1987; 52:267–272.
15. Eppenberger U, Kueng W, Schlaeppi JM, et al. Markers of tumor angiogenesis and proteolysis independently define high- and low-risk subsets of node-negative breast cancer patients. J Clin Oncol 1998; 16:3129–3136.

16. Conover W. Practical Nonparametric Statistics. 2nd ed. New York: Wiley, 1998.
17. Breimann L, Friedman J, Olsen R, et al. Classification and Regression Trees. Wadsworth: Belmont, 1984.
18. Kaplan E, Meier P. Nonparametric estimation from incomplete observations. J Am Stat Assoc 1958; 457–481.
19. Harrington DP, Fleming TR. A class of rank test procedures for censored survival data. Biometrika 1982; 69:553–566.
20. Wild D, Chester D, Perren T. Endocrine aspects of clinical management of breast cancer. Endocr Rel Cancer 1998; 97–110.
21. Yue W, Santen RJ. Aromatase inhibitors: rationale for use following antiestrogen therapy. Semin Oncol 1996; 23:21–27.
22. Miller WR, Mullen P, Telford J, et al. Clinical importance of intratumoral aromatase. Breast Cancer Res Treat 1998; 49:S27–32; see Discussion S33–37.
23. Miller WR, Hawkins RA, Forrest AP. Significance of aromatase activity in human breast cancer. Cancer Res 1982; 42:3365s–3368s.
24. van Landeghem A, Portman J, Mabauurs M. Endogenous concentration and subcellular distribution of estrogens in normal and malignant human breast tissue. Cancer Res 1985; 2900–2906.
25. Lippmann, Dickson. Prog Hormone Res 1989; 383–440.
26. MacDonald PC, Grodin JM, Edman CD, et al. Origin of estrogen in post-menopausal woman with a non endocrine tumor of the ovary and endometrial hyperplasia. Obstet Gynecol 1976; 60:174–177.
27. Bulun SE, Simpson ER. Competitive reverse transcription-polymerase chain reaction analysis indicates that levels of aromatase cytochrome P450 transcripts in adipose tissue of buttocks, thighs, and abdomen of women increase with advancing age. J Clin Endocrinol Metab 1994; 78:428–432.

12

Induction and Suppression of Aromatase by Inhibitors

William R. Miller, R. Vidya, P. Mullen, and J. M. Dixon
Western General Hospital
Edinburgh, Scotland

I. ABSTRACT

Results are presented from in vitro assays with particulate fractions of breast cancer and cultures of breast adipose tissue fibroblasts that new-generation type I aromatase inhibitors (such as exemestane and formestane) and type II agents (such as anastrozole and letrozole) are highly potent drugs (with IC_{50} values in the lower nanomolar range). Studies of in situ aromatase activity in breast cancers performed in women given neoadjuvant letrozole also demonstrate that the drug effectively blocks the local biosynthesis of estrogen within the breast. However, ex vivo studies of breast specimens from women treated with inhibitors and parallel investigations in which fibroblasts were preincubated with inhibitors showed first that in vitro assays may underestimate the inhibitory effects of type II inhibitors (but not type I) and second that type II inhibitors may enhance aromatase activity, an effect not seen with type I agents. The reason for these paradoxical observations and the clinical implications for the use of different inhibitors is discussed.

II. INTRODUCTION

Because the growth of many breast cancers may be dependent upon hormones (most noticeably estrogens), endocrine-deprivation therapy is a major treatment modality for the disease (1). In premenopausal women, this may take the form of ovarian ablation or the use of luteinizing hormone–releasing hormone (LHRH) analogues, but for postmenopausal women, other strategies need to be adopted, as estrogens are no longer produced by the ovaries but by peripheral tissues such as fat, muscle, and breast tumors themselves (2).

Among these concepts, inhibiting estrogen synthesis and blocking the action of estrogen at its receptor by the use of specific drugs have been translated into clinical practice (3). In terms of biosynthesis, estrogens lie at the end of a metabolic pathway in which blockade of any reaction may, theoretically, result in its decreased production. However, the most specific effects are achieved by inhibiting the last step in the pathway by which androgens are converted to estrogens via aromatization. This transformation, like other steps in the pathway, involves steroid hydroxylation using a cytochrome P450 electron-transfer system. The challenge has been to produce potent drugs that have the specificity to inhibit aromatase without affecting other steroid-metabolizing enzymes. In the event, two separate groups of agents have been developed. These are type I inhibitors which interact with the androgen substrate binding site of the aromatase enzyme (and are steroidal structurally) and type II inhibitors which interfere with cytochrome P450–mediated electron transfer and are usually azoles, which are structurally nonsteroidal (4). The objectives of this chapter are to describe the comparative efficacy of new-generation aromatase inhibitors and, in the case of type II agents, the paradoxical data that aromatase activity may be increased following exposure to the drugs. Observations have been made on aromatase within the breast using results obtained from homogenates of breast cancer, cultures of mammary adipose tissue cells, and in situ measurements following administration of radiolabeled steroids to breast cancer patients.

III. RESULTS

A. In Vitro Assays Using Particulate Fractions of Breast Cancer

A relatively simple screen to determine the relative efficacy of aromatase inhibitors is to prepare particulate fractions from large breast cancers and assay aromatase activity in the absence and presence of differing concentrations of aromatase inhibitors. Results are shown in Figure 1a from studies performed in

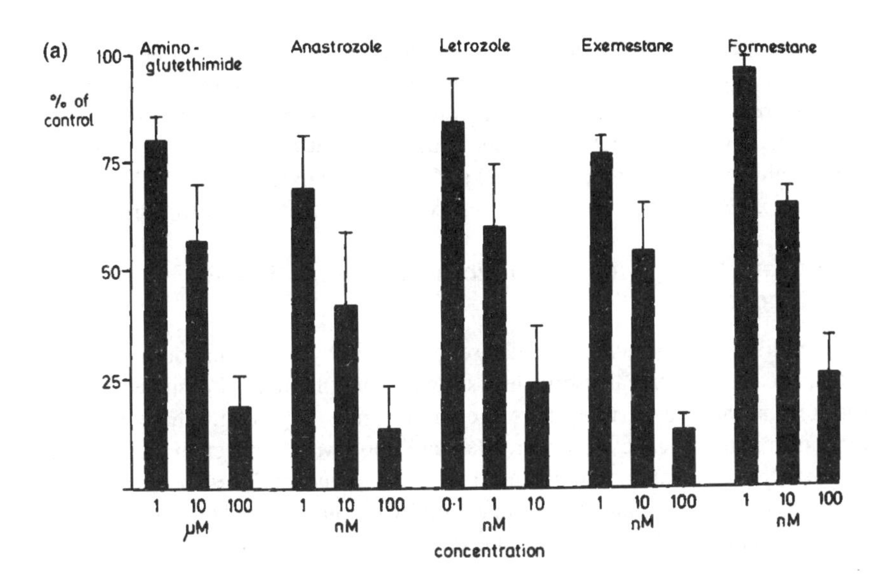

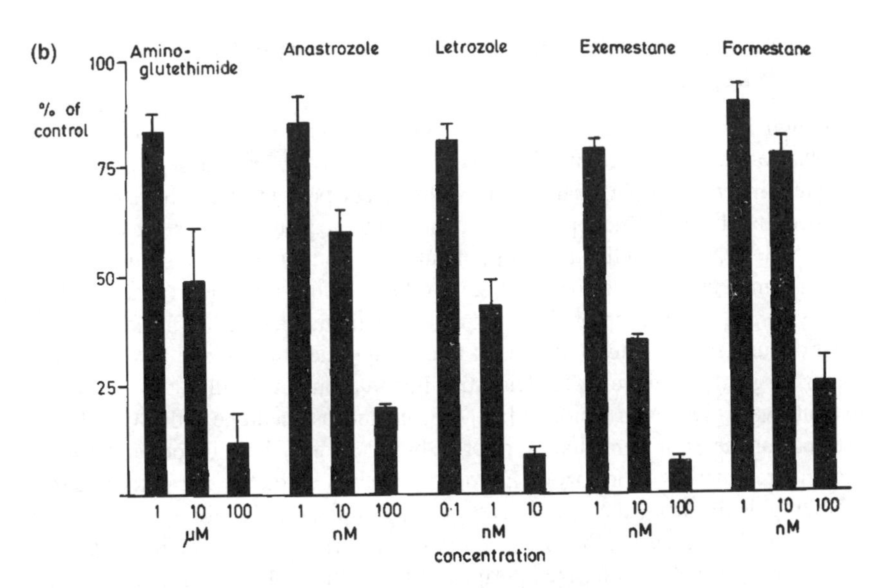

Figure 1 Effects of inhibitors on aromatase activity (as a percentage of control systems without inhibitors) in (a) particulate fractions of breast cancers and (b) cultures of breast adipose tissue fibroblasts. Columns are means and bars are standard errors of the means of either (a) four separate tumors or (b) three different systems.

four separate tumors by adding aminoglutethimide, letrozole, anastrozole, formestane, and exemestane to replicate incubates. All agents inhibited aromatase in a dose-related manner, although micromolar concentrations were required for aminoglutethimide, whereas all other drugs were effective in the nanomolar range. Calculated IC_{50} values were 20 μM for aminoglutethimide, 8 nM for anastrozole, 2 nM for letrozole, 15 nM for exemestane, and 30 nM for formestane.

B. Assays Using Cultures of Fibroblasts from Breast Adipose Tissue

Cultured fibroblasts in which aromatase is induced by preincubation with dexamethazone also represent useful test systems for inhibitors. Combined results from four separate experiments are shown in Figure 1b. As with the particulate fraction from breast cancers, dose-related inhibition was produced by micromolar concentrations of aminoglutethimide and the nanomolar doses of anastrozole, letrozole, exemestane, and formestane. Approximate IC_{50} values were 10 μM for aminoglutethimide, 15 nM for anastrozole, 0.8 nM for letrozole, 5 nM for exemestane, and 30 nM for formestane.

C. In Situ Aromatase Assays Before and During Treatment with Letrozole

Although in vitro and ex vivo studies provide evidence for the potential of aromatase inhibitors, the crucial issue of how effectively aromatase is blocked in vivo requires more sophisticated methodology. The present studies have utilized a protocol in which postmenopausal women with large primary breast cancers have received letrozole neoadjuvantly for 3 months. The advantage of this study design is that tumor is available for investigation both before treatment (as a result of biopsy for estrogen receptor [ER] status—all patients included in the study have ER-rich tumors) and after 3 months of treatment (when the patients have residual tumor surgically removed). To determine in situ aromatase within the breast, the patients were given an infusion of [^{3}H]-labeled androstenedione and [^{14}C]-labeled estrone for the 18 h immediately prior to breast surgery both before and after 3 months treatment with letrozole. Estrogen was then extracted and purified chromatographically from the tumors and peripheral plasma. In situ aromatase was determined by measuring the radioactivity in the purified tumor fractions (correcting for the estrogen synthesized peripherally and then taken up into the tumor). Twenty-four women (12 treated with 2.5 mg daily and 12 with 10.0 mg letrozole daily) were entered into the study, but because one patient (from the 2.5-mg group) experienced a complete pathological response with treatment, only 23 tumor pairs were available for comparison.

Three tumors had no evidence of aromatase activity either before or after treatment, with radioactive estrogen within the tumors being accounted for solely by uptake from the circulation. Results are shown in Figure 2: twenty tumors (10 each at the 2.5- and 10-mg doses), displayed aromatase activity before treatment. This in situ activity was markedly reduced with treatment in 9 of 10 patients treated with the 2.5-mg dose and in all patients receiving the 10-mg dose. The difference between paired pretreatment and 3-month values was statistically significant by sign test ($P = .022$ for the 2.5-mg and $P = .004$ for 10-mg doses).

D. Ex Vivo Studies in Patients Treated with Aromatase Inhibitors

It was of interest to determine in vitro aromatase activity in excised breast tissue from the patients treated neoadjuvantly with letrozole. Sufficient paired material was available in four tumors and seven samples of nonmalignant breast from the cases described above. Treatment was associated with a reduction in aromatase activity in all four tumors (Fig. 3a). However, in each individual tumor, the degree of inhibition was less ex vivo than in situ. Effects of letrozole on in vitro aromatase measured in nonmalignant breast were less marked and, in some instances, particularly when activity was low before treatment, a paradoxical increase in aromatase activity was seen after 3 months' therapy (Fig. 3b).

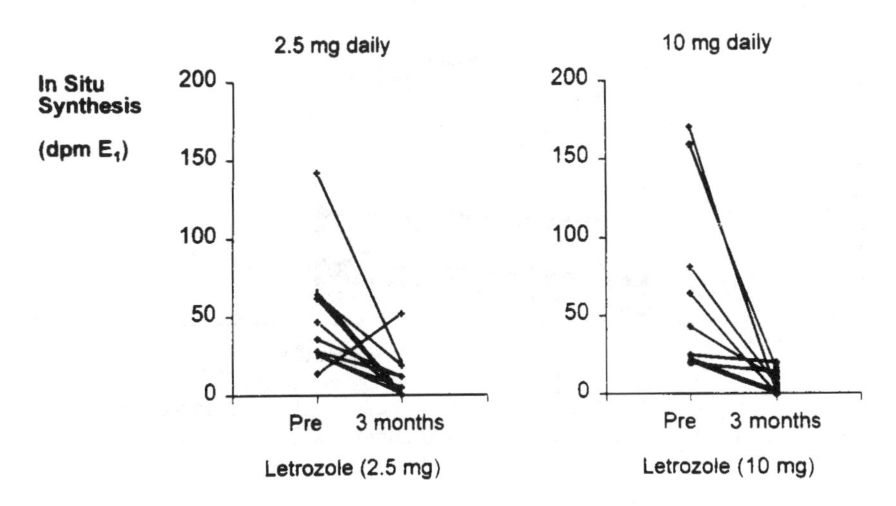

Figure 2 Effects of treatment with letrozole on in situ synthesis of estrogen within the breast.

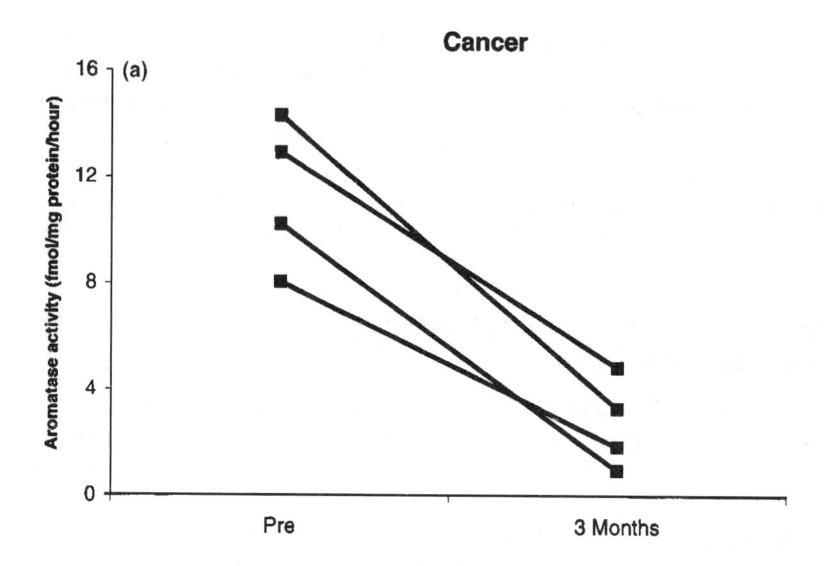

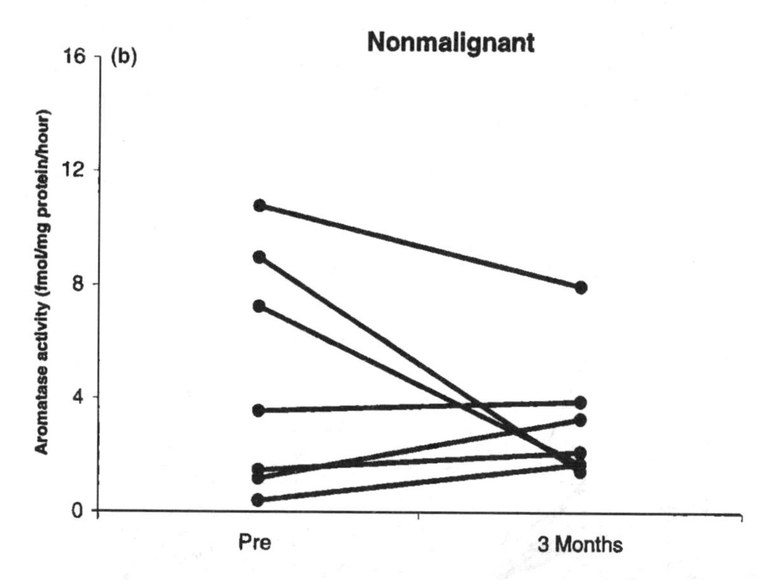

Figure 3 In vitro aromatase in (a) breast cancers and (b) nonmalignant breast tissue taken before and after 3 months of treatment with neoadjuvant letrozole.

E. Preincubation of Cultured Fibroblasts with Aromatase Inhibitors

An important difference between in situ and in vitro measurements following letrozole treatment is that in situ measurements are made while the drug is still present in the body in contrast to in vitro assays where the tumor is removed from the presence of the drug. To mimic this comparison in an experimental system, the study design for testing aromatase inhibitors in cultures of fibroblasts was modified, as in Figure 4. The sequence of preincubating cultures with dexamethazone (to induce aromatase) and then adding inhibitors during the assay procedure (this parallels in situ studies in patients in whom assays are performed in the presence of the drug) was used to obtain the results presented in Figure 1b. Additional cultures were set up using the design of preincubating with inhibitors but assaying for aromatase activity in their absence (mimicking the condition of the ex vivo studies described above).

Results using letrozole as inhibitors in this preincubation system are shown in Figures 5a and b. Concentrations of drugs in excess of 100 nM produced inhibitory effects, but comparison with Figure 1b shows that the effects are magnitudes less than would have been achieved by including letrozole during the assay period. Interestingly, drug doses between 2 and 20 nM produced a consistent increase in activity compared with that in control cultures performed in the absence of letrozole. Figure 5 also shows the effects of preincubation with letrozole in the absence of dexamethasone. Letrozole produced a fivefold increase in

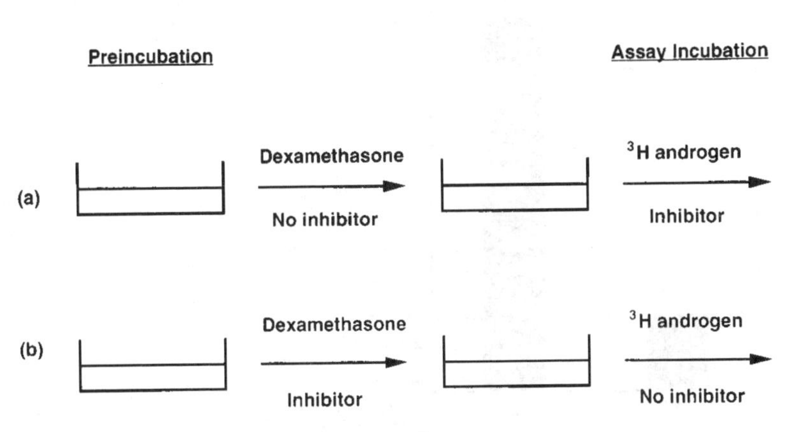

FIGURE 4 Study designs for assay of aromatase activity in cultures of breast adipose tissue fibroblasts (a) aromatase is induced by dexamethasone in the absence of inhibitions which are added only during the incubation with [^{3}H]-androgen; (b) inhibitors are present during the dexamethasone induction phase but not during the period of incubation with [^{3}H]-androgen.

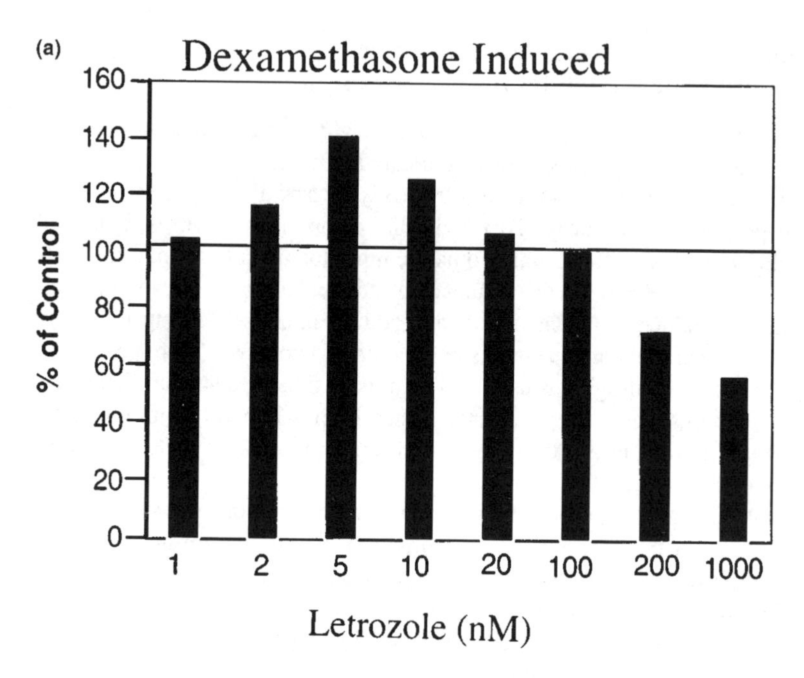

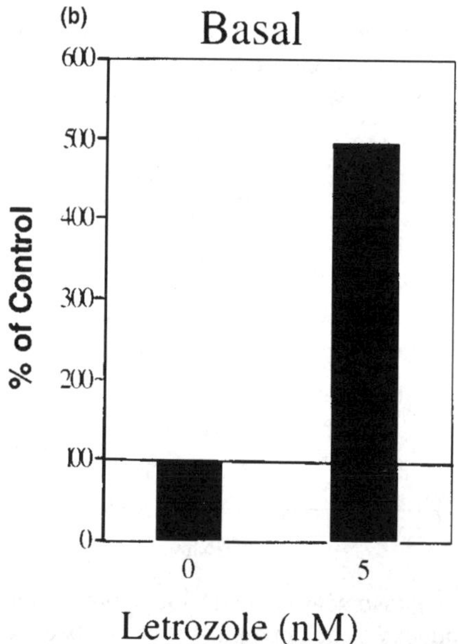

FIGURE 5 Levels of aromatase activity in fibroblasts from mammary adipose tissue preincubated for 18 h and assayed for activity in the absence of letrozole (control) and the presence of inhibitors. (a) Dexamethasone included in the preincubation phase; (b) without dexamethasone.

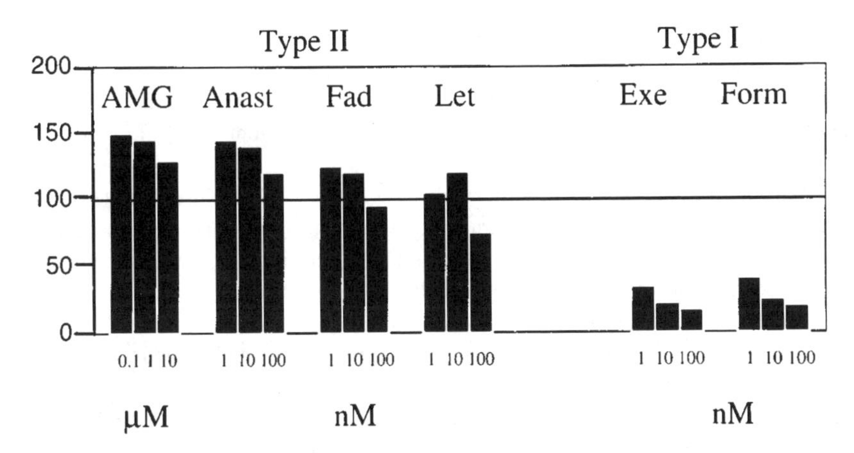

FIGURE 6 Level of aromatase activity in fibroblasts from mammary adipose tissue preincubated for 18 h and assayed for activity in the absence of inhibitors (control) and the presence of inhibitors. AMG, aminoglutethimide; Anast, anastrozole; Fad, fadrozole; Let, letrozole; Exe, exemestane; Form, formestane.

basal levels (which were low). A comparison of several type II inhibitors (aminoglutethimide, anastrozole, fadrozole, and letrozole) showed that all the type II inhibitors were capable of increasing aromatase activity when used at concentrations which would be inhibitory if the drug was added during the aromatase assay (Fig. 6). In contrast, the type I inhibitors (exemestane and formestane) markedly reduced aromatase activity in a dose-related manner; comparison with Figure 1b indicates that the degree of inhibition produced by preincubating with formestane and exemestane was greater than observed when the drugs are added during the assay.

IV. DISCUSSION

The results from incubations of particulate fractions of breast cancers and culture of breast adipose tissue fibroblasts confirm the remarkable potency of newly developed aromatase inhibitors. Thus, anastrozole, letrozole, exemestane, and formestane were able to inhibit aromatase at nanomolar concentrations. This contrasts with the doses needed for aminoglutethimide, which were in the micromolar range and therefore orders of magnitude higher. Although these results are similar to those observed in placental microsomes (5–7), there are consistent differences between drugs and test systems. Thus, in breast cancer particulate fractions, the type II inhibitors, anastrozole and letrozole, were more potent than the two type I drugs, exemestane and formestane (as is the case in placental microsomes), but in cultured fibroblasts, letrozole and exemestane had increased potency. The reason for this is not immediately apparent but may relate to differences between disrupted and whole cell preparations. For example, it has been suggested that letrozole accumulates to a greater extent in whole cell systems (8).

Although observations in experimental test systems are informative, they do not necessarily reflect events occurring within the body. To determine the effects of aromatase inhibitors on in situ aromatase activity within breast cancers, use has been made of a protocol in which patients with large primary tumors have been offered neoadjuvant treatment in an attempt to shrink tumors. These investigations allow not only measurements of tumor size but also the monitoring of endocrinological effects within the breast by biopsying breast tissue before treatment and by surgery after drug treatment. To assess in situ aromatase, patients were infused with radioactively labeled androgen and estrogen before and after treatment. These studies confirmed that breast tumors obtained estrogen both by uptake from the circulation and by local biosynthesis from androgen precursors. In many tumors, local biosynthesis seemed to be the major source of estrogen. In situ aromatase activity was markedly reduced after letrozole treatment in 19 of the 20 tumors with the potential for in situ production of estrogen. The reason for a failure to demonstrate a decrease with treatment in the remaining tumor is not apparent—in this patient therapy produced a marked decrease in amounts of ''whole-body'' conversion of androgen to estrogen, a suppression of endogenous concentrations of estrogen within the breast, and a clinical response that would be compatible with inhibition of aromatase. However, the general inhibitory effects of letrozole clearly illustrate its potential to act as a potent endocrine agent within the breast. Consistent with inhibition of in situ aromatase, it was possible to demonstrate that endogenous tumor estrogen levels were markedly reduced (9). These actions are compatible with the clinical benefits, in terms of tumor shrinkage, which were achieved with letrozole treatment (10).

The use of radioactively labeled infusions to measure in situ aromatase activity is an invasive and extra procedure in patient management. A more acceptable approach would have been to assess in vitro aromatase activity in surgically excised breast material. To determine whether in vitro measurements accurately reflected those within the breast, in vitro assays were performed in paired specimens of both malignant and nonmalignant breast tissues before and after treatment. The results showed the expected inhibition in malignant tissue, but the degree of effect was less than that elicited from in situ measurements. In the nonmalignant breast, the degree of inhibition with letrozole treatment was even less marked. Indeed, increases in activity were seen in cases in which aromatase activity was initially low. We have reported similar paradoxical increases in aromatase activity in tumors from patients treated with aminoglutethimide (11);

this activity was shown still to be sensitive to aminoglutethimide by demonstrating inhibition after in vitro incubation with the drug. In contrast, in vitro incubation of paired tumor samples taken before and after treatment with the type I inhibitor formestane showed the expected decrease in activity. It is clear therefore that, at least with type II agents, in vitro estimates of treated specimens may underestimate the decrease of inhibition produced in vivo; this makes the use of protocols that accurately measure in situ activity essential when assessing the putative efficacy of individual inhibitors.

The reason in vitro assays may underestimate the potency of type II inhibitors is likely to be because the assays are performed without the addition of inhibitors. Even though aromatase activity may be blocked by the drug in situ, the reversible nature of the association of type II drugs may mean that aromatase activity is released from its inhibition (this would not be the case for irreversible type I inhibitors). To explore this concept further, the nature of these effects, cultures of mammary fibroblasts were preincubated with aromatase inhibitors (to simulate patient treatment) and then assayed in the absence of drugs (to mimic ex vivo studies). Interestingly, under these conditions, all three type II inhibitors (aminoglutethimide, letrozole, and anastrozole) failed to realize their full inhibitory potential and, at certain concentrations, were associated with enhanced activity. In contrast to these effects, the type I inhibitor formestane has always been associated with decreased aromatase activity; indeed, inclusion in the 18-h preincubation period tended to produce greater effects than were achieved in the 5-h assay. This probably reflects the irreversible ''suicide'' mechanism of action of this inhibitor (12). The result mirrors those observed in the ex vivo studies described above and draws attention to the different mechanism of actions of the different types of aromatase inhibitors. Although the reversible nature of type II inhibitors accounts for the decreased level of inhibition seen in these studies, it does not immediately explain the increased levels of activity following treatment. However, type II inhibitors might induce aromatase mRNA/stabilize aromatase protein, as has been reported by Harada et al. (13) and Chen et al. (14).

These observations may have clinical relevance in that, although in the short-term type II inhibitors may effectively cause estrogen blockade (see Fig. 2), chronic administration might increase the aromatase enzyme to such an extent that the drugs may no longer be efficient. Compatible with this there is the suggestion that estrogen concentrations may increase at relapse in patients treated with aminoglutethimide (15). Under these circumstances, it might be expected that type I inhibitors such as formestane may produce beneficial effects in patients relapsing while receiving type II drugs, as has been observed clinically (16). Such non-cross resistance to aromatase inhibitors points to a role for type I inhibitors following type II inhibitors despite the inherently lower potency of the former in experimental systems.

V. CONCLUSIONS

The new generation of aromatase inhibitors, which include anastrozole, letrozole, exemestane, and formestane, are extremely potent agents capable of profoundly influencing endocrine events within the breast. It seems likely that they will secure a place when treating postmenopausal women with hormone-deprivation therapy.

REFERENCES

1. Miller WR. Estrogens and endocrine therapy for breast cancer. In: Miller WR, ed. Estrogen and Breast Cancer. Austin, Texas: Landes, 1996:125–150.
2. Miller WR. Aromatase inhibitors. Endocr Rel Cancer 1996; 3:65–79.
3. Miller WR. Aromatase inhibitors—where are we now? Br J Cancer 1996; 73:415–417.
4. Miller WR. Aromatase inhibitors and breast cancer. Cancer Treat Rev 1997; 23:171–187.
5. di Salle E, Ornati G, Giudici D, Lassus M, Evans TRJ, Coombes RC. Exemestane (FCE 24304), a new steroidal aromatase inhibitor. J Steroid Biochem Mol Biol 1992; 43:137–143.
6. Plourde PV, Dyroff M, Dukes M. Arimidex: a potent and selective fourth-generation aromatase inhibitor. Breast Cancer Res Treat 1994; 30:103–111.
7. Bhatnagar AS, Hausler A, Schieweck K, Lang M, Bowman R. Highly selective inhibition of estrogen biosynthesis by CGS20267. A new non-steroidal aromatase inhibitor. J Steroid Biochem Mol Biol 1998; 37:1021–1027.
8. Bhatnagar A, Miller WR. Pharmacology of inhibitors of estrogen biosynthesis. In: Oertel M, Schillinger E, eds. Estrogens and Anti-Estrogens II: Pharmacology and Clinical Applications of Estrogens and Anti-Estrogens. Berlin: Springer-Verlag, 1999:223–230.
9. Miller WR. Biology of aromatase inhibitors: pharmacology/endocrinology within the breast. Endocr Rel Cancer 1999; 6:187–195.
10. Dixon JM, Love CDB, Renshaw L, Bellamy C, Cameron DA, Miller WR, et al. Lessons from the use of aromatase inhibitors in the neoadjuvant setting. Endocr Rel Cancer 1999; 6:227–230.
11. Miller WR, O'Neill J. The importance of local synthesis of estrogen within the breast. Steroids 1987; 50:537–540.
12. Brodie AMH, Garnett WM, Hendrickson JR, Tsai-Morris CH, Marcotte PA, Robinson CH. Inactivation of aromatase in vitro by 4-hydroxyandrostene-3,17-dione and 4-acetoxy-4-androstene-3,17-dione and sustained effects *in vivo*. Steroids 1981; 38:693–702.
13. Harada N, Honda S-I, Hatano O. Aromatase inhibitors and enzyme stability. Endocr Rel Cancer 1999; 6:211–218.
14. Chen S, Zhou D, Okubo T, Kao Y-C, Yang C. Breast tumor aromatase; functional role and transcriptional regulation. Endocr Rel Cancer 1999; 6:149–156.
15. Dowsett M, Harris AL, Smith UIE, Jeffcoate SL. Endocrine changes associated with

relapse in advanced breast cancer patients on aminoglutethimide therapy. J Clin Endocrinol Metab 1984; 58:99–104.
16. Coombes RC, Goss P, Dowsett M, Gazet J-C, Brodie A. 4-Hydroxyandrostenedione in treatment of postmenopausal patients with advanced breast cancer. Lancet 1984; 2:1237–1239.

13

Relevance of Animal Models to the Clinical Setting

Angela M.H. Brodie, B. Long, Q. Lu, and Y. Liu
University of Maryland School of Medicine
Baltimore, Maryland

I. ABSTRACT

A model has been developed to evaluate the effects of aromatase inhibitors and antiestrogens on aromatase activity and tumor proliferation in the same system. Human breast cancer cells transfected with the human aromatase gene (MCF-7_{CA}) are used and can be studied in in vitro cultures and in tumors grown in ovariectomized mice. This model simulates the patient with hormone-dependent tumors expressing aromatase. Tumor cell proliferation is stimulated by estrogen produced in situ via aromatization of androstenedione. The MCF-7_{CA} cells were used in culture to determine the IC_{50} values of aromatase inhibitors. The relative potencies of the compounds were found to be consistent when compared in two other types of human cells expressing aromatase: JEG-3 human choriocarcinoma cells and normal breast fibroblasts.

Letrozole was the most potent inhibitor in all cell types. In mice with tumors of MCF-7_{CA} cells treated with aromatase inhibitors and antiestrogens, tumor growth was inhibited to a greater extent by letrozole than tamoxifen. Also, aromatase inhibitors had no stimulatory effects on the uterus. Estrogen receptor

(ER) levels in the tumors and the uterus were increased by aromatase inhibitors and decreased by tamoxifen. Progesterone receptor (PgR) levels were increased by tamoxifen, which is consistent with its partial agonist properties. In animals treated with letrozole, no PgR could be detected, suggesting that this compound may prevent induction of other estrogen-induced genes, such as those involved in proliferation. The combination of antiestrogens and aromatase inhibitors was found to be no more effective than aromatase inhibitor treatment alone in preventing tumor growth. This suggests that the sequential use of agents with different mechanisms of action, as currently employed in the clinic, is likely to be more advantageous by extending the period of effective treatment when resistance has developed to first-line therapy. Studies in this model may be helpful in guiding the use of these agents in clinical practice.

II. INTRODUCTION

Two strategies are now being used to inhibit the stimulatory effects of estrogen on breast cancers. These are the inhibition of estrogen production and action. The antiestrogen tamoxifen competes with estrogen for ER in the tumor, thereby blocking the actions of the hormone and reducing tumor proliferation. Tamoxifen therapy has been shown to be significantly more beneficial than chemotherapy for postmenopausal patients with hormone-dependent breast cancer (1). Tamoxifen increases survival and is associated with lower toxicity (2). For these reasons, tamoxifen is currently the first-line agent of choice for postmenopausal breast cancer patients. In the last few years, highly selective inhibitors of estrogen synthetase (aromatase) have become available and some have been approved as second-line agents for patients who have relapsed from first-line treatment (usually tamoxifen) (3). These new aromatase inhibitors have low toxicities and are now preferred in this context to other agents, such as megestrol acetate and aminoglutethimide, which resulted in more side effects for the patients. Despite the success of tamoxifen in improving the treatment of breast cancer over the last decade, its safety in long-term use have become a concern (4). Tamoxifen therapy increases the risk of developing endometrial cancer, and the incidence is correlated with the duration of treatment (5,6). This effect is thought to be due to the partial agonist action of tamoxifen (7). In older patients, there is an increase in the incidence of stroke, which may similarly be related to the partial agonist effect. Aromatase inhibitors are highly effective in blocking estrogen production, and they are without any agonist actions, offering the possibility of being safer and more effective.

In order to investigate the effects of aromatase inhibitors and to compare them to antiestrogens in different treatment strategies, a mouse model has been developed in which hormone-responsive tumors synthesize estrogens (8,9). Thus, this model represents an intratumoral aromatase system. Although several animal models were extensively utilized during the development of aromatase inhibitors (10–12), we believed that a more relevant model was needed to address the issues arising from recent clinical studies of aromatase inhibitors and antiestrogens. Some of the limitations of the previously used rat models are that the tumors are induced by carcinogens, either 7,12-dimethylbenzanthracene (DMBA) or nitrosourethrane (NMU), and are dependent on ovarian steroids (13). In addition, the rat tumors have a number of characteristics differing from human breast cancers, such as significant dependence on prolactin as well as estrogen.

After menopause, estrogens are no longer produced in the ovaries and plasma levels are reduced. Estrogen synthesis in peripheral tissues, such as fat and muscle, increases, becoming the main source of circulating estrogens in postmenopausal women. Aromatase activity in breast cancers has been reported for many years, since the early work of Miller and colleagues (14). However, controversy has surrounded the question of whether the enzyme activity was sufficient to produce enough estrogen to activate ER (15). Some recent studies have reported that the concentrations of estrogen in breast tissue are four- to sixfold higher than in serum and similar to those in premenopausal patients (16). Moreover, estrogen concentrations in tumors are higher than in breast fat (17,18). This suggests that estrogen production within the breast and by breast cancers may have a role in stimulating tumor proliferation in postmenopausal patients. Although the source of estrogen could be due to an increased gradient of steroids from the circulation into breast tissue (19), local synthesis may account for the major proportion of estrogen concentrations within the breast (20). In addition to evidence of aromatase activity (21,22) and mRNA (23) in breast tumors, aromatase expression has been detected in sections of breast tumors by immunocytochemistry using monoclonal (24) and polyclonal antibodies (25,26). Recently, aromatase activity and $mRNA_{arom}$ were detected in tumor cells isolated from ascites fluid from metastatic breast cancer, thus confirming that the enzyme is expressed by the tumor (27).

To demonstrate that intratumoral aromatase activity has functional significance, we carried out histocultures of several breast cancers. Proliferation of some tumors in histoculture was found to be enhanced by testosterone as well as estrogens, and the stimulation by testosterone, but not by estrogen, could be blocked by aromatase inhibitors (24). This suggests that androgens are aromatized to estrogens by the tumors and the concentrations are sufficient to stimulate tumor proliferation. In addition, we found that there was a correlation between aromatase activity and the level of proliferating cell nuclear antigen (PCNA). These results indicate that aromatase within the breast and in the tumor has functional consequences and that effective inhibition of locally produced estrogen might be important in determining the outcome of aromatase inhibitor treatment.

Based on the above evidence that breast tumors are stimulated by estrogens in a paracrine or autocrine fashion, we developed an intratumoral aromatase model in which estrogen made by the tumor cells stimulates their own proliferation. The human breast carcinoma cell line, MCF-7, is the only available cell line that is responsive to the proliferative effects of estrogen. Although the T47D human breast cancer cell line is responsive to estrogen to a lesser extent (28), the cells are more dependent on progesterone. In addition, the effects of estrogen on MCF-7 cells have been extensively studied in the nude mouse (29). Unfortunately, for the purpose of an intratumoral model, aromatase activity in MCF-7 cells is very low. This may be due to the cells being passaged many times or due to being grown under conditions containing estrogen. If the cells are maintained in very low concentrations of estrogen, aromatase activity is significantly increased (20), demonstrating that aromatase is normally expressed in this cell line. In order to have a consistent and higher level of aromatase activity, we have used cells stably transfected with the human aromatase gene and designated MCF-7_{CA}.

III. HUMAN BREAST CARCINOMA CELLS (MCF-7) TRANSFECTED WITH THE AROMATASE GENE (MCF-7_{CA})

The MCF-7_{CA} cells have proved useful for examining the effects of aromatase inhibitors in vitro. In addition to proliferating in response to estrogens, the aromatase-transfected cells will respond to androstenedione, whereas wild-type cells will not. Inhibition of aromatase activity and cell proliferation is correlated to the extent that both aromatase activity and proliferation can be blocked by the addition of aromatase inhibitor to the cell culture. Thus, when letrozole is used in a wide range of concentrations, it is evident that growth inhibition is occurring within the range of aromatase inhibition. Further studies are necessary to determine whether there is a direct correlation between these two effects.

The IC_{50} values of a number of aromatase inhibitors were determined and compared in several different cell types that expressed aromatase. In addition to MCF-7_{CA} breast cancer cells, we used JEG-3 human choriocarcinoma cells and breast fibroblasts. Breast fibroblasts were obtained from women without malignant disease who had undergone mammary reduction surgery. Although both JEG-3 cells and fibroblasts express aromatase, they are devoid of ER and therefore do not show increased proliferation in response to estrogens or androstenedione.

MCF-7 human breast cancer cells stably transfected with the aromatase gene (MCF-7_{CA}) were kindly provided by Dr. S. Chen (City of Hope, Duarte, CA) (30) and were maintained in Eagle minimal essential medium (EMEM) with 5% fetal bovine serum (FBS), 1% penicillin/streptomycin, 1% nonessential amino acids, and 600 μg/mL geneticin (G418). JEG-3 cells were obtained from the American Type Culture Collection and were cultured in the same medium as MCF-7_{CA} cells with 10% FBS and without the G418. Cells were plated into six-well plates (50,000 cells/well) and left overnight to attach. For determination of IC_{50} values, cells were treated with letrozole, anastrozole, and 4-hydroxyandrostenedione (4-OHA) for 2 h in the presence of 0.5 μCi of [1β^{3}H]-androstenedione. Then, 300 μL of trichloroacetic acid (TCA) was added to the medium to precipitate the proteins. After centrifugation, 1 mL of media was mixed with 2 mL of chloroform to extract unconverted substrate and other steroids. An aliquot of 0.7 mL of the aqueous phase was treated with 2.5% activated charcoal suspension (0.7 mL) to remove any residual steroids. Tritiated water [^3H_2O] formed during the aromatization of [$^3H\Delta^4A$] to estrogen was measured by counting the radioactivity in the aqueous supernatant. Aromatase activity levels were determined as fmol/100,000 cells/6 h. All experiments were repeated in triplicate and the results are expressed as mean ± SEM. Aromatase activity levels were then deter-

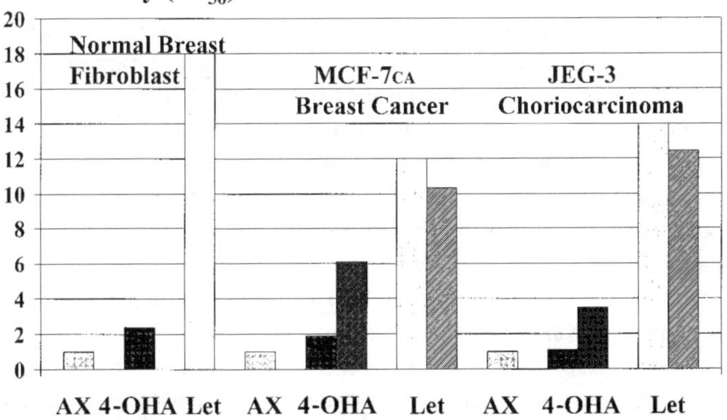

FIGURE 1 The relative potencies of aromatase inhibitors in human cell cultures. IC_{50} values in breast fibroblasts were obtained by cotreating the cells with dexomethasone (1 μM) plus inhibitors for 30 h followed by an 18-h assay with 0.5 μCi [1β ^{3}H]-androstenedione in the presence of aromatase inhibitors. The IC_{50} values in MCF-7_{CA} and JEG-3 cells were obtained by incubating with 0.5 μCi [1β ^{3}H]-androstenedione for 2 h in the presence of inhibitors. Each assay was performed in triplicate. Results are expressed as relative potency to anastrozole. The experiment was performed twice with different samples of compounds and cells, MCF-7_{CA} and JEG-3. The relative potencies of 4-OHA and letrozole (LET) is compared to anastrozole (Ax).

mined as described above. IC_{50} values for inhibitors were calculated from the linear regression line in a plot of percentage of enzyme activity versus log inhibitor concentration.

For the normal breast fibroblasts, semiconfluent flasks of cells from a 22-year-old female patient were washed with DPBS and grown for 48 h in routine medium. Cells (100,000–300,000) were then plated into six-well plates and allowed to attach. Cells were washed and treated with dexamethasone (1 μM) in the same medium and coincubated with vehicle or inhibitors at various concentrations for 30 h followed by incubation with 0.5 μCi of [1β ^{3}H]-androstenedione for 18 h. Aromatase activity was determined as above.

The results of two experiments are shown in Figure 1. In cultures of all three cell types, letrozole was the most potent aromatase inhibitor with IC_{50} values ranging from 0.14 to 0.45 nM. Aromatase was inhibited 2- to 10-fold more effectively by 4-OHA than anastrozole in all of the cell cultures. These results suggest that letrozole, anastrozole, and 4-OHA have consistent effects on aromatase activity in the different types of cells (31).

IV. INTRATUMORAL AROMATASE MODEL

In order to investigate the antitumor effects of aromatase inhibitors in vivo, the MCF-7_{CA} cells were inoculated into ovariectomized, athymic, immunosuppressed mice. Although tumors will result from inoculation of MCF-7_{CA} into intact animals (8), tumor growth in ovariectomized mice is comparatively slow, as the adrenal glands of these animals secrete reduced levels of adrenocortical hormones. We found that supplementing the mice with androstenedione, the substrate for aromatase, improved the growth rate of tumors (8).

Ovariectomized female BALB/c mice (aged 4–6 weeks) were inoculated subcutaneously (sc) in four sites, each with 0.1 mL of a suspension of MCF-7_{CA} cells. The cell suspension of 3×10^7 cells/mL was prepared from subconfluent MCF-7_{CA} resuspended in Matrigel (10 mg/mL). Animals were injected throughout the course of the experiment with 0.1 mg/mouse/day sc androstenedione. Tumor growth was measured with calipers weekly and tumor volumes calculated according to the formula $4/3 \times \pi \times r_1^2 \times r_2$ ($r_1 < r_2$) (10). The animals were housed in a pathogen-free environment under controlled conditions of light and humidity and received food and water ad libitum (8,9,32,33).

When all tumors reached a measurable size (~500 mm^3), usually 28–35 days after inoculation, animals were assigned to groups of four or five mice. They were then treated daily with injections sc and tumors measured weekly. In the following studies, letrozole (CGS 20,267) and fadrozole (CGS 16949A) (kindly provided by Dr. Ajay Bhatnagar, Novartis, Basel, Switzerland), anastrozole (ZD 1033) (kindly provided by Dr. Michael Dukes, Zeneca Pharmaceuticals, Macclesfield, UK), and the antiestrogen, tamoxifen (Sigma), were prepared for injection in 0.3% hydroxypropyl cellulose (HPC). The pure antiestrogen, faslodex (ICI 182,780) (kindly provided by Dr. A. Wakeling, Zeneca Pharmaceuticals, Macclesfield, UK) was injected in oil once per week. Control animals received vehicle (0.3% HPC, 0.1 mL/mouse/day) sc daily. Groups of mice were also injected with a combination of an aromatase inhibitor and an antiestrogen or a combination of tamoxifen and ICI 182.780. The doses administered in combination were the same as for each agent used alone. The treatments lasted 5–6 weeks, as indicated in each figure. Animals were autopsied 4–6 h after the last injection. Tumors and uteri were removed from the mice, cleaned, weighed, and stored at −80°C until assayed. The data from the tissue weights were analyzed by one-way analysis of variance (ANOVA) followed by Newman-Kiels multiple range test.

V. EFFECTS OF AROMATASE INHIBITORS AND ANTIESTROGENS ON TUMOR GROWTH

Aromatase inhibitors and antiestrogens were effective in reducing tumor growth in the intratumoral aromatase model (8,9,32,33) (Table 1). Dose-response effects were evident with tamoxifen and letrozole. Tamoxifen at 60 μg/day almost completely suppressed tumor growth, whereas tumor regression occurred with letrozole at the same dose (Fig. 2). This was evident from tumor weights of a group of mice sacrificed at the beginning of the experiment and compared with tumor weights of mice treated with 10 μg/day and 60 μg/day letrozole (32).

As the uterus is an estrogen target tissue and because of concerns that it is stimulated by the agonist actions of tamoxifen in patients, we examined the effects of tamoxifen and aromatase inhibitors on the uterus. In our model, estrogen produced by the tumor enters the circulation and is sufficient to maintain uterine weight at the level of the intact mouse in metestrus. When mice were treated with tamoxifen (3 μg/day), the weight of the uteri were not significantly different from those of the vehicle-treated animals even though tumor weights were reduced. In contrast, letrozole significantly decreased uterine weights, suggesting that the aromatase inhibitor does not have agonist effects on the uterus and reduces the amount of estrogen stimulating it (32).

In a separate experiment, we investigated the effects of aromatase inhibitors letrozole and fadrozole and tamoxifen on ER and PgR in the tumors and uteri (9). Both inhibitors increased ER levels in these tissues in a dose-dependent manner (Fig. 3). Tamoxifen, on the other hand, decreased ER levels in tumors and uteri. Furthermore, tamoxifen tended to increase PgR levels in the uterus, although not in the tumors. Since the progesterone receptor is an estrogen-responsive gene, induction of PgR in the uterus is consistent with the agonist effects of this compound on the uterus. The aromatase inhibitors significantly reduced PgR levels in tumors and uteri. Interestingly, there was no detectable PgR in the tumors of animals treated with letrozole (60 μg/day and 250 μg/day). This sug-

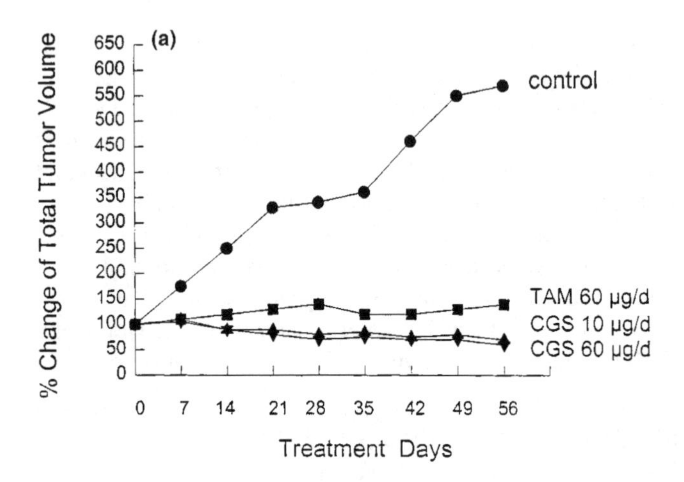

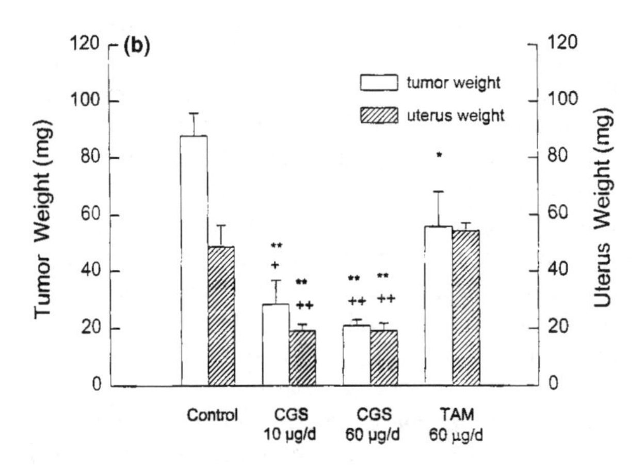

FIGURE 2 The effect of the antiestrogen tamoxifen and aromatase inhibitor letrozole on tumor and uterine wet weight in the nude mouse model. Groups of four mice were injected sc daily with letrozole CGS 10 μg/mouse/day or 60 μg/mouse/day, tamoxifen 60 μg/mouse/day, or vehicle. (a) Tumors were measured weekly and the percentage change in volume calculated. (b) After 56 days of treatment, mice were sacrificed and tumors and uteri were weighed. Values, mean ± SE, are significantly different from control. $^{*}P < .05$; CGS 10 μg vs. TAM^{+} $P < .05$; CGS 60 μg vs. TAM^{2+} $P < .01$ (32).

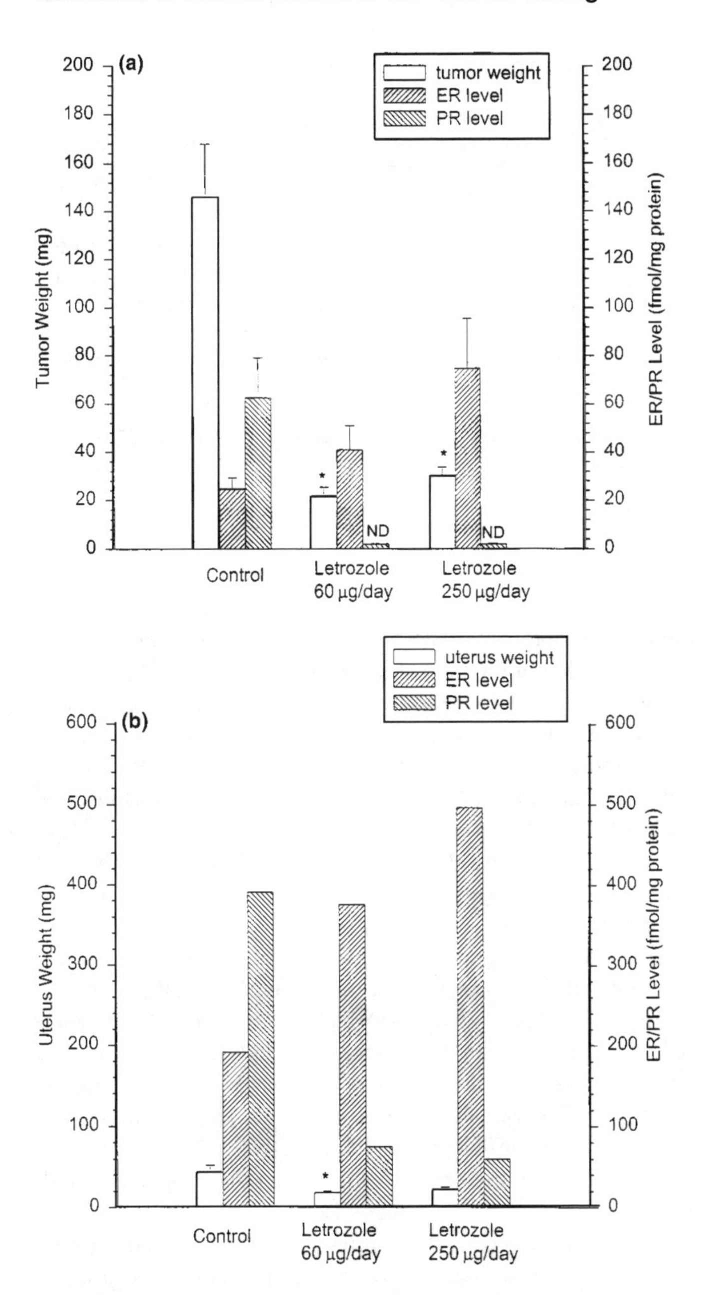

FIGURE 3 (a) Effects of letrozole on ER/PgR levels and tumor weight of MCF-7_{CA} tumors grown in OVX mice. (b) Effects of letrozole on ER/PgR levels and tissue weight of uteri of OVX mice with MCF-7_{CA} tumors. ER and PgR were measured in frozen tissues. Bars, mean ± SE of three determinations; ND, no PgR detected. $^*P < .05$ vs. control (9).

Table 1 The Effect of Letrozole and Tamoxifen on Tumor Weight

Treatment	Treatment days	Mice (n)	Tumors (n)	Tumor (mg wet weight[b])
Vehicle	1	4	21	53.54 ± 7.51[c]
Vehicle	56	2	17	226.3 ± 31.10
Tamoxifen	56	4	26	55.88 ± 12.20[c]
Letrozole	56	4	22	20.58 ± 2.09[c]
Letrozole[a]	35–56	2	14	74.64 ± 9.10[b]

Groups of four mice were injected sc with tamoxifen or letrozole (60 μg/mouse/day) or vehicle. One group of vehicle-treated mice were autopsied on day 1, and tumors were removed and weighed.

[a] Two mice in the control group were crossed over to letrozole treatment on Day 35. All other mice were autopsied on day 56 of treatment.

[b] Mean ± SE.

[c] Values are significantly different from control ($P < .01$) (32).

gests that this compound may prevent induction of other estrogen-induced genes, such as those involved in proliferation (9) (Fig. 3a & b).

As both antiestrogens and aromatase inhibitors are effective in treating breast cancer patients, combining these agents with different modes of action might result in greater antitumor efficacy than either alone. As a guide to future clinical strategies, we used the intratumoral aromatase model to address this question. When fadrozole (250 μg/day) was combined with tamoxifen (3 μg/day), tumor weights were not significantly different from those with the aromatase inhibitor alone, whereas uterine weights were similar to tamoxifen alone (9). The ER and PgR levels reflected these tumor weights (Fig. 4a & b). Thus, it is apparent that when estrogen levels are reduced by the aromatase inhibitor, tamoxifen has direct agonist actions on the uterus but not on the tumor. Similar results to those with fadrozole were obtained with 4-OHA (9). In order to determine whether greater reduction in antitumor effects could be achieved by combining these two types of agents, we used doses of the compounds which resulted in partial tumor suppression. However, the dose of tamoxifen (3 μg/day) in these experiments may not have been sufficient to block effectively the actions of the high local level of estrogen in the tumors, as evident from the induction of PgR by tamoxifen alone.

In further studies, we investigated the combination of other aromatase inhibitors with antiestrogens (32,33). Since 10 μg/mouse/day letrozole caused almost complete regression of tumors, a dose of 5 μg/day letrozole was used in the combined treatments. This was compared with the same dose of anastrozole. All compounds alone, or in combination at these doses, were effective in sup-

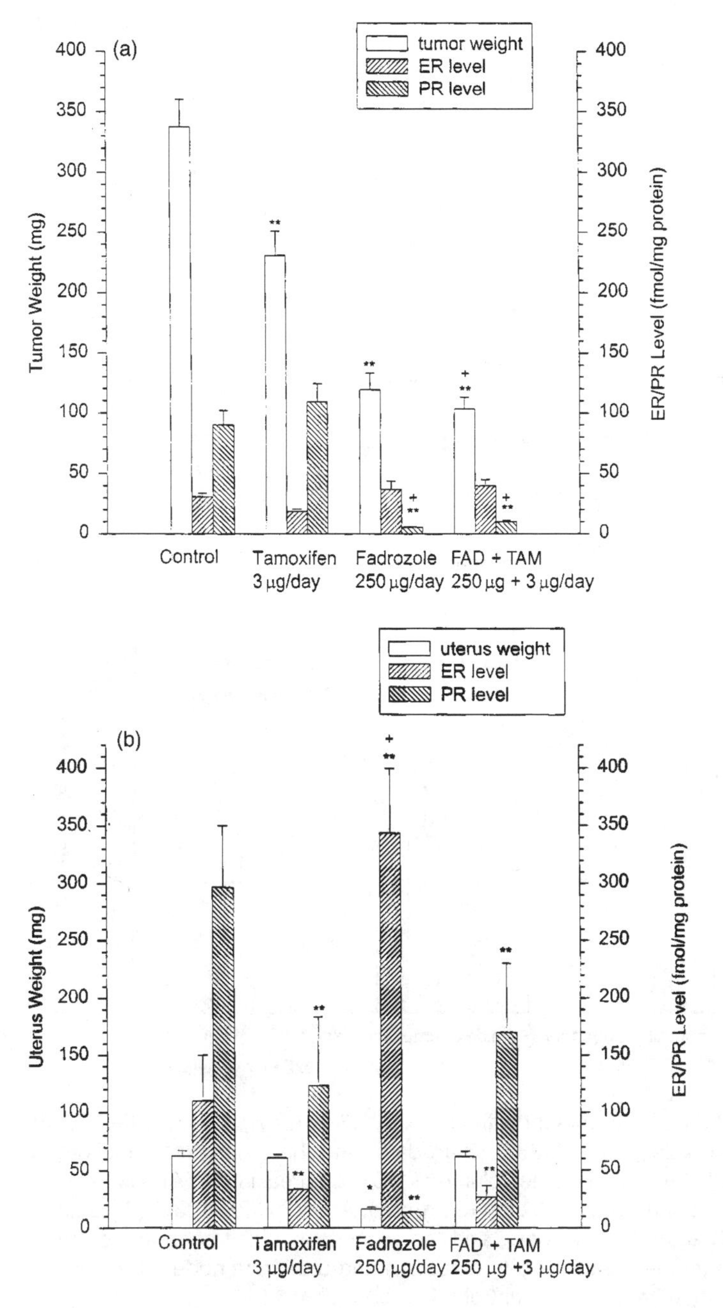

FIGURE 4 (a) Effects of combining an aromatase inhibitor and TAM on ER/PgR levels and tumor weight of MCF-7_{CA} tumors grown in OVX mice. (b) Effects of aromatase inhibitor and TAM on ER/PgR levels and tissue weight of uteri of OVX mice with MCF-7_{CA} tumors. Estrogen receptors and PgR were measured in frozen tissues. Bars, mean ± SE of three determinations; TAM, tamoxifen; FAD, fadrozole (CGS 16949); $^{*}P < .05$ vs control; $^{**}P < .01$ vs control; $^{+}$ $P < .01$ vs TAM (9).

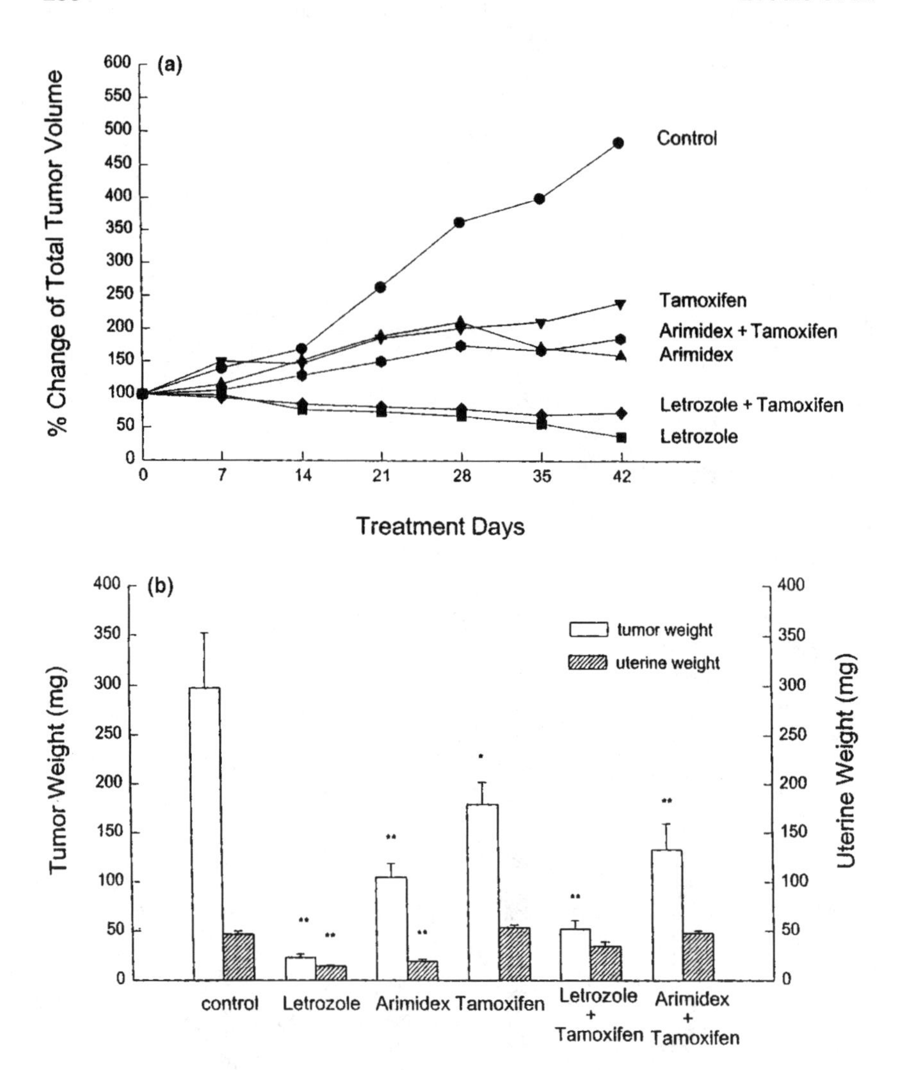

FIGURE 5 (a) Effect of treatment with tamoxifen (3 μg/day), Arimidex (ZD 1033) (5 μg/day), letrozole (CGS 20,267) (5 μg/day), and the aromatase inhibitors in combination with tamoxifen on the volume of MCF-7_{CA} breast tumors in nude mice. (b) Effect of treatment with tamoxifen (3 μg/day), Arimidex (ZD 1033) (5 μg/day), letrozole (CGS 20,267) (5 μg/day), and the aromatase inhibitors in combination with tamoxifen on the mean weights of tumors and uteri from nude mice. Values were significantly different from control $^{*}P < .05$ (33).

pressing tumor growth in comparison to that of the control mice. Weights of tumors removed at the end of treatment were significantly reduced by treatment with the aromatase inhibitors letrozole and anastrozole compared to tamoxifen ($P < .05$) (Fig. 5) (33). Treatment with either anastrozole or letrozole together with tamoxifen did not produce greater reductions in tumor growth, as measured by tumor weight, than either aromatase inhibitor treatment alone. Estrogen concentrations measured in tumor tissue of the letrozole treated mice were markedly reduced from 460 to 20 pg/mg tissue. The combination of aromatase inhibitor and tamoxifen tended to be less effective than either aromatase inhibitor alone, but this was not statistically significant for either letrozole or anastrozole (Fig. 5) (32,33). The pure antiestrogen ICI 182.780 has been reported to have significant antitumor effects in the nude mouse (28). When it was combined with tamoxifen, the mean tumor weight was also slightly greater than after treatment with ICI 182.780 alone (33). We did not investigate the possibility of interactions between tamoxifen and the aromatase inhibitors which might reduce the plasma levels of these compounds. However, a reasonable conclusion, as indicated above, is that tamoxifen may have a partial agonistic action on the tumors, overriding the reduction in estrogen concentrations—even in the tumors—achieved by aromatase inhibitors. The agonist action of tamoxifen may also counteract the effect of the pure antiestrogen. This effect was evident in the tumors and uteri, as treatment with the combination of tamoxifen and ICI 182.780 was less effective in reducing uterine weight than the pure antiestrogen alone ($P < .05$) (33).

In summary, letrozole is the most potent agent of those investigated in the mouse model. Our studies reveal the agonist effects of tamoxifen on the tumors, particularly when estrogen levels were reduced. Our findings suggest that combining any of the four aromatase inhibitors with antiestrogens did not result in any greater antitumor efficacy than occurs with these agents alone. Although these agents are used in higher doses in patients and may have different pharmacokinetic properties from those in mice, our results do not suggest that combining aromatase inhibitors with antiestrogens will be of greater benefit to patients compared to their sequential use, as currently applied. This strategy of using agents with different mechanisms of action is likely to be more advantageous by extending effective treatment when resistance has developed to first-line therapy.

ACKNOWLEDGMENT

This research was supported by the NIH grant CA 62483.

REFERENCES

1. Early Breast Cancer Trials Collaborative Group. Systemic treatment of early breast cancer by hormonal, cytotoxic, or immune therapy. 133 randomized trials involving

31,000 recurrence and 24,000 deaths among 75,000 women. Lancet 1992; 339:1–15.
2. Jordan VC. Tamoxifen: toxicities and drug resistance during the treatment and prevention of breast cancer. Ann Rev Pharmacol Toxicol 1995; 35:195–211.
3. Njar CVO, Brodie AMH. Comprehensive pharmacology and clinical efficacy of aromatase inhibitors. Drugs 1999; 58:233–255.
4. Fornander T, Rutqvist LE, Cedermark B, Mattsson A, Silfversward C, Skoog L, Somell A, Theve T, Wilking N, Askergren J, Hjalmar ML. Adjuvant tamoxifen in early breast cancer: occurrence of new primary cancer. Lancet 1989; 1:117–119.
5. Fornander T, Hellstrom AC, and Moberger B. Descriptive clinicopathologic study of 17 patients with endometrial cancer during or after adjuvant tamoxifen in early breast cancer. J Natl Cancer Inst 1993; 815:1850–1855.
6. Killackey MA, Hakes TB, Pierce VK. Endometrial adenocarcinoma in breast cancer patients receiving anti-estrogens. Cancer Treat Rep 1985; 69:237–238.
7. Wakeling AE, Valcaccia B, Newboult E, Green LR. Non-steroidal antioestrogens—receptor binding and biological response in rat uterus, rat mammary carcinoma, and human breast cancer cells. J Steroid Biochem 1984; 20:111–120.
8. Yue W, Zhou DJ, Chen S, Brodie AMH. A new nude mouse model for postmenopausal breast cancer using MCF-7 cells transfected with the human aromatase gene. Cancer Res 1994; 54:5092–5095.
9. Yue W, Wang J, Savinov A, Brodie A. Effect of aromatase inhibitors on growth of mammary tumors in a nude mouse model. Cancer Res 1995; 55:3073–3077.
10. Brodie AMH, Schwarzel WC, Shaikh AA and Brodie HJ. The effect of an aromatase inhibitor, 4-hydroxy-4-androstene-3,17-dione, on estrogen dependent processes in reproduction and breast cancer. Endocrinology 1977; 100:1684–1695.
11. Schieweck K, Bhatnagar AS, Batzl CH, Lang M. Anti-tumor and endocrine effects of non-steroidal aromatase inhibitors on estrogen-dependent rat mammary tumors. J Steroid Biochem Mol Biol 1993; 44:633–636.
12. Wouters W, Ginckel RV, Krekels M, Bowden C, De Coster R. Pharmacology of vorozole. J Steroid Biochem Mol Biol 1993; 44:617–621.
13. Wing LY, Garrett WM, Brodie AMH: The effect of aromatase inhibitors, aminogluthimide and 4-hydroxyandrostenedione on cyclic rats with DMBA-induced mammary tumors. Cancer Res 1985; 45:2425–2428.
14. Miller WR, Forrest APM. Estradiol synthesis from C19 steroids by human breast cancer. Br J Cancer 1974; 33:16–18.
15. Bradlow HL. A reassessment of the role of breast tumor aromatization. Cancer Res 1982; 42:3382S–3386S.
16. Szyczak J, Milewicz A, Thijssen JHH, Blankenstein MA, Daroszewski J. Concentration of sex steroids in adipose tissue after menopause. Steroids 1998; 63:319–321.
17. Blankenstein MA, van de Ven J, de Jong PC, Thijssen JHH. Intratumoral levels of estrogens in breast cancer. Xth International Congress on Hormonal Steroids, Quebec City, Canada, 1998.
18. van Landegham AAJ, Portman J, Nabauurs M. Endogenous concentration and subcellular distribution of estrogens in normal and malignant human breast tissue. Cancer Res 1985 45:2900–2906.
19. Masamura S, Santner SJ, Gimotty P, Santen RJ. Mechanism for maintenance to high

breast tumor estradiol concentrations in the absence of ovarian function: role of very high affinity tissue uptake. Breast Cancer Res Treat 1997; 42:215–226.
20. Yue W, Santen RJ, Wang JP, Hamilton CJ, Demers LM. Aromatase within the breast. Endocr Rel Cancer 1999; 6:157–164.
21. Lipton A, Santen RJ, Santen SJ, Harvey HA, Sanders SI, Mathews YL. Prognostic value of breast cancer aromatase. Cancer 1992; 70:1951–1955.
22. Li K, Adams JB. Aromatization of testosterone and oestrogen receptor levels in human breast cancer. J Steroid Biochem Mol Biol 1981; 14:269–272.
23. Koos RD, Banks PK, Inkster SE, Yue W, Brodie AMH. Detection of aromatase and keratinocyte growth factor expression in breast tumors using reverse transcription-polymerase chain reaction. J Steroid Biochem Mol Biol 1993; 45:217–225.
24. Lu Q, Nakamura J, Savinov A, Yue W, Weisz J, Dabbs DJ, Wolz G, Brodie A. Expression of aromatase protein and messenger RNA in tumor epithelial cells and evidence of functional significance of locally produced estrogen in human breast cancers. Endocrinology 1996; 137:3061–3068.
25. Sasano H, Nagura H, Harada N, Goukon Y, Kimura M. Immunolocalization of aromatase and other steroidogenic enzymes in human breast disorders. Hum Pathol 1994; 25:530–535.
26. Esteban JM, Warsi Z, Haniu M, Hall P, Shively JE, Chen S. Detection of intratumoral aromatase in breast carcinomas. Am J Pathol 1992; 140:337–343.
27. Tanino H, Bellavance E, Long B, Lu Q, Brodie A: Aromatase activity in cancer cells derived from a patient with recurrent breast cancer. AACR Annual Meeting, Philadelphia, 1999.
28. Dowsett M, Macaulay V, Gledhill J, Ryde C, Nicholls J, Ashworth A, McKinna JA, Smith IE. Control of aromatase in breast cancer cells and its importance for tumor growth. J Steroid Biochem Mol Biol 1993; 44:605–609.
29. Osborne CK, Coronado-Heinsohn EB, Hilsenbeck SG, McCue BL, Wakeling AE, McClelland RA, Manning DL, Nicholson RI. Comparison of the effects of a pure steroidal anti-estrogen with those of tamoxifen in a model of human breast cancer. J Natl Cancer Inst 1995; 87:746–750.
30. Zhou D, Pompon D, Chen S. Stable expression of human aromatase complementary DNA in mammalian cells: a useful system for aromatase inhibitor screening. Cancer Res 1990; 50:6949–6954.
31. Long B, Tilghman S, Yue W, Thiantanawat A, Grigoryev D, Brodie A. The steroidal anti-estrogen ICI 182,780 is an inhibitor of cellular aromatase activity. J Steroid Biochem Mol Biol 1998; 67:293–304.
32. Lu Q, Yue W, Wang J, Liu Y, Long B, Brodie A. The effects of aromatase inhibitors and anti-estrogens in the nude mouse model. Breast Cancer Res Treat 1998; 50:63–71.
33. Lu Q, Liu Y, Long BJ, Grigoryev D, Gimbel M, Brodie A. The effect of combining aromatase inhibitors with anti-estrogens on tumor growth in a nude mouse model for breast cancer. Breast Cancer Res Treat 1999; 57:183–192.

Panel Discussion 5

New Directions from Aromatase Research

List of Participants

Ajay Bhatnagar Basel, Switzerland
Angela Brodie Baltimore, Maryland
Serdar Bulun Chicago, Illinois
Mitch Dowsett London, England
Urs Eppenberger Basel, Switzerland
Dean Evans Basel, Switzerland
Nobuhiro Harada Toyoake, Japan
Harold Harvey Hershey, Pennsylvania
S. Holmberg Gothenberg, Sweden
William Miller Edinburgh, Scotland
Hironobu Sasano Sendai, Japan
Arnold Verbeek Basel, Switzerland

Dean Evans: In terms of animal models that are currently available for use in aromatase research, there are the NMU (nitrosourethrane) and DMBA (7,12-dimethylbenzanthracene) carcinogen-induced models, your MCF-7 aromatase xenograft model, and the aromatase transgenic model of Raj Tekmal. What are your thoughts on the need for additional models? For example, in the use of the MCF-7 aromatase cells as an orthotropic model where you actually implant the tumors or inoculate the cells directly into the cleared fat pad rather than administering them subcutaneously or even to look for or develop estrogen receptor (ER)-positive/aromatase-positive metastatic breast models.

Angela Brodie: Yes, we haven't used the metastatic model. The reason being that when we've been looking at inhibitor effects, we wanted to measure changes in growth. If tumors metastasize, the control animals may die before we finish the experiment. Also, there are multiple sites involved. We found it more convenient to look at the tumors growing subcutaneously. However, it is certainly a model that we will be looking at in the future. I think that it will give us even more information about how the tumors are going to respond to the treatment.

Serdar Bulun: Those were outstanding presentations, thank you very much. I have a question for Angela. What is the estrogen levels or the concept of estrogen production in ovariectomized mice—have you looked into that?

Angela Brodie: The estrogen level is extremely low in ovariectomized mice unlike the patients. There is really no peripheral conversion, so we are using the tumor to produce estrogen.

Serdar Bulun: I see, and I have a comment for Dr. Eppenberger's talk. Three different groups, including ours, Dr. Harada's and Chen's, have found increased aromatase mRNA levels in tumors and also in tissues around the majority of tumors. What was interesting was in terms of promoter use. We thought that when we would get the promoters in adipose tissue taken 1 cm and 10 cm from the tumor or within the tumor, we were thinking that we would see a grading effect. In other words, promoter I.4 or 1d would be used away from the tumor, but as you get near the promoter, use would be changed. We were surprised to find out that it was uniform—it was all promoter II or I.3, or in other terminology, it was 1c/1d, the cAMP inducible promoters. So we might be looking at a systemic effect as opposed to a local tumor effect. In terms of promoter usage, we still do not have a ready explanation, and I would like to hear your thoughts about that.

Urs Eppenberger: We found similar data for promoter distribution. We were thinking that there was a distinct discrimination between normal and tumor tissue, but we couldn't detect that. These findings seem to be similar to studies that we carried out on ER variants where we were able to detect the same variants in normal tissue as well. In other words, there is a kind of polymorphism in terms of promoters.

One thing we are aware of is that high levels of aromatase expression are not only found in tumors with positive ER levels but also with negative ER levels.

Dean Evans: Before we move on from this last point, one question which comes up from time to time is whether the patients with ER-negative/aromatase-positive breast tumors are actually candidates for aromatase inhibitor therapy. If we consider the local and direct mechanism of action of the aromatase inhibitors in the breast, then it does not really make sense that these would be candidates, since the ER is absent. So considering this from a local mechanistic viewpoint, are these patients with ER-negative/aromatase-positive breast tumors candidates for aromatase inhibitors?

Urs Eppenberger: Yes, but there is this subset which is ER/progesterone receptor (PgR)-negative, but with high levels of aromatase, and that is one group I would suggest to treat with letrozole in the future as a first-line therapy if possible, since this subset might represent an aggressive tumor type.

William Miller: I was just going to say that Mike Dixon yesterday indicated a selection procedure for neoadjuvant treatment with letrozole and indeed other forms of endocrine therapy which is clearly based on ER. So potentially, in clinical practice, the decision has already been made that the population of tumors which is likely to respond will be ER rich. Maybe we might have to reassess that, but at the moment that is the situation.

S. Holmberg: I have a question for Dr. Brodie. Could you speculate (in a mechanistic fashion) why the combination group did so poorly? It seems that it should be an ideal treatment.

Angela Brodie: I think that tamoxifen is acting as a partial agonist as well as an antagonist. I think that when we lower the level of estrogen with the aromatase inhibitor we are seeing more of the agonist effect of tamoxifen on the tumor. The inhibitor is giving much better suppression of tumor growth.

Dean Evans: Angela, you showed one figure using the MCF-7 aromatase xenograft model in which you have administered letrozole as a long-term treatment where, after the reduction and maintenance of a reduced tumor volume, that at about 18 to 20 weeks of treatment, you begin to observe a small increase in tumor volume. Although the tumor size is still well below the initial starting tumor volume before the treatment commenced, is this observed small increase in tumor volume related purely to the fact that maybe the dose of letrozole that was used in this experiment was suboptimal?

Angela Brodie: Yes, a very low dose of letrozole was used. It could be that there is a loss of inhibition after several weeks, but that is something that we need to investigate further.

Ajay Bhatnagar: Just to comment on what Angela said about the combination of tamoxifen and aromatase inhibitors. It is very important that this reason be documented in these proceedings, because it often tends to be forgotten that the only clinical trial that is going on right now that I know of is the ATAC (Anastrozole, Tamoxifen, and Combined) trial, where there is such a combination using an antiestrogen and aromatase inhibitor. And the effect that Angela mentioned, which is that the partial agonists tend to become even more agonistic when the level of estrogen is lower occurs not only in animals, but also occurs in the human situation. And thus this will be something that needs to be taken into account when ATAC results are finally analyzed.

Mitch Dowsett: Can I respond to that first? ATAC is by far the biggest situation assessing the combination, but I think Ian Smith outlined our IMPACT trial, which is a neoadjuvant trial with exactly the same randomization as ATAC. So, there is the opportunity there to see whether this Arimidex, tamoxifen, aromatase inhibitor, tamoxifen interaction is negative as you are suggesting actually on the primary tumor.

Ajay Bhatnagar: I think, Bill, that the data you presented, i.e., your hypothesis, along with the data about induction of enzyme activity with the type II inhibitors could have enormous potential impact as we move the aromatase inhibitors into early breast cancer and longer therapeutic times. I would like to give a little bit of information, which has already been published, and to ask you for your comment. About 3 years ago, we published results we obtained with diurnal variation of estradiol in both healthy postmenopausal women and in adult males. This was done as part of a Phase I study for letrozole and fadrozole. When you look at the placebo group, you will find that there is a diurnal variation in estradiol levels. They tend to go up and peak at about 10:00 to 12:00 o'clock in the morning, and then start to go down and start coming up again in the evening. And they very much pattern the cortisol values with a lag time of 2 to 3 hours, which are also diurnal variant under normal situations. There is very much a biological human clinical correlate to the fact that we are using glucocorticoids. One can have an effect on estradiol levels which come through the induction of the aromatase enzyme. Now, in your studies, since dexamethasone is also a steroid—and one would have to make some sort of assumption that the effect of dexamethasone on the stability of the enzyme has something to do with binding to the active site—it would seem quite reasonable to expect that steroidal aromatase inhibitors that also bind at the substrates site using their steroid structure would then antagonize the effects of dexamethasone. And maybe one could explain what you have shown in those terms that the steroidal inhibitors antagonize the effects of dexamethasone. This does not explain what the type II are doing in terms of increasing the effects of dexamethasone. Do you think that the model you have presented may be something that is not as solidly based in terms of induction of aromatase by the type II inhibitor as compared to the type I inhibitors?

William Miller: Well, thank you very much for that information. Maybe I can take a step back. When we first presented the data on the induction of aromatase activity by aminoglutethimide we were aware that this regime was not simply aromatase inhibitor but aromatase inhibitor and hydrocortisone. And when we first presented, we were afraid to attribute the effect to aminoglutethimide. We were more driven towards the idea that, in fact, it was the hydrocortisone components of the regime that actually did the induction. And we said that against a background that we also knew that it produced some result in culture situations. What we were able to show was that, if you change the pattern of exposure of target cells to corticoids, then you change the results and expression of aromatase. My feeling is that the results that are now shown with letrozole would suggest that what we are seeing is a generic effect of a type II inhibitor. Hence, the other reason that we showed on one of the slides data, where in culture we incubated with dexamethasone to induce synthesis, and then put the letrozole in. And then we did the experiments where we left the dexamethasone out, and we simply added the type II inhibitors. Under these conditions, we still observed enhanced aromatase activity.

It may also have clinical implications, if in fact we do believe there is a real clinical effect of inducing aromatase, then the thing to do to get round this is to combine type I and type II, so that the induction by the type II is inhibited by the type I. But, we have not done the experiment. It is clearly an important experiment to do.

Mitch Dowsett: These were exceptionally good presentations this morning. Bill, with the nonsteroidal inhibitors the data you have in vitro on the stabilization or induction of aromatase enzyme showed changes by about 50% above baseline. With respect to aminoglutethimide, you get up to a 10-fold increase in vivo. Could the hydrocortisone have some impact there? Is that the explanation, or do you have another explanation for the differential quantitative effects?

William Miller: No, all I can do is to make the interesting observation which I did during the talk. The inductive effect we actually saw in the patients with aminoglutethimide/hydrocortisone is quite marked, whereas with letrozole, of course, you still get an inhibition of tumor aromatase, but it is not to the same degree that we would have expected. I simply made the correlation that, if you actually look in the model system, then the induction which is produced by aminoglutethimide is substantially more than that produced by letrozole. So these two do go parallel; whether this is coincidence or not I do not know.

Angela Brodie: Can I just add a comment to that? We have also seen the same effect in the cell culture systems as have others. The induction with the aminoglutethimide is considerably more than it is with letrozole or fadrozole. These results are published in W. Yue and A. Brodie. Steroid Biochem 1997; 63:317–385.

Mitch Dowsett: Thanks, Angela. There is another point to Bill. The model you have where you are looking at the measurement of aromatase after 3 months—we have talked a lot about the importance of timing of the biological observations in these neoadjuvant models depending on which endpoint one is looking at. What are your thoughts about the cell population you are left with after 3 months and what the aromatase activity, which you have measured, is therefore showing us after 3 months?

William Miller: I think you are pointing to a limitation in the experimental system that we have. We study events at the end of 3 months, and the assumptions we have made that we are studying the same tumor. It is very clear from Mike Dixon's results that, if you are actually looking at the residual tumor, it may have changed. I think one of the questions that could have come forward is to ask, if you are getting these wonderful responses as we do get with letrozole, is there going to be substantially less tumor cells in the biopsy? And, if in fact the tumor cells are the major site of aromatase activity, are you actually seeing an inhibition of aromatase, but just a decellularization of the tumor? Whilst this could be true in some of these tumors (and we do not identify the components of the tumor responsible for aromatase activity), I do not think that that is true of all of them. In fact, when you look at some of these tumors histologically, which we have done, although the tumor has shrunk in size, if you look at a section of the tumor, it has not changed morphologically from before.

Perhaps I may just take a step back. We have one outlyer where in fact we seem to have an increase in aromatase activity, which I cannot really explain. I think what is interesting is that that tumor was actually one of the tumors that showed the greatest responses. There was still tumor left in the biopsy, but only microscopic residual disease. I have tried to indicate that there are differences when you look at activities in nonmalignant breast and in tumor. It may just be associated with that. Just to complicate these things a little further, our methodology repeats radioactive steroid infusion 3 months after the first infusion in order to show that there is a change in aromatase activity. We make the assumption that all the radioactivity that we pick up after that second infusion comes from that second infusion. And there is just the possibility that this radioactivity may reside from the first infusion. If then you have a complete response of the tumor, and what is left is largely fat, and fat acts as the reservoir for steroid hormones, what we might be seeing there artifactually is material that is left from the first infusion.

Mitch Dowsett: Thanks Bill, that was a very full explanation. Angela, during your talk I picked up one particular comment, which I think was almost an aside. When you were going through the MCF-7 experiments in vitro, I think you made the comment that some of the antiproliferative effects were perhaps greater than would have been anticipated from just aromatase inhibition. As I made the argument yesterday that letrozole might, at some dosages, have some impact on other pathways, I was sensitive to that particular comment. I just wondered whether you examined that in more detail. For example, have you looked at MCF-7 cells, which when they are stimulated with estradiol, would not be expected to be growth dependent on any aromatase activity, and whether letrozole has some impact in that circumstance?

Angela Brodie: Actually no, we haven't. I think that is certainly a very good suggestion. We will look into that further in the future.

Dean Evans: A question to Hironobu in regard to the topic of intratumoral aromatase and one which is maybe of more academic importance in relation to the precise cellular localization of aromatase in the breast. Based on the immunohistochemical data obtained using the two previously existing aromatase antibodies, differential cellular staining patterns have been observed between them even when used on sequential tissue sections. What are you thoughts on the importance of the precise sublocalization of aromatase in the breast in regard to the use of aromatase inhibitors; because whether aromatase is localized within the stromal cell compartment or the carcinoma cells themselves, there would be still aromatase activity present within the breast to actually contribute to the local supply of estrogens.

Hironobu Sasano: Of necessity there are controversies of whether it is the tumor cells or the stroma cells that express aromatase. In our own study, desmoplastic stromal cells and adipocytes around the tumoral invasion sites are primary sites of aromatase; i.e., estrone production. Some tumor cells may also convert androgens to estrone by aromatase, but one must also consider the number of each cell type contained in human breast cancer. It may depend on histological types of the tumor, but stromal cells and adipocytes comprise the bulk of human breast cancer tissue. Therefore, I believe that predominant sources of intratumoral estrogens by aromatase are these desmoplastic fibroblasts and reactive adipocytes.

William Miller: Can I ask a generic question about strategy with regard to these antibodies. We have already heard that two reasonably characterized antibodies, which in model systems or in systems expressing large amounts of aromatase, seem to give positive answers. But when you actually look at breast cancers they may give different answers both in terms of subcellular distribution or in the predominant type of cells in which you are staining. If then the new antibodies which you are going to characterize also show this heterogeneity of staining, what will the criteria be for determining which antibody is actually measuring aromatase in tumors and which is not?

Dean Evans: I am going to refer this question in part to Hironobu. These aspects are very important in the decision-making strategy for the selection and development of the new monoclonal antibodies that we are generating. My own feeling on this point is that we should not reject any antibody based on our prior knowledge or expectations that has been obtained using the previous polyclonal and monoclonal antibodies. At least biochemically the new monoclonal antibodies are initially selected based on reactivity to recombinantly expressed aromatase protein that was then purified to homogeneity using sequential affinity column chromatography and gel filtration procedures as well as being assessed for reactivity on control mock-infected cell lysates and other recombinant purified proteins. This should initially narrow the possibilities for cross reactivities. At the histological level, these initial antibodies are then screened sequentially on several different aromatase-expressing tissues based on immunoreactivity of samples prepared using procedures routinely employed in pathology labs. Localization patterns are important selection criteria that need to be included to help in guiding the selection, but at the same time this also needs to be handled with some flexibility and especially in the early histological rounds of screening to avoid eliminating antibodies based on any preset expectations. That is why these new monoclonal antibodies will not be initially released too early to investigators outside of the core group of collaborators involved in this project. However, once these monoclonals are fully validated biochemically and histologically, it is the intention that these antibodies will then be made generally available on request to interested investigators.

Hironobu Sasano: I think the first criterion is the exclusion of inappropriate antiaromatase monoclonal antibodies in the presence of immunoreactivity in the nuclei of the cells. I believe that everyone agrees with this point. With respect to the other exclusion criteria, we should be open minded whether it is stroma cells or carcinoma cells. Ideally, immunoreactivity detected by these monoclonal antibodies is correlated with biochemical activity of the enzyme, on the degree of aromatase expression determined by other methods, and possible response of aromatase inhibitors in the patients with breast carcinoma. However, as I mentioned in my presentation, a good and reproducible screening system is absolutely required for this purpose.

Dean Evans: This is a collective question to the speakers. One of the basic aspects which is initially an important consideration is whether aromatase is being expressed or not in the breast tissue and breast tumor cells. To what importance is the gradation in aromatase expression levels to the actual local biology of the breast? In the talk of Urs Eppenberger, data were shown where there was quite a wide variation in aromatase mRNA levels in the primary breast tumors. Now whether this mRNA is then translated to protein and the protein is functionally active awaits further study. Is it the presence or absence of aromatase expression in the breast or the gradation in level of expression that is important?

Urs Eppenberger: We will address this question in the context of the BIG trial 1-98, which is a Phase III study evaluating letrozole in postmenopausal women with ER/PgR–positive tumors.

Hironobu Sasano: I believe that the variations observed reflect marked diversity of histological subtypes of breast carcinoma examined, such as invasive ductal carcinoma, lobular carcinoma, ductal carcinoma in situ cirrhus type, or medullary type. You have totally different cell populations in the specimens that you examined. For instance, some carcinoma specimens may contain 80% carcinoma on epithelial cells. This is the primary reason of such diversity. Correlation of the findings with morphological findings on in situ approach is therefore considered as a "must" in the analysis of clinical materials in this case.

Urs Eppenberger: I was mentioning or comparing in my lecture something about erbB-2 overexpression. In this context, we started to realize that histological subsets have to be discriminated, since they have different prediction with respect to adjuvant chemotherapy. A similar situation is possible with high levels of aromatase expression, but this has to be proved first different prediction.

Harold Harvey: I have two related questions to Dr. Miller, and they concern the kinetics of the aromatase enzyme. Bill, what is known about the rate of the resynthesis of this enzyme, for example, after a stop at a type I inhibitor? Secondly, when you talk about stabilization of the enzyme, of a protein, are you referring to lack of the protein, and if so, what is known about the catabolic pathways for aromatase, because presumably this could be a target for therapy as well?

William Miller: Can I defer these to other people on the panel, because I feel they have done the basic work on stabilization and know much about it? Professor Harada and Dr. Brodie have actually done the stabilization work I presented.

Angela Brodie: As far as the resynthesis of the enzyme, we have done that experiment in JEG cells, choriocarcinoma cells, which have a comparatively high expression of aromatase. It seems that it takes about 24 hours for resynthesis. If you inhibit it completely with formestane, which inactivates the enzyme, resynthesis occurs in about 24 hours. I think that Dr. Brueggemeier has done similar experiments. So, that seems to be in those cells at least the time for resynthesis. The stabilization experiments have been done with cycloheximide, and really represent degradation of the enzyme. We do not think we have looked at the time it takes to degrade aromatase.

Dean Evans: At present, there are two schools of thought relating to explaining the phenomenon of why aromatase inhibitors increase aromatase levels, and both have supportive evidence. One explanation is that it occurs via stabilization of the protein and the other is via the induction of mRNA. Which of these possibilities so far has the strongest supportive data? What Angela has just said is that what we are maybe actually getting is a feedback loop where aromatase promoter activity is being activated. There is some data from Shiuan Chen on mRNA measurements suggesting that transcriptional activity of the aromatase promoter is the regulatory component that is being affected.

William Miller: Maybe I could respond on that and tie in one of your earlier questions. My feeling is that until recently we did not have the tools to do things properly, and certainly not in clinical material, to see whether experimental observations translate into clinical material. For this reason, it is so important that Eppenberger's measurements on messenger RNA are actually characterized to see whether they are meaningful, because up until now we really have not been able to measure accurately the messenger RNA in clinical specimens of breast cancer. And now we have what seems to be very exquisite real-time polymerase chain reaction (PCR) technology. I think that this is important. It is the same with regard to the aromatase protein—Part of the problem is that the level of aromatase activity in the majority of breast cancers is really very small, although it might really be important, because it produces estrogens locally. And again, I do not think we have the tools to measure with confidence the aromatase activity or protein in individual tumor compartments. I think that is why it is so important that we do characterize with well-defined criteria the new antibodies that are produced. If I then tie that into the question you asked about quantification of aromatase—I think it is important to quantitate. Mitch Dowsett showed us a slide yesterday which showed a relationship between immunochemical staining and response to aromatase inhibitors, but there is a huge overlap between groups. The reason that you get a statistically relevant difference is that the high levels of aromatase are in tumors which tend to respond. The problem is that we are not in bad shape when looking at high levels of aromatase, but I do not think we can very accurately quantitate low levels.

Arnold Verbeek: Question to Bill. There was some discussion going on this morning of what the proper sequence should be of treatment with a steroidal and nonsteroidal aromatase inhibitor, and whether they should be given in combination. I would just like to mention there is some quite nice clinical proof available that the steroidal aromatase inhibitor like exemestane works very well after nonsteroidal aromatase inhibitors. I think clinically that the sequence is one of activity. Could you please comment?

William Miller: I am grateful for that question. I did quote the Coombes study, which suggested that you could get response with formestane after the nonsteroidal inhibitor. And you are absolutely right. I think that Steve Johnston showed a very nice slide yesterday from a review by Per Lønning which showed very clearly that you do get responses with nonsteroidal inhibitors, such as exemestane after a use of a nonsteroidal. I think that is a very important point to make.

Mitch Dowsett: Could I just ask Bill to summarize a quarter of century of his activity in the area? Do you think that the absence of aromatase in a breast carcinoma excludes the possibility of responding to an aromatase inhibitor?

William Miller: No, absolutely not. It is very clear, if you look at the type of studies where you infuse patients with both radioactively labeled androgen and estrogen, you can look at the relative proportions of radioactivity— whether ^{14}C or ^{3}H— and purify the estrogen fractions. You can then show that certain tumors do not get their estrogen from their own biosynthesis but rather from the circulation. And in those situations, it is clear that an aromatase inhibitor working peripherally will cause a response.

Nobuhiro Harada: This may be a more general question, so I do not know whom I should ask. Probably Dr. Miller or Dr. Dowsett would best be able to answer it. Recently, highly potent aromatase inhibitors were developed causing complete suppression of aromatase activity and resulting in estrogen deprivation for a long period. This complete suppression could make ERs become hypersensitive, so that trace levels of estrogen could promote tumor proliferation, as supported by data from Drs. Masamoto and Santen and as mentioned by Dr. Dowsett in yesterday's session (Panel Discussion 1). Is it likely that this could be a drug-resistance mechanism of tumor cells to a long-term treatment with an aromatase inhibitor?

Mitch Dowsett: I do think that we have some consistent data now from Dick Santen's and my lab that you can get this hypersensitivity. We are doing very detailed molecular analysis of the cells during that acquisition of hypersensitivity to try to explain why this might occur. We have not distinguished anything very different about ER. Our favored hypothesis was that coactivators might be overexpressed in this circumstance. But those coactivators we have looked at have not yet revealed any real difference. So I think we have a consistent but unexplained phenomenon which is recapitulated in the clinical scenario as well, and I think it is therefore relevant.

In terms of can you avoid it or can you utilize strategies to get patients to respond thereafter, one of the rather interesting aspects which I did not go into yesterday, but people may have picked up was that you have this bell-shaped curve with estrogen stimulation. The wild-type cells there have reduced stimulation above 10^{-9} molar, and with the long-term estrogen-deprived cells, the curve tends to go over above about 10^{-11} molar. These higher doses of estrogen are actually those that are achievable by just straightforward hormone-replacement therapy (HRT), and are consistent with the hormone levels in a premenopausal woman.

Professor Lønning and Professor Howell are doing some studies in which they are utilizing diethylstilbestrol after aromatase inhibitors to determine whether or not you can get responses to that particular mode. My only concern about that is that it is actually still a rather high dose of estrogen above that which we are seeing in premenopausal women. The real test would be in premenopausal women to treat initially with a gonadotropin-releasing hormone (GnRH) agonist, like Zoladex, and then perhaps add in a compound like formestane, and to extend it further by using a compound like letrozole. At resistance, withdraw all treatment and allow the estrogen levels to reassert a premenopausal level, then determine whether the tumor is actually sensitive to premenopausal estrogen levels. I think that is the route to test the concept, and it is not only a test, it is an opportunity.

Part V

NEW CLINICAL INDICATIONS FOR AROMATASE INHIBITORS

C. Rose, Chair
A. Bhatnagar, Chair

14

Aromatase Inhibitors as Therapy for Pubertal Gynecomastia

Paul B. Kaplowitz
Virginia Commonwealth University School of Medicine
Richmond, Virginia

I. ABSTRACT

Pubertal gynecomastia is a common condition which probably reflects subtle alterations in the balance of testosterone and estrogen production in teenage boys. Studies to date have failed to identify a consistent hormonal abnormality in these teenage boys. Most cases are mild and transient, but in a minority, the gynecomastia is sufficiently marked and persistent to be distressing to the boy. Even in these cases, one study reported that only 7 of 60 had a defined etiology. Surgical reduction is not an option for most of these patients, and there has been relatively little experience with medical therapy with antiestrogens and nonaromatizable androgens. The importance of aromatase in the pathogenesis of gynecomastia is supported by studies of rare cases of aromatase excess, and in vitro studies showing increased aromatase activity in skin fibroblasts from men with gynecomastia versus controls. In addition, a single study published in 1986 indicated that the drug testolactone taken three times a day was moderately effective in reducing breast size in a small group of boys with pubertal gynecomastia. The case is made that a controlled trial should be undertaken with one of the newer, more potent long-acting aromatase inhibitors such as letrozole in boys with marked persistent pubertal gynecomastia.

II. INTRODUCTION

Gynecomastia refers to an abnormal amount of glandular breast tissue in males. It must be distinguished from lipomastia in obese individuals, in which what appears to be enlarged breast is actually all adipose tissue. ''Physiological'' gynecomastia is seen in newborns, adolescents, and the elderly. Pathological gynecomastia is seen mostly in adults, and it can generally be attributed to either a decrease in testosterone formation or action, or enhanced estrogen production, or drugs that alter the testosterone/estrogen balance (1).

Pubertal gynecomastia is defined as the presence of breast development in otherwise normal boys, usually starting in mid-puberty. Two studies have shown that a small amount of breast tissue can be palpated in approximately half of normal boys (2,3). This benign pubertal gynecomastia rarely causes problems and often regresses within a year. However, a smaller proportion of boys have a more marked increase in breast size, which tends to persist and may cause emotional distress, anxiety, and even depression (4). Although firm data are not available on the history of this more severe form of pubertal gynecomastia, anecdotal experience suggests it can persist for years, and that spontaneous regression is much less likely to occur. The incidence of this more severe form of pubertal gynecomastia is unknown, but a recent study from Johns Hopkins University included 60 boys with marked breast development (diameter of greater than 4 cm) seen over a 10-year period (5). Only 7 of the 60 boys had a defined etiology: Klinefelter's syndrome in two and one each with familial increased aromatase activity, partial androgen insensitivity, estrogen-secreting hepatocarcinoma, and 46 XX maleness. The remainder were considered idiopathic.

It is generally assumed that the physiological rise in serum estradiol during male puberty due to aromatization of testosterone is required for the development of breast enlargement. However, comparison of sex steroid levels between boys with gynecomastia and control boys has often failed to detect consistent or significant differences (4). One study found modest transient elevations in serum estrogen levels (6), another reported a decrease in the ratio of adrenal androgens to estrone or estradiol, but normal ratios of testosterone to estrogens (7), and another found decreases in free testosterone in boys with breast tissue versus controls (3). One of the difficulties in interpreting such studies is that some looked at all boys with any palpable breast tissue, whereas others examined only boys with marked gynecomastia. Another difficulty is that the critical time for detecting hormonal imbalances may be when breast development first starts; however, at the time of evaluation in these boys, they may have had breast development for a variable number of months or years.

III. INVOLVEMENT OF AROMATASE IN GYNECOMASTIA

Evidence that increased aromatase activity can be linked to gynecomastia comes from several sources. Rare families have been described in which increased aromatase activity in peripheral tissues results in the prepubertal onset of gynecomastia. One 8-year-old boy was able to convert half of his circulating androstenedione to estrone with a production rate of 780 μg per day, about 50-fold above normal (8). Two additional reports described two brothers (9) and a brother and sister (10) who had similar severe prepubertal breast enlargement, suggesting a single gene mutation resulting in increased levels of P450 aromatase gene transcription (10). Prepubertal gynecomastia has also resulted from an aromatase-producing sex cord tumor in a 4-year old boy; aromatase activity was found to be about 100-fold greater than in normal adult testicular tissue (11). A recent study found strong aromatase immunoreactivity in the breast tissue of 37% of boys and men with idiopathic gynecomastia, indicating that, in some cases, local increases in estrogen production within breast tissue itself may favor growth in males (12).

Finally, the possibility that idiopathic gynecomastia might be related to increased aromatase activity in peripheral tissues was investigated. Cultured pubic skin fibroblasts from eight males with gynecomastia (five were 16–20 years old) were evaluated for their ability to convert labeled androstenedione to estradiol and compared with similar fibroblasts from five control men (13). After 4-h of incubation, seven of eight gynecomastia versus none of five control fibroblasts produced detectable amounts of labeled estradiol. The authors speculate that such an increase in aromatase activity in peripheral tissues could cause local (but perhaps not systemic) increases in estradiol production and favor development of gynecomastia.

IV. THERAPY OF GYNECOMASTIA

A. Surgery

The only widely accepted treatment for pubertal gynecomastia is surgical reduction (1). However, insurance companies consider the use of surgery for this condition as cosmetic surgery, and so rarely cover it. Many refinements of the basic technique have been made in the past 30 years. Very good results with minimal scarring have been obtained by suctioning the nonglandular tissue and having the patient wear a Velcro chest binder for 8–12 weeks (14). Given the expense and potential morbidity of a surgical procedure, medical alternatives to surgery should have attracted considerable interest. It is therefore surprising that very few studies have been published which describe the results of hormonal therapy for pubertal gynecomastia. Those studies are summarized below.

B. Drug Therapy for Pubertal Gynecomastia: Sex Steroid–Like Drugs

The first agent used for pubertal gynecomastia was clomiphene citrate, an anti-estrogenic drug which acts in the hypothalamic-pituitary axis to induce a surge of gonadotropins and has been used for ovulation induction. In a 1983 study, 12 boys with persistent pubertal gynecomastia were treated with 50 mg/day for 1–3 months (15). Since only five boys had a reduction of breast diameter of more than 20%, the outcome of this therapy was not considered satisfactory.

A gel form of dihydrotestosterone (DHT) was applied directly to the skin over the breast tissue in a group of 40 men with idiopathic gynecomastia in a study reported in 1983 (16). The rationale was that since DHT cannot be aromatized to estradiol, there would be a favorable shift in the DHT/estradiol ratio, which was in fact observed. Complete or partial disappearance of breast tissue was noted in 29 of 40 subjects, although no objective measurements were reported. In 1986, the use of an intramuscular preparation of DHT given every 2–3 weeks in four boys with pubertal gynecomastia was reported. After 16 weeks, the area measured at the base of the breast decreased from a mean of 24.3 to 7.0 cm^2, and this reduction persisted for 2 months posttherapy (17). Despite the promise of this form of therapy, no follow-up studies have been published and neither preparation of DHT has ever been marketed.

The use of the estrogen antagonist tamoxifen would seem to offer a safe and effective means of reducing breast tissue mass in patients with gynecomastia. There are two reported studies of its use in adult men with benign or painful gynecomastia at a dose of 10 mg twice daily (18,19). The majority of patients experienced pain reduction and at least partial reduction in breast size. One study showed a doubling of both luteinizing hormone (LH) and estradiol levels during therapy (19), suggesting that the antagonism of estrogen effects may be partly overcome by reduced negative feedback on the pituitary and increased stimulation of estrogen secretion. To date, no reports of tamoxifen use in pubertal gynecomastia have been published.

C. Aromatase Inhibitors

The appearance in the 1980s of agents for use in advanced estrogen-dependent cancer, which act specifically to inhibit aromatase, offered a new potential therapy for gynecomastia. In 1986, Zachmann et al. described the use of testolactone (150 mg three times a day for 2–6 months) in 22 boys with pubertal gynecomastia (20). Median breast diameter decreased from 3.8 cm at baseline to 3.0 cm at 2 months, 2.8 cm at 4 months (n = 14), and 1.5 cm at 6 months (n = 4). There was little change in testosterone and estradiol but a marked increase in serum androstenedione. There were no undesirable side effects, and the psychological effects were said to be very favorable. Surprisingly, there have been no additional published reports on the use of testolactone or other aromatase inhibitors for pubertal gynecomastia. Zachmann reported that he continued treating patients after the manuscript was published with equally good results. However, his free source of drug was discontinued, and since testolactone was expensive and not an established treatment, the insurance companies in Switzerland did not want to pay for it (M. Zachmann, personal communication, 1998).

A 1994 report described the favorable results of testolactone on gynecomastia developing in an adult after unilateral orchiectomy for a Leydig cell tumor. They documented an initial lowering of the elevated serum estradiol, which returned to baseline after 3 months of treatment, but serum testosterone nearly doubled, increasing the T/E2 ratio (21). In the family with aromatase excess discussed above, 3 years of testolactone resulted in estradiol levels becoming undetectable and a slowing of bone maturation (10).

One reason for the lack of studies on aromatase inhibitors for gynecomastia may be evidence from animal models of the effects on testicular function. In dogs, two different inhibitors, formestane and letrozole, caused Leydig cell hypertrophy and hyperplasia, presumably due to a decrease of estradiol-mediated feedback suppression of LH (22). Disturbed spermatogenesis was also observed. In five adult male bonnet monkeys, long-term treatment with a new potent long-acting aromatase inhibitor resulted in a large increase in serum testosterone from 10 to 80 days with a return to close to baseline after 120–180 days. There was also an acute suppression of sperm counts and sperm motility, which was maximal between 55 and 85 days, with partial recovery in some of the monkeys thereafter (23). This suggests that estradiol may have a critical role in sperm development. Thus, the potential benefits of aromatase inhibitors in boys with gynecomastia will need to be balanced against the potential of alterations in testicular function, which may or may not be completely reversible with discontinuation of therapy.

V. SUMMARY AND PROPOSAL

Marked persistent pubertal gynecomastia is a relatively common disorder in teenage boys, the hormonal basis of which is still poorly understood. However, most evidence points to the elevated production of estradiol or alteration in the testosterone to estrogen ratio during male puberty as being required for gynecomastia to develop. A small number of uncontrolled studies have shown some clinical benefit from hormonal therapies designed to increase the effective androgen/estrogen ratio. The most impressive of these is a 1986 study showing significant breast size reduction in a group of 22 boys treated with testolactone. No studies have been done with the newer aromatase inhibitors such as letrozole, which are more potent than testolactone and can be taken daily instead of three times a day. Such a study should be placebo controlled and involve in the range of 20–30 actively treated patients to be recruited at a minimum of six pediatric endocrinology sites. It should be designed to examine the following parameters.

1. *Effect of the Drug on Breast Size*. Although measurement of breast diameter is not very precise, it is probably the easiest measure to obtain and has been used in most previous studies. With proper technique, the glandular tissue can be separated from adipose tissue. In addition, because breasts with similar diameter at the base may differ in how much they protrude, hemicircumference should also be measured, although it must be recognized that this can be affected by overlying adipose tissue. The use of ultrasound objectively to measure breast volume is a consideration, but no reports of either a standard method for this or normative data could be found.

2. *Time Course of This Effect*. Based on the testolactone data, it is proposed to monitor efficacy of the drug at either 2, 4, and 6 months or 3 and 6 months, with a 6 month follow-up after discontinuation. Because of the possibility of effects on testicular function, longer treatment periods should be avoided until there is more human male safety data.

3. *Evaluation of Changes in Hormone Levels During Therapy*. This should include assessment of testosterone, estradiol, estrone, androstenedione, LH, and follicle-stimulating hormone levels. It should be noted that although estradiol and estrone levels may not show large or sustained decreases, the ratios of testosterone or androstenedione to estradiol or estrone should increase.

4. *Safety and Tolerability of the Medication*. Since drug-related adverse reactions have been uncommon in women with advanced breast cancer, intolerance to the drug is not expected to be a major problem in healthy adolescent boys.

5. *Drug Level*. At steady state 2–3 months after initiation of therapy.

6. *The Effect of the Medications on the Emotional State and Quality of Life of Boys with Gynecomastia*. There is currently no information in the literature on the topic; so this study should be used as an opportunity to collect this type of data by developing a questionnaire to address general mood, self-esteem, and sensitivity about the size of the breasts. Showing positive psychological benefits as well as a decrease in breast size will help in the process of getting approval of the drug for this indication.

REFERENCES

1. Frantz AG, Wilson JD. Endocrine disorders of the breast In: Wilson JD and Foster DW, eds. Williams Textbook of Endocrinology. 8th ed. Philadelphia: Saunders, 1992:953–976.

2. Nydick M, et al. Gynecomastia in adolescent boys. JAMA 1961; 178:449–454.
3. Biro FM, Lucky AW, Huster GA, Morrison JA. Hormonal studies and physical maturation in adolescent gynecomastia. J Pediatr 1990; 116:450–455.
4. Marynick SP, Nisula BC, Pita JC, Loriaux DL. Persistent pubertal macromastia. J Clin Endocrinol Metab 1980; 50:128–130.
5. Sher ES, Migeon CJ, Berkowitz GD. Evaluation of boys with marked breast development at puberty. Clin Pediatr 1998; 37:367–372.
6. La Franchi SH, Parlow AF, Lippe BM, et al. Pubertal gynecomastia and transient elevation of serum estradiol level. Am J Dis Child 1975; 129:927–931.
7. Moore DC, Schlaepfer LV, Paunier L, Sizonenko PC. Hormonal changes during puberty: V. Transient pubertal gynecomastia: abnormal androgen: estrogen ratios. J Clin Endocrinol Metab 1984; 58:492–499.
8. Hemsell DL, Edman CD, Marks JF, Siteri P, MacDonald PC. Massive extraglandular aromatization of plasma androstenedione resulting in feminization of a prepubertal boy. J Clin Invest 1977; 60:455–464.
9. Berkovitz GD, Guerami A, Brown TR, MacDonald PC, Migeon CJ. Familial gynecomastia with increased extraglandular aromatization of plasma $carbon_{19}$-steroids. J Clin Invest 1985; 75:1763–1769.
10. Stratakis CA, Vottero A, Brodie A, et al. The aromatase excess syndrome is associated with feminization of both sexes and autosomal dominant transmission of aberrant P450 aromatase gene transcription. J Clin Endocrinol Metab 1998; 83:1348–1357.
11. Coen P, Kulin H, Ballantine T, Zaino R, Frauenhoffer E, Boal D, Inkster S, Brodie A, Santen R. An aromatase-producing sex-cord tumor resulting in prepubertal gynecomastia. N Engl J Med 1991; 324:317–322.
12. Sasano H, Kimura M, Shizawa S, Kimura N, Nagura H. Aromatase and steroid receptors in gynecomastia and male breast cancer: an immunohistochemical study. J Clin Endocrinol Metab 1996; 81:3063–3067.
13. Bulard J, Mowszowicz I, Schaison G. Increased aromatase activity in public skin fibroblasts from patients with isolated gynecomastia. J Clin Endocrinol Metab 1987; 64:618–623.
14. Cohen K, Pozez AL, McKeown JE. Gynecomastia. In Courtiss EH, ed: Male Aesthetic Surgery, 2nd ed. St. Louis: Mosby-Year Book, 1991:373–394.
15. Plourde PV, Kulin HE, Santner SJ. Clomiphene in the treatment of adolescent gynecomastia. Am J Dis Child 1983; 137:1080–1082.
16. Kuhn JM, Roca R, Laudat M-H, Rieu M, Luton J-P, Bricaire H. Studies on the treatment of idiopathic gynecomastia with percutaneous dihydrotestosterone. Clin Endocrinol 1983; 19:513–520.
17. Eberle AJ, Sparrow JT, Keenan BS. Treatment of persistent pubertal gynecomastia with dihydrotestosterone heptanoate. J Pediatr 1986; 109:144–149.
18. Parker LN, Gray DR, Lai MK, Levin ER. Treatment of gynecomastia with tamoxifen: a double-blind crossover study. Metabolism 1986; 35:705–708.
19. McDermott MT, Hofeldt FD, Kidd GS. Tamoxifen therapy for painful idiopathic gynecomastia. South Med J 1990; 83:1284–1286.
20. Zachmann M, Eiholzer U, Muritano M, et al. Treatment of pubertal gynecomastia with testolactone. Acta Endocrinol 1986; 279(Suppl):218–224.

21. Auchus RJ, Lynch SC. Treatment of post-orchioectomy gynecomastia with testolactone. Endocrinologist 1994; 4:429–432.
22. Junker Walker U, Nogues V. Changes induced by treatment with aromatase inhibitors in testicular Leydig cells of rats and dogs. Exp Toxicol Pathol 1994; 46:211–213.
23. Shetty G, Krishnamurthy H, Krishnamurthy HN, Bhatnagar AS, Moudgal NR. Effect of long-term treatment of aromatase inhibitor on testicular function of adult male bonnet monkeys. Steroids 1998; 63:414–420.

Panel Discussion 6

Aromatase Inhibitors and Gynecomastia November 12, 1999

List of Participants

S. Holmberg Gothenberg, Sweden
Paul Kaplowitz Richmond, Virginia
William Miller Edinburgh, Scotland
Matthew Smith Boston, Massachusetts
G. Stathopoulos Athens, Greece
D. Tsiftsis Athens, Greece
Adrian Weiss Basel, Switzerland

Matthew Smith: Do these boys have breast pain in addition to gynecomastia, and if so, did you plan to measure that in your studies?

Paul Kaplowitz: Some of them do, and I think that that is a very good point. I would say about half of them in my experience have breast pain, and I think you are right that we should figure out at least, if not an accurate way, then a subjective way to assess diminution of tenderness of the breast tissue.

D. Tsiftsis: I wonder, as you mentioned, this is a self-limiting disease that sometimes lasts less than 6 months, so you must control your study very carefully. And the other thing is, I am a bit skeptical about exposing young boys to an endocrine treatment in contrast to surgery, which is considered minimal but does not last more that 10 minutes, and gives a permanent, very good result. Would you like to comment on that?

Paul Kaplowitz: Regarding the first question, I think you are right—We need a control group. Anecdotal experience has been that when the gynecomastia becomes of a sufficient size to concern the boy that it does not regress within 6 months, I think the time course is much longer. On the other hand, I don't keep seeing these patients back, so I don't accurately know how long it lasts. The question of surgery is a good one. I think that surgery is a bit more than a 10-minute surgery—I think it is a little more complicated. If there is a mechanism for boys to get the surgery paid for, I think it would be a more viable treatment, but I think a lot of families don't want to go through that. Looking for another alternative, I think that with as many antiestrogens and aromatase inhibitors as we have available, it is time that some company took the step of being the first one to study this.

S. Holmberg: I see in my clinic quite a few patients that have been taking anabolic steroids as they go to gyms or in prison. Quite a few of them tell me that they take tamoxifen at the same time to avoid developing what they call ''bitch tits.'' Do you have any experience in this field?

Paul Kaplowitz: No, I am a little surprised, because most of the anabolic steroids are nonaromatizable. You would think they would not cause gynecomastia, whereas testosterone obviously could, as it is converted to estrogen. So I don't have any experience with boys wanting tamoxifen to try to counteract the effects of anabolic steroids.

G. Stathopoulos: What happens after some years in those young boys? Do they get a recurrence? If so, do you have to repeat the initial successful treatment, do you put them on long term treatment, and then what happens in the future?

Paul Kaplowitz: I think you would have to follow these boys up for ideally 2 years, because the question is, if all we are doing is causing a temporary shrinkage and then by 6 months it is back to where it was before, then perhaps this is not the best treatment. But I agree with you, the key is good documentation of effects and long-term follow-up. I think that there is reason to hope that once you get the breasts to shrink, that you may—assuming that the hormonal process has quieted down—keep that reduction for longer time periods. I know that in a lot of young girls I have evaluated, who have what we call premature thelarche (a little breast enlargement due to a transient increase in estrogens). Even though the increased circulation in estrogen from a transiently functioning ovarian cyst may only last a brief period, once the breast tissue is there, it stays for a long time. So if we can shrink it, perhaps we can maintain that reduction. But I think your point is well taken.

William Miller: Can I add two further planks in your argument in the use of aromatase inhibitors in gynecomastia. You mentioned data of treating patients with dihydrotestosterone, and presented as if this was a systemic effect—and it probably could be—but 5α-dihydrotestosterone (5α-DHT) is probably the best aromatase inhibitor amongst the naturally occurring steroids. And indeed if I remember correctly, John McIndoe, at the first aromatase symposium way back in 1981, presented evidence that 5α-dihydrotestosterone would inhibit aromatase activity in MCF-7 breast cancer cells, so you may wish to interpret it that way. The second thing is perhaps I can also draw your attention to some of our own data, where in fact we took gynecomastic breast tissue and did some steroid incubations. And what we actually found was that they apparently had no 5α-reductase activity. So the level of 5α-reduced steroids that came out of that breast tissue was in fact negligible. Whereas if you take female breast tissue and do the same incubation, you can find it. So maybe a reduction in the 5α-DHT is associated with a local effect.

Paul Kaplowitz: I think that your point is very interesting, and I don't know why the 5α-DHT has not been pursued. So it is certainly another avenue.

Adrian Weiss: Male breast cancer is quite rare. But I remember I have seen several patients in clinics with male breast cancer, and they had gynecomastia, and I remember one of them where the diagnosis of male breast cancer was made during the operation for gynecomastia. And that is one question, the relationship between gynecomastia and male breast cancer. The second question—I think that the study you are proposing may be better indicated for bilateral gynecomastia, since surgery may be more complicated in that case.

Paul Kaplowitz: I don't have any information on gynecomastia and male breast cancer. Male breast cancer is very rare and obviously is seen only by physicians in adults. But I am not aware of any studies in the literature showing whether there is or is not a relationship between that and patients with gynecomastia. Interestingly, when patients come to you with breast development, that is one of the things they may be worried about. First of all the boys worry whether they are turning into girls, and then some are worried whether the breasts are malignant. Concerning the second question you had, in the very early cases of gynecomastia, you may see some asymmetry, but when they get to the point as we saw in the slide that I showed, it is almost always bilateral. So those would be the cases that we would be interested in treating.

15

Aromatase Inhibition and Prostate Cancer

Matthew R. Smith
Massachusetts General Hospital
Boston, Massachusetts

I. ABSTRACT

Estrogens and other nonandrogenic growth factors may contribute to the development and progression of prostate cancer. Preclinical studies support an important role of estrogens in prostate cancer growth. Phase II studies of men with androgen-independent prostate cancer, however, suggest that antiestrogens have minimal activity. In contrast, a Phase I/II study of the first-generation aromatase inhibitor rogletimide reported promising early results. Future studies of newer and more potent aromatase inhibitors are warranted.

II. ROLE OF ESTROGEN AND ESTROGEN RECEPTORS IN PROSTATE CANCER

Androgen deprivation by either orchiectomy or chronic administration of gonadotropin-releasing hormone (GnRH) agonists is the mainstay of treatment for advanced prostate cancer (1). Androgen deprivation results in responses in the majority of patients, but the median duration of response is only about 2 years. Adrenal androgens and other nonandrogenic hormones, including estrogens, may contribute to the growth of prostate cancer following androgen deprivation (2,3).

Estrogen receptors (ER) are expressed in normal prostate epithelium (4,5) and most primary prostate cancers (6). Androgen deprivation, the mainstay of treatment for metastatic prostate cancer, appears to increase prostate ER density (7). Estrogen receptor β is the predominant ER expressed in the prostate (8).

III. ANTIESTROGENS

Antiestrogens have activity in preclinical models of prostate cancer. Antiestrogens inhibit the growth of human prostate cancer cell lines in vitro (9) and also have activity in a variety of rodent prostate cancer models (9–13).

Previous studies of the antiestrogen tamoxifen reported minimal activity in men with metastatic prostate cancer (14–20). The results of these clinical trials are summarized in Table 1. Patient characteristics, prior hormonal therapy, tamoxifen dose, and criteria for response varied among the studies. Response rates between 0 and 23% were reported.

The tamoxifen studies may have underestimated the activity of antiestrogens for two reasons. First, these studies were conducted prior to the routine use of prostate-specific antigen (PSA), and responses were defined by insensitive clinical and radiographic criteria. Second, the potential antineoplastic activity of tamoxifen may have been masked by treatment-related increases in serum testosterone levels. None of the men in the tamoxifen studies was treated with GnRH agonists, and only about half were treated with orchiectomy prior to study entry. Treatment of noncastrate men with antiestrogens increases serum testosterone levels by increasing the release of GnRH (21). In a study of men with advanced prostate cancer reported by Torti and collaborators, treatment with tamoxifen increased serum testosterone levels in most men (19). In the other tamoxifen studies, the effects of treatment on testosterone levels were not reported.

TABLE 1 Phase II Studies of Tamoxifen for Metastatic Prostate Cancer

Study	Reference	n	Prior orchiectomy (%)	Tamoxifen dose (mg/d)	Response rate (%)
Philadelphia	14	31	55	20	23
Antwerp	15	10	50	20–40	0
Bronx	16	10	Not reported	20	10
Multicentered	17	51	37	20	7
Multicentered	18	41	82	40	5
Palo Alto	19	17	12	20–100	0
ECOG	20	19	53	10–30	0

Toremifene is a triphenylethylene-derivative antiestrogen, related chemically and pharmacologically to tamoxifen. Like tamoxifen, toremifene binds with high affinity to cytoplasmic ER (22) and has both antiestrogenic and partial estrogenic activity (23–27). In a Phase II study of 12 men with androgen-independent prostate cancer, there were no responses using sensitive PSA response criteria [response rate = 0%; 95% confidence interval (CI) = 0–22%] (28). Median time to treatment failure was 16 weeks (range 8–19 weeks). Treatment did not significantly change serum testosterone levels.

The limited activity of antiestrogens for advanced androgen-independent prostate cancer may be explained by decreased ER expression. In a recent study, 57 of 73 (78%) primary prostatectomy specimens expressed functional ER (6). In contrast, none of the 22 (0%) metastatic prostate cancers expressed ER, suggesting that transition to the metastatic phenotype is accompanied by loss of ER expression (29). The effect of antiestrogens in androgen-dependent prostate cancer has not been prospectively evaluated.

The results of these studies do not exclude the possibility that high-dose antiestrogens have activity in androgen-independent prostate cancer. Tamoxifen and other antiestrogens inhibit the growth of prostate cancer cells in vitro by mechanisms independent of ER. In micromolar concentrations, tamoxifen inhibits protein kinase C, induces Rb dephosphorylation, and increases p21 expression in vitro (30,31). These alternative mechanisms of growth inhibition may contribute to the clinical activity of high-dose antiestrogens. In a Phase II trial, 13 men with androgen-independent prostate cancer were treated with high-dose tamoxifen (160–200 mg/m^2/day), resulting in a >50% PSA decrease in one of the men (8%) and stable disease reported in four (31%) men (32). Treatment resulted in grade 3 cardiovascular toxicity (QTc prolongation) in three of 13 (23%) men. Two of the six (33%) men treated with tamoxifen 200 mg/m^2/day experienced grade 3 cerebellar toxicity.

IV. STEROID SYNTHESIS INHIBITORS

Inhibitors of steroid synthesis have been evaluated as secondary hormonal treatment for androgen-independent prostate cancer. Aminoglutethimide inhibits several steroidogenic enzymes, including the aromatase enzyme system responsible for the conversion of androgens to estrogens. Ketoconazole inhibits C17,20-lyase and other steroidogenic enzymes. Toxicity and marginal efficacy limit the use of these agents for prostate cancer (1).

V. AROMATASE INHIBITION

Selective inhibitors of the aromatase enzyme system may have activity in androgen-independent prostate cancer. In a Phase I/II study, 23 men with androgen-independent prostate cancer were treated with the aromatase inhibitor rogletimide (33). Among 13 men treated at the highest dose level (800 mg orally daily), there were two subjective responses and two PSA responses (>80% PSA decrease). The effects of rogletimide on sex hormone levels were not reported. The activity of newer and more potent aromatase inhibitors has not yet been evaluated in men with androgen-independent prostate cancer.

Letrozole is a nonsteroidal competitive inhibitor of the aromatase enzyme system. Letrozole is indicated for the treatment of advanced breast cancer in postmenopausal women with disease progression following antiestrogen treatment. Letrozole is approximately 10,000-fold more potent than aminoglutethimide in vivo. In women, treatment with letrozole significantly lowers serum estrogens without significant effects on serum levels of androgens, adrenal steroids, aldosterone, or thyroid hormones. Letrozole treatment in women appears to increase serum insulin-like growth factor (IGF)-I levels without significant effects on IGF-binding protein-3 (IGFBP-3) levels (34).

We recently initiated a Phase II study of letrozole for men with androgen-independent prostate cancer. Men with histologically confirmed prostate cancer and a rising PSA after androgen deprivation and antiandrogen withdrawal will be treated with letrozole (2.5 mg orally daily) until disease progression or toxicity. The primary study endpoint will be PSA response. Study accrual will be completed in the first quarter of the year 2000 and preliminary results will be available in the third quarter of the same year.

VI. CONCLUSIONS

Preclinical studies suggest that estrogens play an important role in prostate cancer development and progression. Clinical trials of men with androgen-independent prostate cancer, however, suggest that antiestrogens have minimal activity. Aromatase inhibition represents an attractive alternative approach for the treatment of men with androgen-independent disease. A Phase I/II study of the first-generation aromatase inhibitor rogletimide demonstrated promising activity. A Phase II study of the third-generation aromatase inhibitor letrozole was recently initiated at our institution. Additional studies of aromatase inhibition for prostate cancer are warranted.

REFERENCES

1. Moull JW. Contemporary management of advanced prostate cancer. Oncology 1998; 12:499–507.
2. Culig Z, Hobisch A, Cronauer MV, Radmayr C, Hittmair A, Zhang J, Thurnher M, Bartsch G, Klocker H. Regulation of prostatic growth and function by peptide growth factors. Prostate 1996; 28(6):392–405.

3. Ware JL. Growth factors and their receptors as determinants in the proliferation and metastasis of human prostate cancer. Cancer Metas Rev 1993; 12(3–4):287–301.
4. Wagoner RK, Schulze KH, Jungblut PW. Estrogen and androgen receptors in human prostate and prostatic tumor tissue. Acta Endocrinol 1975; 193(suppl):52.
5. Hawkins EF, Nijs M, Brassinne C. Steroid receptors in the human prostate. Biochem Biophys Res Commun 1976; 70:854–861.
6. Nativ O, Umehara T, Colvard DS, Therneau TM, Farrow GM, Spelsberg TC, Lieber MM. Relationship between DNA-ploidy and functional estrogen receptors in operable prostate cancer. Eur Urol 1997; 32(1):96–99.
7. Kruithof-Dekker IG, Tetu B, Janssen PJ, Van der Kwast TH. Elevated estrogen receptor expression in human prostatic stromal cells by androgen ablation therapy. J Urol 1996; 156(3):1194–1197.
8. Kuiper GG, Enmark E, Pelto-Huikko M, Nilsson S, Gustafsson JA. Cloning of a novel receptor expressed in rat prostate and ovary. Proc Natl Acad Sci USA 1996; 93(12):5925–5930.
9. Lanoit Y, Veilleux R, Dufour M, Simard J, Fabrie F. Characterization of the biphasic action of androgens and of the potent antiproliferative effects of the new pure antiestrogen EM-139 on cell cycle kinetic parameters in LNCaP human prostatic cancer cells. Cancer Res 1991; 51:5165–5170.
10. Ip MM, Milholland RJ, Rosen F. Functionality of the estrogen receptor and tamoxifen treatment of R3327 Dunning rat prostate adenocarcinoma. Cancer Res 1980; 40: 2188–2193.
11. Schneider MR, von Angerer E, Hohn W, Sinowatz F. Antitumor activity of antiestrogenic phenylindoles on experimental prostate tumors. Eur J Cancer Clin Oncol 1987; 23:1005–1015.
12. Schneider MR, Schiller CD, Humm A, von Angerer E. Effect of zindoxifene on experimental prostatic tumors of the rat. J Cancer Res Clin Oncol 1991; 117:33–36.
13. Neubauer BL, Best KL, Counts DF, Goode RL, Hoover DM, Jones CD, Sarosdy MF, Shaar CJ, Tanzer LR, Merriman RL. Raloxifene (LY156758) produces antimetastatic responses and extends survival in the PAIII rat prostatic adenocarcinoma model. Prostate 1995; 27:220–229.
14. Glick JH, Wein A, Padavic K, Negendank W, Harris D, Brodovsky H. Tamoxifen in refractory metastatic carcinomas of the prostate. Cancer Treat Rep 1980; 64:813–818.
15. Arnold DJ, Hallbridge E, Rosen N, Usher S. Tamoxifen therapy for metastatic prostate cancer. Proc Am Soc Clin Oncol 1981; 21:468.
16. Denis L, Dedlercq DL. Anti-estrogens in the treatment of prostatic cancer. Acta Urol Belg 1980; 48:106–109.
17. Glick JH, Wein A, Padavic K, Negendank W, Harris D, Brodovsky H. Phase II trial of tamoxifen in metastatic carcinoma of the prostate. Cancer 1982; 49:1367–1372.
18. Spremulli E, Desimone P, Durant J. A phase II study of Nolvadex tamoxifen citrate in the treatment of advanced prostatic adenocarcinoma. Am J Clin Oncol 1982; 5: 149–153.
19. Torti FM, Lum BL, Lo R, Freiha F, Shortliffe L. Tamoxifen in advanced prostatic carcinoma. Cancer 1984; 54:739–743.

20. Horton J, Rosenbaum C, Cummings FJ. Tamoxifen in advanced prostate cancer: an ECOG pilot study. Prostate 1988; 12:173–177.
21. Schill WB. Recent progress in pharmacological therapy for male subfertility: a review. Andrologia 1979; 11(2):77–107.
22. Kallio S, Kangas L, Blanco G, Johansson R, Karjalainen A, Perila M, Pippo I, Sundquist H, Sodervall M, Toivola R. A new triphenyethylene compound, Fc-1157a II. hormonal effects. Cancer Chemother Pharmacol 1986; 17:103–108.
23. Kangas L, Nieminen AL, Blanco G, Gronroos M, Kallio S, Karjalainen A, Perila M, Sodervall M, Toivola R. A new triphenylethylene compound, Fc-1157a II. antitiumor effects. Cancer Chemother Pharmacol 1986; 17:109–113.
24. Robinson SP, Mauel DA, Jordan VC. Antitumor actions of toremifene in the 7,12-dimethylbenzenthracene (DBMA)–induced rat mammary tumor model. Eur J Cancer Clin Oncol 1988; 24:1817–1821.
25. Robinson SP, Parker CJ, Jordan VC. Preclinical studies with toremifene as an antitumor agent. Breast Cancer Res Treat 1990; 16(suppl):9–17.
26. DiSalle E, Zaccheo T, Ornati G. Anti-estrogenic and antitumor properties of the new triphenylethylene derivative in toremifene in the rat. J Steroid Biochem 1990; 36:203–206.
27. Homesley HD, Shemano I, Gams RA, Harry DS, Hickox PG, Rebar RW, et al. Antiestrogenic potency of toremifene and tamoxifen in postmenopausal women. Am J Clin Oncol 1993; 16:117–122.
28. Smith MR, Kantoff PW, Oh W, Elson G, Manola J, McMullin M, Jacobsen J, Brufsky A, Kaufman D. Phase II trial of the anti-estrogen toremifene for androgen-independent prostate cancer. Prostate J 1999;
29. Hobisch A, Hittmair A, Daxenbichler G, Wille S, Radmayr C, Hobisch-Hagen P, et al. Metastatic lesions from prostate cancer do not express estrogen and progesterone receptors. J Pathol 1997; 182(3):356–361.
30. Rohlff C, Blagosklonny MV, Kyle E, Kesari A, Kim IY, Zelner DJ, Hakim F, Trepel J, Bergan RC. Prostate cancer cell growth inhibition by tamoxifen is associated with inhibition of protein kinase C and induction of p21 (waf1/cip1). Prostate 1998; 37(1):51–59.
31. Rohlff C, Lee S, Genieser H, Bergan R. Multiple signal pathways mediate tamoxifen-induced cell death in hormone refractory PC3M prostate carcinoma cells (meeting abstr). Proc Annu Meet Am Assoc Cancer Res 1997; 38:A3353.
32. Bergan RC, Blagosklonny M, Dawson NA, Headlee D, Figg WD, Neckers L, Reed E, Myers CE. Significant activity by high dose tamoxifen in hormone refractory prostate cancer (meeting abstr). Proc Annu Meet Am Soc Clin Oncol 1995; 14: A637.
33. Dearnaley DP, Shearer RJ, Gallagher C, Oliver T, Gadd J, Boschoff C, Meely K. A Phase (Ph) I/II study with rogletimide (ROG) in patients (pt) with hormone resistant (HR) prostatic cancer (CaP) (meeting abstr). Ann Oncol 1995;(suppl 8):179.
34. Bajetta E, Ferrari L, Celio L, Mariani L, Miceli R, Di Leo A, Zilembo N, Buzzoni R, Spagnoli I, Martinetti A, Bichisao E, Seregni E. The aromatase inhibitor letrozole in advanced breast cancer: effects on serum insulin-like growth factor (IGF)-I and IGF-binding protein-3 levels. J Steroid Biochem Mol Biol 1997; 63(4–6):261–267.

Panel Discussion 7

Aromatase and Prostate Cancer November 12, 1999

List of Participants

Ajay Bhatnagar Basel, Switzerland
Matthew Smith Boston, Massachusetts
G. Stathopoulos Athens, Greece
Carsten Rose Odense, Denmark

Ajay Bhatnagar: Dr. Smith, you said that for the letrozole studies the patient recruitment takes place when the prostate-specific antigen (PSA) levels rise after androgen deprivation. Is this androgen deprivation a form of castration or is it medical androgen deprivation?

Matthew Smith: Either form of androgen deprivation is acceptable. For cosmetic and psychological reasons, most men elect to undergo chronic treatment with a gonadotropin-releasing hormone (GnRH) agonist. Androgen deprivation will continue during letrozole treatment.

Carsten Rose: May I ask you, have you any knowledge about the use of aminoglutethimide in prostate cancer treatment?

Matthew Smith: Aminoglutethimide has significant side effects, so our local preferences may be to use ketoconazole. Ketoconazole is the most active form of secondary hormonal therapy we have for prostate cancer. Like aminoglutethimide, its use is limited by side effects. About one-third of men are unable to continue therapy because of abdominal pain, diarrhea, or fatigue. The response rates are about one in three. Some of these responses are durable. It is unclear whether that is due to aromatase inhibition or by its other mechanisms of action.

G. Stathopoulos: Do you see in the future any combination of letrozole with antiandrogens?

Matthew Smith: In our clinical trials of antiandrogen monotherapy, we treat these men with prophylactic breast radiation to prevent the predictable and often severe gynecomastia that accompanies treatment. Another approach would be to evaluate other pharmacological interventions to prevent the gynecomastia, and certainly letrozole represents a sensible approach. Tamoxifen improves the gynecomastia associated with antiandrogen monotherapy. Antiandrogen monotherapy increases estrogen levels, and some of the potential benefits of antiandrogen monotherapy may be related to higher circulating estrogen levels. Letrozole may eliminate this potential advantage.

Ajay Bhatnagar: This model of yours, Dr. Smith, actually raises some very fascinating questions that are very basic to endocrinology, because I do not think in all the literature I followed over all of these years one has really ever well explained negative feedback. And I think that in these patients of yours you use a GnRH clamp. So one would imagine that the entire system from the hypothalamus downward is clamped out. And yet we are going to have a situation here where the reduction of estrogen is going to have negative feedback influence on the pituitary. And it will be fascinating to know whether this feedback influence is within the clamp or is something which escapes the clamp, and it will be interesting to know whether you plan to measure any estrogen or any androgen levels during your study.

Matthew Smith: Administration of antiestrogens to intact men results in increased testosterone levels. In our toremifene study, total testosterone levels and SHBG (sex hormone–binding globulin) increased, but free testosterone levels remained unchanged. We will be measuring gonadal steroid levels in the letrozole study.

Ajay Bhatnagar: Because when you give letrozole to normal males, testosterone levels rise by about four- to fivefold.

Matthew Smith: Androgen deprivation for prostate cancer is the most common cause of hypogonadism in men worldwide. Hypogonadal men with prostate cancer are an important population for fundamental endocrinology investigation.

16

Aromatase in Endometriosis: Biological and Clinical Application

Serdar E. Bulun
University of Illinois
Chicago, Illinois

Khaled M. Zeitoun
Columbia University College of Physicians and Surgeons
New York, New York

Kazuto Takayama and Hironobu Sasano
Tohoku University School of Medicine
Sendai, Japan

Evan R. Simpson
Prince Henry's Institute of Medical Research
Monash Medical Centre
Clayton, Victoria, Australia

I. ABSTRACT

Aromatase activity is not detectable in normal endometrium. In contrast, aromatase is expressed aberrantly in endometriosis and is stimulated by prostaglandin E_2 (PGE_2). This results in local production of estrogen, which induces

PGE_2 formation and establishes a positive-feedback cycle. Another abnormality in endometriosis—deficient 17β-hydroxysteroid dehydrogenase (17β-HSD) type 2 expression—impairs the inactivation of estradiol to estrone. These molecular aberrations collectively favor the accumulation of increasing quantities of estradiol and PGE_2 in endometriosis. The clinical relevance of these findings was demonstrated by the successful treatment of an unusually aggressive case of postmenopausal endometriosis using an aromatase inhibitor.

II. INTRODUCTION

Endometriosis is linked to pelvic pain and infertility, and it is defined as the presence of endometrial glands and stroma within the pelvic peritoneum and other extrauterine sites. It is estimated that 2–10% of women are affected in the reproductive age group (1,2). Endometriosis is viewed to be a polygenically inherited disease of complex multifactorial etiology (3). Sampson's theory of transplantation of endometrial tissue on the pelvic peritoneum via retrograde menstruation is the most widely accepted explanation for the development of pelvic endometriosis because of convincing circumstantial and experimental evidence (4). Since retrograde menstruation is observed in almost all cycling women, endometriosis is postulated to develop as a result of the coexistence of a defect in clearance of the menstrual efflux from pelvic peritoneal surfaces, possibly involving the immune system (5). Alternatively, intrinsic molecular aberrations in pelvic endometriotic implants were proposed to contribute significantly to the development of the endometriosis. Some of the molecular abnormalities include the aberrant expression of aromatase, deficiency of 17β-hydroxysteroid dehydrogenase (17β-HSD) type 2, and resistance to the protective action of progesterone and certain cytokines by tissues (6–12). Since endometriosis is an estrogen-dependent disorder, aromatase expression and 17β-HSD type 2 deficiency are of paramount importance in the pathophysiology of endometriosis. In this chapter, aberrant mechanisms of estrogen biosynthesis and metabolism in women with endometriosis are reviewed, with emphasis on identifying targets for new treatment strategies.

III. DISCUSSION

A. Estrogen Biosynthesis and Metabolism in Humans

The conversion of androstenedione and testosterone to estrone and estradiol is catalyzed by aromatase, which is expressed in a number of human tissues such as ovarian granulosa cells, placental syncytiotrophoblast, adipose tissue and skin fibroblasts, and the brain. In reproductively active women, the ovaries are the most important sites of estrogen biosynthesis, which takes place in a cyclic fashion. The binding of follicle-stimulating hormone (FSH) to its G-protein–coupled receptor in granulosa cell membranes stimulates a rise in intracellular cAMP levels. This is turn enhances the binding of two critical transcription factors—steroidogenic factor-1 (SF-1) and cAMP response element–binding protein (CREB)—to the classically located proximal promoter II of the aromatase gene (13,14). This action activates aromatase expression and consequently estrogen secretion from the preovulatory follicle (14,15).

In postmenopausal women, the most important sites of estrogen formation are in extraglandular tissues such as adipose tissue and skin fibroblasts (16–18) (Fig. 1). In contrast to ovarian aromatase regulation by cAMP, this is controlled primarily by cytokines (interleukin-6 [IL-6], IL-11, tumor necrosis factor-α [TNF-α]) and glucocorticoids via the I.4 promoter (15). The major substrate for aromatase in adipose tissue and skin is androstenedione of adrenal origin. In postmenopausal women, approximately 2% of circulating androstenedione is converted to estrone, which is further converted to estradiol in these peripheral

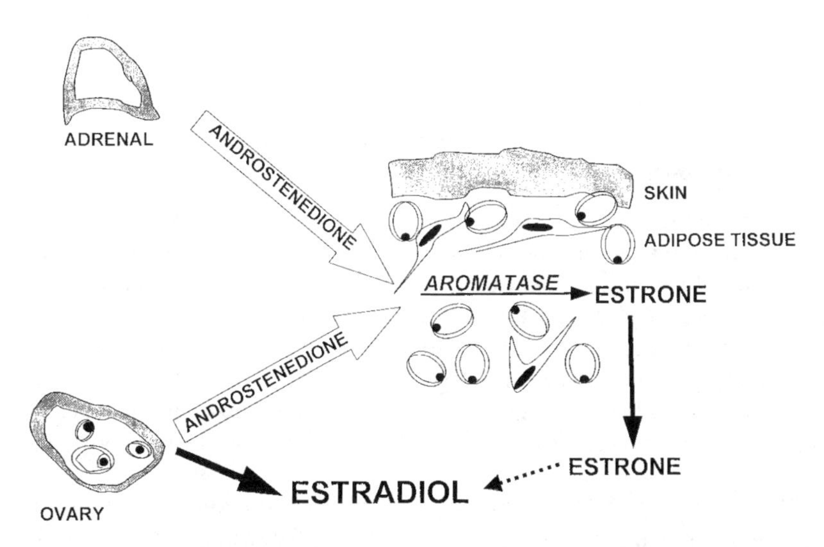

FIGURE 1 Extraovarian estrogen synthesis in women. Estradiol in women is either directly secreted by the ovary or produced in extraglandular sites (adipose tissue and skin). The principal substrate for extraglandular aromatase activity in ovulatory women is ovarian and adrenal androstenedione. In women receiving GnRH agonists or postmenopausal women, however, adrenal androstenedione becomes primary substrate. Androstenedione is converted by aromatase to estrone in adipose tissue and skin fibroblasts. Estrone is further converted to estradiol by 17β-HSD type 1 activity in these peripheral tissues. Thus, peripheral aromatization is the major source for circulating estradiol in the postmenopausal period or during ovarian suppression.

tissues. This may give rise to significant serum levels of estradiol capable of causing endometrial hyperplasia or even carcinoma (17,18).

B. Aromatase Expression in Müllerian-Derived Tissues

Müllerian tissues are known targets of estrogen action. Until recently, estrogen action had been viewed as occurring via an ''endocrine'' mechanism only. It was believed that only circulating estradiol, whether secreted by the ovary or the adipose tissue, could exert an estrogenic effect after delivery to target tissues via the bloodstream. However, studies on aromatase expression in breast cancer demonstrated that paracrine mechanisms play an important role in estrogen action in these tissues (19). Estrogen produced by aromatase activity in breast adipose tissue fibroblasts has been demonstrated to promote the growth of adjacent malignant breast epithelial cells (20). Finally, we have demonstrated an ''intracrine'' effect of estrogen in uterine leiomyomas and endometriosis. Estrogen produced by aromatase activity in the cytoplasm of leiomyoma smooth muscle cells or endometriotic stromal cells can exert its effects by readily binding to its nuclear receptor within the same cell (6,21,22). Disease-free endometrium and myometrium, on the other hand, lack aromatase expression (21,22).

C. Significance of Aromatase Expression in Endometriosis

Among estrogen-responsive pelvic disorders, aromatase expression has been studied in greatest detail in endometriosis (6,7,22,23). Extremely high levels of aromatase mRNA were found in extraovarian endometriotic implants and endometriomas. In addition, cultured endometriosis-derived stromal cells incubated with a cAMP analogue displayed extraordinarily high levels of aromatase activity; comparable to that in placental syncytiotrophoblast (22). These exciting findings led us to test a battery of growth factors, cytokines, and other substances that might induce aromatase activity via a cAMP-dependent pathway in endometriosis.

PGE_2 was found to be the most potent inducer of aromatase activity in endometriotic stromal cells known (22). This PGE_2 effect was found to be mediated via the cAMP-inducing EP_2 receptor subtype (our unpublished observations). Moreover, estrogen was reported to increase PGE_2 formation by stimulating cyclooxygenase type 2 (COX-2) enzyme in endometrial stromal cells in culture (24). Thus, a positive-feedback loop for continuous local production of estrogen and PG is established, favoring the proliferative and inflammatory characteristics of endometriosis (Fig. 2). Additionally, aromatase mRNA was also detected in the eutopic endometrial samples of women with moderate to severe endometriosis (but not in those of disease-free women) albeit in much smaller quantities compared with endometriotic implants (6). This may be suggestive of a genetic defect

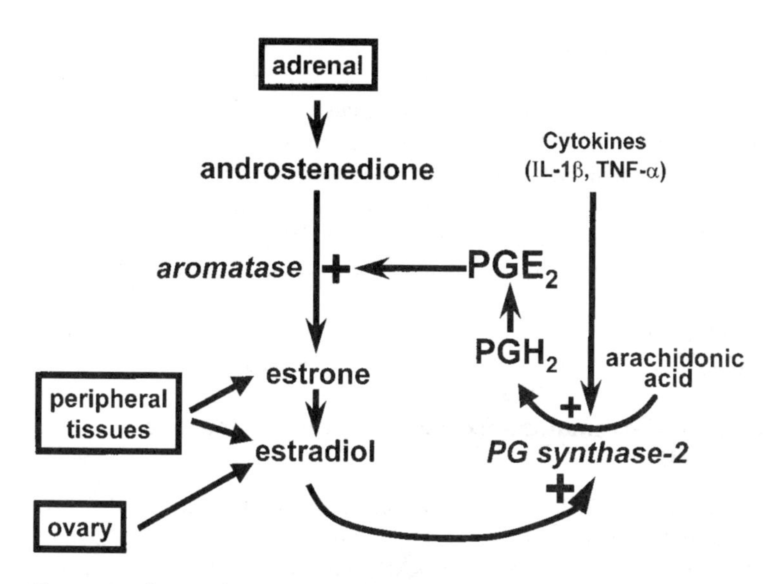

FIGURE 2 Origin of estrogen in endometriotic lesions. Estradiol in an endometriotic lesion arises from several sites. In an ovulatory woman, estradiol is secreted directly from the ovary in a cyclic fashion. In the early follicular phase and after menopause, peripheral tissues (adipose and skin) are the most important sources to account for the circulating estradiol. Estradiol is also produced locally in the endometriotic implant itself in both ovulatory and postmenopausal women. The most important precursor, androstenedione of adrenal origin, becomes converted to estrone that in turn is reduced to estradiol in the peripheral tissues and endometriotic implants. We demonstrated significant levels of 17β-HSD type 1 expression in endometriosis, which catalyzes the conversion of estrone to estradiol (12). Estradiol both directly and indirectly (through cytokines) induces COX-2, which gives rise to elevated concentrations of PGE_2 in endometriosis (24). PGE_2, in turn, is the most potent stimulator of aromatase known in endometriotic stromal cells (22). This establishes a positive feedback loop in favor of continuous estrogen formation in endometriosis.

in women with endometriosis, which is manifested by this subtle finding in the eutopic endometrium. We propose that when defective endometrium with low levels of aberrant aromatase expression reaches the pelvic peritoneum by retrograde menstruation, it causes an inflammatory reaction that exponentially increases local aromatase activity, that is, estrogen formation, induced directly or indirectly by PG and cytokines (22).

It would be naive to propose that aberrant aromatase expression is the only important molecular mechanism in the development and growth of pelvic endometriosis. There may be many other molecular mechanisms that favor the development of endometriosis, such as abnormal expression of proteinase type enzymes that remodel tissues or their inhibitors (matrix metalloproteinases, tissue inhibitor of metalloproteinase-1), certain cytokines (IL-6, RANTES) and growth factors (epidermal growth factor, EGF) (8–11). Alternatively, a defective immune system failing to clear peritoneal surfaces of the retrograde menstrual efflux could be the cause of the endometriosis (5,25). The development of endometriosis in an individual woman probably requires the coexistence of a threshold number of these aberrations.

Nonetheless, aberrant aromatase expression is clinically relevant, since aromatase inhibitors suppress postmenopausal endometriosis (26).

D. Regulation of Aromatase Expression in Endometriotic Stromal Cells

As emphasized earlier, PGE_2 was found to be the most potent inducer of aromatase activity known through increasing cAMP levels in endometriotic stromal cells (22). On the other hand, neither cAMP analogues nor PGE_2 was capable of stimulating any detectable aromatase activity in cultured eutopic endometrial stromal cells. Attention focused on the potential molecular differences that give rise to aromatase expression in endometriosis and its inhibition in eutopic endometrium.

To address this, we first identified that cAMP-inducible promoter II stimulated in vivo aromatase expression in endometriotic tissue (7). Stimulating SF-1 and inhibiting chicken ovalbumin upstream promoter transcription factor (COUP-TF) were then found to compete for the same binding site in aromatase promoter II. Although COUP-TF was ubiquitously expressed in both eutopic endometrium and endometriosis, SF-1 was specifically expressed in endometriosis and not the eutopic endometrium, binding to the aromatase promoter more avidly than COUP-TF (7). Thus, although SF-1 and other transcription factors (e.g., CREB) activate transcription in endometriosis, whereas COUP-TF, although occupying the same DNA site in eutopic endometrium, inhibits this process (7) (Fig. 3). In summary, one of the molecular alterations leading to local aromatase expression in endometriosis but not in normal endometrium is the aberrant production of SF-1 in endometriotic stromal cells, which overcomes the protective inhibition maintained normally by COUP-TF in the eutopic endometrium.

E. Interconversions of Estrone and Estradiol in Endometriosis

Aromatase's primary substrate in endometriosis in premenopausal women is adrenal or ovarian androstenedione, whereas in postmenopausal women, it is adrenal androstenedione alone. The major product of aromatase activity in endometri-

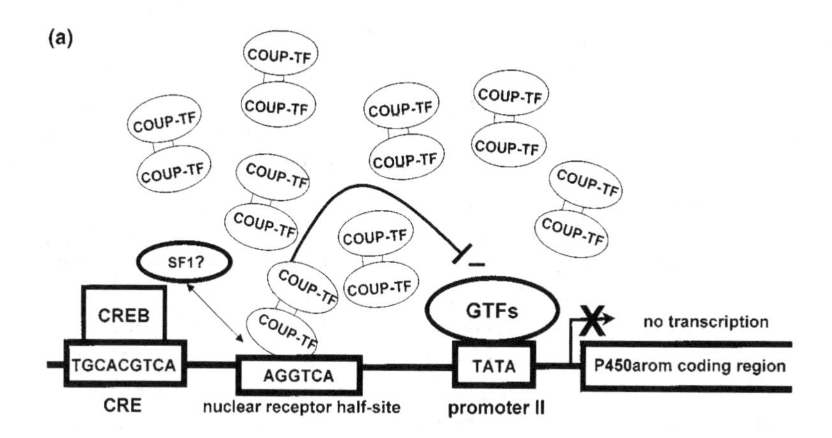

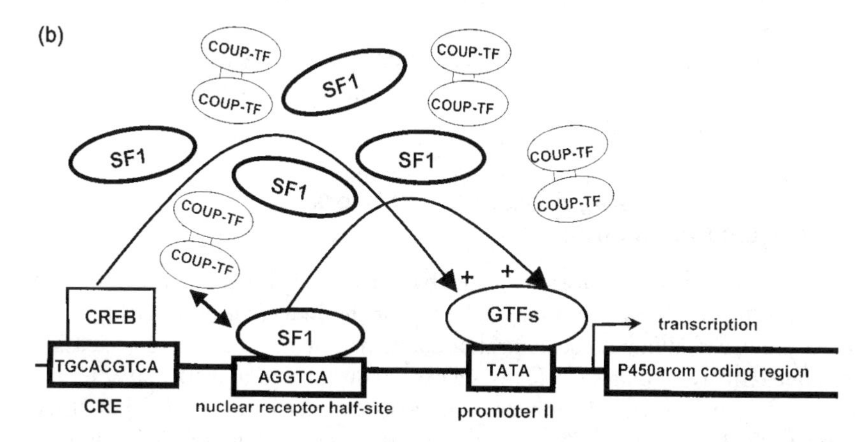

FIGURE 3 Proposed mechanism of regulation of aromatase (P450arom) expression by SF-1 and COUP-TF in eutopic endometrium and endometriosis. (a) Binding of COUP-TF (dimer) to a specific DNA site (nuclear receptor half-site) upstream of aromatase promoter II in eutopic endometrial stromal cells. In the eutopic endometrium, COUP-TF binds to nuclear receptor half-site practically in the absence of any competition by SF-1, since SF-1 expression is not detected in the majority of eutopic endometrial samples. Thus, COUP-TF exerts its inhibitory effect on the complex of general transcription factors (GTFs) that bind to TATA box. (b) In the endometriotic stromal cell, where SF-1 is also present, SF-1 binds as a monomer to the nuclear receptor half-site with a higher affinity compared with that of COUP-TF, which binds to this site relatively loosely as a dimer. Upon replacing COUP-TF, SF-1 synergizes with cAMP response element binding protein (CREB, bound to upstream CRE) and other factors to activate transcription of the aromatase gene in response to cAMP (7).

osis is estrone, which is only weakly estrogenic and must be converted to the potent estrogen estradiol to exert a full estrogenic effect. We demonstrated that the enzyme 17β-HSD type 1, which catalyzes the conversion of estrone to estradiol, is expressed in endometriosis (12,27). In contrast, the enzyme 17β-HSD type 2 (encoded by a separate gene) inactivates estradiol by catalyzing its conversion to estrone in eutopic endometrial glandular cells during the luteal phase (27). Progesterone induces the activity of 17β-HSD type 2 in cultured endometrial glandular cells, making the inactivation of estradiol one of the antiestrogenic properties of progesterone (28). The expression of 17β-HSD type 2 is absent from endometriotic glandular cells, as demonstrated in paired samples of eutopic endometrium and pelvic endometriosis obtained simultaneously during the luteal phase (12). Consequently, the protective mechanism that lowers estradiol levels is lost in endometriotic tissue (12). The aberrant expression of aromatase, the presence of 17β-HSD type 1, and the absence of 17β-HSD type 2 from endometriosis collectively give rise to elevated local levels of estradiol compared with eutopic endometrium. Additionally, 17β-HSD type 2 deficiency may also be viewed as a defective action of progesterone, which fails to induce this enzyme in endometriotic tissue (Fig. 4).

F. Rationale for Using Aromatase Inhibitors to Treat Endometriosis

Endometriosis is successfully suppressed by estrogen deprivation through the use of gonadotropin-release hormone analogues or the induction of surgical menopause. However, although the control of pelvic pain with gonadotropin-releasing hormone (GnRH) agonists is usually successful during and immediately after the treatment, pain associated with endometriosis returns in up to 75% of these women (29,30). There may be multiple reasons for the failure of GnRH agonist treatment of endometriosis. One likely explanation is the presence of significant estradiol production that continues in the adipose tissue, skin, and endometriotic implant per se during the GnRH agonist treatment (Fig. 5). Therefore, blockage of aromatase activity in these extraovarian sites with an aromatase inhibitor may keep larger number of patients in remission for longer periods of time. The most striking evidence for the significance of extraovarian estrogen production is the recurrence of endometriosis after successfully completed hysterectomy and bilateral salpingo-oophorectomy in a number of women (26,31). Endometriotic tissue in one such aggressive case was found to express much higher levels of aromatase mRNA compared with premenopausal endometriosis (26).

We recently reported the treatment of a 57-year-old overweight woman who had recurrence of severe endometriosis after hysterectomy and bilateral salpingo-oophorectomy. Two additional laparotomies were performed owing to persistent severe pelvic pain and bilateral ureteral obstruction leading to left renal

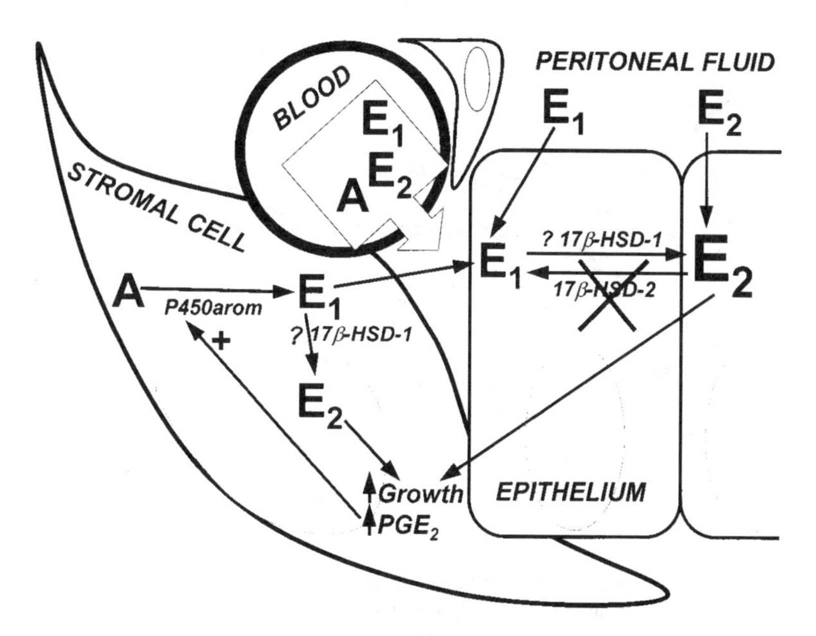

Figure 4 Defective inactivation of estradiol in endometriosis. Estradiol (E_2) reaches the endometriotic lesions via the blood stream (and possibly peritoneal fluid). Additionally, aromatase (P450arom) in the stromal cell catalyzes the conversion of androstenedione (A) to estrone (E_2), which is further reduced to E_2 by 17β-HSD type 1 in the endometriotic tissue. (At this time, the cell type that expresses 17β-HSD type 1 in endometriotic lesions is not known.) E_2 is normally inactivated by conversion to E_1 by 17β-HSD type 2 in epithelial cells of the eutopic endometrium during the secretory phase. In endometriotic tissue, however, E_2 is not metabolized owing to the lack of 17β-HSD type 2 giving rise to increased local concentration of this potent estrogen. Elevated E_2, in turn, will promote the growth of endometriotic tissue and, also, local PGE_2 formation in stromal cells (24). Since PGE_2 is the most potent known inducer of aromatase in endometriosis, this will complete the positive feedback cycle that favors increased levels of E_2 in endometriosis through enhanced biosynthesis and deficient metabolism.

atrophy and right hydronephrosis. Treatment with megestrol acetate was ineffective and a large (3 cm) vaginal endometriotic lesion containing unusually high levels of aromatase mRNA was found. The patient was given anastrozole (an aromatase inhibitor) for 9 months, and dramatic pain and regression of the vaginal endometriotic lesion were observed within the first month of treatment. At the same time, circulating estradiol levels were reduced to 50% of the baseline value. The markedly high pretreatment levels of aromatase mRNA in the endometriotic

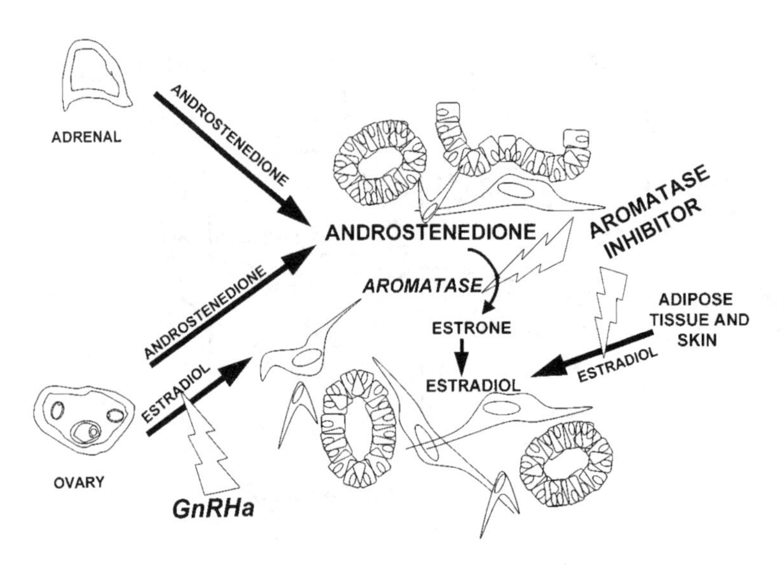

FIGURE 5 Site of action of GnRH agonists and aromatase inhibitors to treat endometriosis. This figure depicts the origin of estrogen in women with endometriosis: (1) delivery from the ovary and adipose tissue/skin via circulation and (2) local biosynthesis in endometriosis. GnRH agonists will eliminate estradiol secreted by the ovary by downregulating the pituitary hypothalamic unit. In cases resistant to treatment with GnRH agonists or in postmenopausal endometriosis, the use of aromatase inhibitors to block estrogen formation in the skin and adipose tissue as well as in endometriotic stromal cells may be critical in controlling the growth of endometriotic tissue. Recurrent endometriosis, especially after surgical removal of the ovaries, may represent lesions that are sensitive to extremely low levels of estradiol, and thus suppression of estradiol production in the periphery (adipose tissue/skin) and in endometriotic tissue may be mandatory for successful treatment of endometriosis.

tissue became undetectable in a repeat biopsy 6 months later, and the lesion nearly disappeared after 9 months of therapy. Despite the addition of calcium and alendronate (a nonsteroidal inhibitor of bone resorption), bone density in the lumbar spine decreased by 6.2%. The occurrence of significant bone loss in this particular case should be studied further.

Two potential mechanisms may account for the strikingly successful result. First, there was evidence of the suppression of peripheral aromatase activity (i.e., skin and adipose tissue), giving rise to a significant decrease in serum estradiol level. Second, unusually high levels of aromatase expression in the endometriotic lesion disappeared after treatment with the aromatase inhibitor anastrozole. The disappearance of aromatase mRNA expression in the lesion may be due to the decrease in local estrogen-stimulated PGE_2 biosynthesis. This reduction in PGE_2, in turn, will have a negative effect on aromatase expression (see Fig. 2).

In summary, the recently developed potent aromatase inhibitors are candidate drugs in the treatment of endometriosis that is resistant to standard regimens. In fact, the use of aromatase inhibitors may be the only available treatment for aggressive postmenopausal endometriosis. It remains to be seen whether aromatase inhibitors alone or together with present lines of therapy in premenopausal women will increase the pain-free interval and time to recurrence after discontinuation (see Fig. 5). Studies are under way to address these questions.

IV. CONCLUSIONS

The development and growth of endometriotic lesions are estrogen dependent. The mechanisms and effectiveness of hormonal treatments for endometriosis should be reevaluated in view of the latest advances in the understanding of estrogen production in women with endometriosis. In addition to ovarian secretion, estradiol is also produced in peripheral sites such as skin, adipose tissue, and endometriotic lesions per se. We suggest that the intracrine and paracrine effects of estradiol produced in the target tissue amplify the estrogenic action of steroid hormones delivered via the circulation. Additionally, this local effect may further enhance defective inactivation of estradiol in endometriosis compared to eutopic endometrium. Aberrant aromatase activity and defective estradiol metabolism in endometriosis are consequences of specific molecular aberrations, such as inappropriate expression of a stimulatory transcription factor or progesterone resistance in this tissue. The clinical relevance of these findings was recently exemplified by the successful treatment of a severe case of recurrent postmenopausal endometriosis with an aromatase inhibitor. Future treatment strategies may be designed to target the signal transduction for aromatase expression in endometriosis or to enhance progesterone action in this tissue and to eliminate bone loss that was an undesirable affect of aromatase inhibition in our otherwise successfully treated case of severe endometriosis.

ACKNOWLEDGMENT

We thank Dee Alexander for providing expert editorial assistance.

REFERENCES

1. Vessey MP, Villard-Mackintosh L, Painter R. Epidemiology of endometriosis in women attending family planning clinics. Br Med J 1993; 306:182–184.
2. Kjerulff KH, Erickson BA, Langenberg PW. Chronic gynecological conditions re-

ported by US women: findings from the National Health Information Survey, 1984 to 1992. Am J Public Health 1996; 86:195–199.
3. Olive DL, Schwartz LB. Endometriosis. N Engl J Med 1993; 328:1759–1769.
4. Sampson JA. Peritoneal endometriosis due to the menstrual dissemination of endometrial tissue into the peritoneal cavity. Am J Obstet Gynecol 1927; 14:422–425.
5. Halme J, White C, Kauma S, Estes J, Haskill S. Peritoneal macrophages from patients with endometriosis release growth factor activity in vitro. J Clin Endocrinol Metab 1988; 66:1044–1049.
6. Noble LS, Simpson ER, Johns A, Bulun SE. Aromatase expression in endometriosis. J Clin Endocrinol Metab 1996; 81:174–179.
7. Zeitoun K, Takayama K, Michael MD, Bulun SE. Stimulation of aromatase P450 promoter (II) activity in endometriosis and its inhibition in endometrium are regulated by competitive binding of SF-1 and COUP-TF to the same cis-acting element. Mol Endocrinol 1999; 13:239–253.
8. Khorram O, Taylor RN, Ryan IP, Schall TJ, Landers DV. Peritoneal fluid concentrations of the cytokine RANTES correlate with the severity of endometriosis. Am J Obstet Gynecol 1993; 169:1545–1549.
9. Sharpe-Timms KL, Penney LL, Zimmer RL, Wright JA, Zhang Y, Surewicz K. Partial purification and amino acid sequence analysis of endometriosis protein-II (ENDO-II) reveals homology with tissue inhibitor of metalloproteinases-1 (TIMP-1). J Clin Endocrinol Metab 1995; 80:3784–3787.
10. Bruner KL, Matrisian LM, Rodgers WH, Gorstein F, Osteen KG. Suppression of matrix metalloproteinases inhibits establishment of ectopic lesions by human endometrium in nude mice. J Clin Invest 1997; 99:2851–2857.
11. Osteen KG, Bruner KL, Sharpe-Timms KL. Steroid and growth factor regulation of matrix metalloproteinase expression and endometriosis. Semin Reprod Endocrinol 1996; 15:301–308.
12. Zeitoun K, et al. Deficient 17β-hydroxysteroid dehydrogenase type 2 expression in endometriosis: failure to metabolize estradiol-17β. J Clin Endocrinol Metab 1998; 83:4474–4480.
13. Michael MD, Michael LF, Simpson ER. A CRE-like sequence that binds CREB and contributes to cAMP-dependent regulation of the proximal promoter of the human aromatase P450 (CYP19) gene. Mol Cell Endocrinol 1997; 134:147–156.
14. Michael MD, Kilgore MW, Morohashi KI, Simpson ER. Ad4BP/SF-1 regulates cyclic AMP-induced transcription from the proximal promoter (PII) of the human aromatase P450 (CYP19) gene in the ovary. J Biol Chem 1995; 270:13561–13566.
15. Simpson ER, et al. Aromatase cytochrome P450, the enzyme responsible for estrogen biosynthesis. Endocr Rev 1994; 15:342–355.
16. Ackerman GE, Smith ME, Mendelson CR, MacDonald PC, Simpson ER. Aromatization of androstenedione by human adipose tissue stromal cells in monolayer culture. J Clin Endocrinol Metab 1981; 53:412–417.
17. MacDonald PC, Rombaut RP, Siiteri PK. Plasma precursors of estrogen. I. Extent of conversion of plasma Δ^4-androstenedione to estrone in normal males and nonpregnant normal, castrate and adrenalectomized females. J Clin Endocrinol Metab 1967; 27:1103–1111.

18. MacDonald PC, Edman CD, Hemsell DL, Porter JC, Siiteri PK. Effect of obesity on conversion of plasma androstenedione to estrone in postmenopausal women with and without endometrial cancer. Am J Obstet Gynecol 1978; 130:448–455.
19. Bulun SE, Price TM, Mahendroo MS, Aitken J, Simpson ER. A link between breast cancer and local estrogen biosynthesis suggested by quantification of breast adipose tissue aromatase cytochrome P450 transcripts using competitive polymerase chain reaction after reverse transcription. J Clin Endocrinol Metab 1993; 77:1622–1628.
20. Yue W, Wang JP, Hamilton CJ, Demers LM, Santen RJ. In situ aromatization enhances breast tumor estradiol levels and cellular proliferation. Cancer Res 1998; 58: 927–932.
21. Bulun SE, Simpson ER. Word RA. Expression of the CYP19 gene and its product aromatase cytochrome P450 in human leiomyoma tissues and cells in culture. J Clin Endocrinol Metab 1994; 78:736–743.
22. Noble LS, et al. Prostaglandin E_2 stimulates aromatase expression in endometriosis-derived stromal cells. J Clin Endocrinol Metab 1997; 82:600–606.
23. Bulun SE, Mahendroo MS, Simpson ER. Polymerase chain reaction amplification fails to detect aromatase cytochrome P450 transcripts in normal human endometrium or decidua. J Clin Endocrinol Metab 1993; 76:1458–1463.
24. Huang JC, Dawood MY, Wu KK. Regulation of cyclooxygenase-2 gene in cultured endometrial stromal cells by sex steroids (meeting abstr). Proc Am Soc Reprod Med 1996.
25. Hill JA. Immunology and endometriosis. Fertil Steril 1992; 58:262–264.
26. Takayama K, Zeitoun K, Gunby RT, Sasano H, Carr BR, Bulun SE. Treatment of severe postmenopausal endometriosis with an aromatase inhibitor. Fertil Steril 1998; 69:709–713.
27. Andersson S, Moghrabi N. Physiology and molecular genetics of 17β-hydroxysteroid dehydrogenases. Steroids 1997; 62:143–147.
28. Satyaswaroop PG, Wartell DJ, Mortel R. Distribution of progesterone receptor, estradiol dehydrogenase, and 20α-dihydroprogesterone dehydrogenase activities in human endometrial glands and stroma: progestin induction of steroid dehydrogenase activities *in vitro* is restricted to the glandular epithelium. Endocrinology 1982; 111: 743–749.
29. Henzl MR, Corson SL, Moghissi K, Buttram VC, Berqvist C, Jacobson J. Administration of nasal nafarelin as compared with oral danazol for endometriosis. N Engl J Med 1988; 318:485–489.
30. Waller KG, Shaw RW. Gonadotropin-releasing hormone analogues for the treatment of endometriosis: long-term follow-up. Fertil Steril 1993; 59:511–515.
31. Metzger DA, Lessey BA, Soper JT, McCarty KS, Jr., Haney AF. Hormone-resistant endometriosis following total abdominal hysterectomy and bilateral salpingo-oophorectomy: correlation with histology and steroid receptor content. Obstet Gynecol 1991; 78:946–950.

Panel Discussion 8

Endometrium: Treatment of Benign and Malignant Disease November 12, 1999

List of Participants

Ajay Bhatnagar Basel, Switzerland
Serdar Bulun Chicago, Illinois
Paul Goss Toronto, Ontario, Canada
Harold Harvey Hershey, Pennsylvania
D. Irlé Geneva, Switzerland
William Miller Edinburgh, Scotland
Matthew Smith Boston, Massachusetts

Matthew Smith: That was a very nice presentation. On your second to last slide you showed the results of aromatase expression in the women who had the dramatic response. Is it not equally possible that you basically killed off the cells that expressed aromatase at high level leaving a population of cells that just has low aromatase expression?

Serdar Bulun: Sure, that is a possibility, but we checked it histologically and the cells did not look necrotic—they were viable, at least histologically. And also we could have looked at the expression of some other gene, but we didn't, as we were satisfied by the histologic examination, but that is a good point. In fact, I think we still have some of that RNA so we could check the expression of some of the other genes.

Ajay Bhatnagar: With respect to the lady that you treated with anastrozole, she had bilateral salpingo-oophorectomy as well so she did not have any ovarian function. Endometriosis is a disease that is more prevalent in the premenopausal women and aromatase inhibitors in the premenopausal women would lead to all sort of disruption in the negative feedback loop. What would be your thoughts about how one would go about then using aromatase inhibitors—you don't want to use a gonadotropin-releasing hormone (GnRH) clamp, so how would you go about using these aromatase inhibitors without inducing cysts in the ovary?

Serdar Bulun: That is a good question. In fact I would use high doses of progestins to shut down the hypothalamus and pituitary. So once the ovaries are quiescent, you wouldn't get into the problems that you would with an intact hypothalamo-pituitary-ovarian axis.

Ajay Bhatnagar: But you could do the same thing with GnRH analogue?

Serdar Bulun: Yes, you can, but that would be very expensive.

Ajay Bhatnagar: So it's a question of cost rather than efficacy?

Serdar Bulun: Possibly. They both give rise to similar results and, in fact, progestins have the advantage of preventing bone loss to a degree. I should bring that point up when I finish answering your question. So that would be a good way of suppressing the hypothalamus, I would think. A lot of investigators ask me the question about the bone loss this patient had despite the bisphosphonate treatment. I continued to treat this patient, as she did not want to stop medication, but after 9 months I halved the anastrozole dose and gave it to her every other day, and she still did well. We also continued with alendronate and, in fact, she gained back the 6% bone loss. Thus, the bone loss was reversible, and these patients could be started on lower doses of aromatase inhibitors and could still do well. We are right at the beginning of these treatments and we should fine-tune it as we go along.

D. Irlé: I have a question regarding low-grade leiomyosarcomas. The data you showed were quite intriguing, and it's a problem in the clinic, because these patients often recur locally and there are some old data about tamoxifen use in sarcomas. I would like your opinion on that particular problem which might or not be a good target as well. Are there any data?

Serdar Bulun: That is a very good point. The slide I showed belonged to a leiomyoma which was a benign disease. Sarcomas are very rare, and the dogma is that they not estrogen dependent. In my private practice, for example, if I see a patient with a myoma, most of the time I do not check the histology because sarcomas are so rare. I mean we are talking about two cases in 100,000 cases or such. So I wasn't referring to sarcomas, I was referring to myomas.

D. Irlé: I asked the question because in the literature there are old data on the use of tamoxifen and there was a regression in the number of these disease even among fibrosarcomas. So I was wondering whether there was some more information about that.

Serdar Bulun: It could be that the stromal tumors of the endometrium may be more responsive to estrogen. I am not aware of leiomyosarcomas, but I am not an oncologist.

William Miller: I wonder if I could ask you two questions. One is a general question at the end that the rest of the audience might like to participate with. The first question I was going to ask was you brought up the control of aromatase in endometrium and quite clearly showed that this control worked off certain promoters, the cyclic-AMP system and prostaglandins. That brings up the possibility that, given that there seems to be some tissue specificity about induction of aromatase, that one could preferentially switch off aromatase because of these different promoters in these different systems. Do you have views about maybe switching off aromatase activity in malignant cells but still leaving it on in bone so it's protected? The second question I was then going to ask, since you have introduced the subject of myomas, and we are therefore straying a bit from your topic, but I think it is important, because the title of this session was "New Indications," what your thoughts are in terms of other potential systems that might be a good target for aromatase inhibitors, and maybe the audience might like to put a few views forward as well.

Serdar Bulun: Sure, that is an excellent point, Bill. I believe that is going to be done. It's just a matter of developing the right techniques. Gene transfer has not been resolved yet, but we should see those developments probably within the next few decades. Regarding other indications for the use of aromatase inhibitors, possibly they could be used in other cancers that may be estrogen dependent, for example, there are a lot of central nervous system tumors, of which there are attempts to treat some of them, like meningiomas, with RU-486. Aromatase may be expressed in some of these estrogen-dependent tumors. This may lead to the development of new indications, and I am sure other people in other fields would have original ideas.

Harold Harvey: Some years ago, through serendipity, we showed that melanoma in fact contains very high levels of aromatase, so we proceeded to treat a small population of patients with aminoglutethimide. We saw no response, but in fact the levels in some of the melanoma tissue with the normal skin as control were extraordinarily high. I think that perhaps with more potent compounds, we should reexamine that tumor.

Ajay Bhatnagar: If I could also comment. I think that we have seen over the last 2 days really good information about very well-tolerated, potent, and effective aromatase inhibitors. We don't just have one now, we have several, and I think that these compounds by their mechanism of action—which is estrogen deprivation—would be potentially of use in any sort of pathology or physiology which is estrogen dependent. We chose these three for the session this afternoon because people had already had thoughts about these areas and some data were available. There is, of course, also the area of bone and short stature on which there are some publications. There is the area of fertility control. The one slide which Dr. Kaplowitz showed of the data on the Bonnet monkeys was an offshoot of work that had been done in terms of contraception in female monkeys using aromatase inhibitors. So we have many areas, but many of them will relate to benign rather than malignant pathophysiology.

Paul Goss: What are your thoughts about benign endometrial hyperplasia with and without atypia?

Serdar Bulun: That is a very good point. Endometrial hyperplasia still contains large numbers of stromal cells which can potentially express aromatase, and they would be important targets, but we haven't looked into the treatment of those.

Paul Goss: How common is endometrial hyperplasia in comparison to endometriosis as a medical issue?

Serdar Bulun: Endometriosis is probably much more common than even endometrial cancer. The epidemiology could not be studied very well because of the difficulties diagnosing the disease. Endometrial hyperplasia is a transient state. It either remains the same and doesn't cause problems or it regresses or it progresses to endometrial cancer. Most of the time the kind of lesions with atypia progress to endometrial cancer, and there are different percentages recorded by different investigators. So it is another unknown type of disease, although endometrial hyperplasia can be easily diagnosed by a simple endometrial biopsy which is performed using a plastic catheter. It wouldn't be as difficult as endometriosis, for example. Usually these are the women who are anovulatory, so I would guess that the prevalence of this condition would be much less than endometriosis.

Index